Scientific Inquiry and the Social Sciences

A Volume in Honor of Donald T. Campbell

Marilynn B. Brewer
Barry E. Collins
Editors

Scientific Inquiry and the Social Sciences

Jossey-Bass Publishers

San Francisco • Washington • London • 1981

Essay Index

SCIENTIFIC INQUIRY AND THE SOCIAL SCIENCES
A Volume in Honor of Donald T. Campbell
by Marilynn B. Brewer, Barry E. Collins, Editors

Copyright © 1981 by: Jossey-Bass Inc., Publishers
433 California Street
San Francisco, California 94104
&
Jossey-Bass Limited
28 Banner Street
London EC1Y 8QE

Library of Congress Cataloging in Publication Data
Main entry under title:

Scientific inquiry and the social sciences.

 Includes bibliographies and indexes.
 Contents: Perspectives on knowing / Marilynn B.
Brewer and Barry E. Collins — Evolution, adapta-
tion, and human understanding / Stephen Toulmin —
Natural selection and other models in the histo-
riography of science / Robert J. Richards — [etc.]
 1. Social sciences—Methodology—Addresses,
essays, lectures. 2. Campbell, Donald Thomas,
1916- . I. Campbell, Donald Thomas, 1916-
II. Brewer, Marilynn B., 1942- . III. Collins,
Barry E.
H61.S444 300'.72 81-6023
ISBN 0-87589-496-8 AACR2

Manufactured in the United States of America

JACKET DESIGN BY WILLI BAUM

FIRST EDITION

Code 8110

The Jossey-Bass
Social and Behavioral Science Series

Special Adviser
Methodology of Social and
Behavioral Research
DONALD W. FISKE
University of Chicago

Preface

This book is about scientific inquiry—both its logical foundations and its application to the social sciences. It was conceived and written in honor of Donald T. Campbell and includes contributions from a number of former students as well as colleagues who have been directly influenced by Campbell's thinking. The chapters, all written expressly for this volume, build on Campbell's work, extend it, apply it, or critically examine it.

The contributors represent several disciplines: philosophy, history of science, anthropology, and psychology (including social psychology and methodology). These fields reflect the wide range of areas within which Donald Campbell has worked and to which he has made major contributions. The book is intended for all those involved in these and related areas and has been written at a level that will be useful for students, while at the same time providing new ideas and perspectives for knowledgeable professionals.

Chapter One provides a general introduction to the book and identifies six main themes in the thinking of Donald Campbell. The subsequent chapters, which explore these themes in depth, have been organized into three parts—Models of Knowing, Methodological Applications, and Substantive Applications—each with its own introductory comments.

We would like to thank all those students and friends of Donald T. Campbell who encouraged the production of this volume and, in particular, Donald W. Fiske, who gave the project its initial impetus and was helpful and supportive throughout the whole process of bringing it to fruition.

July 1981   Marilynn B. Brewer
Santa Barbara, California

Barry E. Collins
Los Angeles, California

Contents

Contents

Contributors

Marilynn B. Brewer is professor and chairperson of psychology at the University of California, Santa Barbara. Following receipt of a B.A. degree in Social Science from North Park College in Chicago, she did her graduate work in social psychology under the direction of Donald Campbell at Northwestern University, where she was awarded the M.A. degree in 1966 and the Ph.D. degree in 1968. Before joining the faculty at the University of California, Santa Barbara, Brewer was an assistant professor of psychology at Loyola University of Chicago. She is coauthor, with Donald Campbell, of *Ethnocentrism and Intergroup Attitudes: East African Evidence* (1976) and has published other books and articles in the areas of research methodology, social perception, and intergroup relations. She has served on the Board of Scientific Affairs of the American Psychological Association and on the editorial boards of

Journal of Personality and Social Psychology, the *Journal of Social Issues,* and other journals.

Barry E. Collins is professor of psychology at the University of California, Los Angeles. He was a student at Northwestern University, where he received his B.S. degree in speech in 1959, an M.A. degree in speech in 1960 and his Ph.D. degree in social psychology in 1963 as a student of Donald Campbell. Collins held his first job as an assistant professor of psychology at Yale University and has also taught at University of Wisconsin and Stanford University. He is coauthor with Harold Guetzkow of *A Social Psychology of Group Processes for Decision Making* (1964) and with Norman Miller and Charles Kiesler of *Attitude Change: A Critical Analysis of Theoretical Approaches* (1969). He has published a number of experimental studies on attitude change induced by forced compliance and, most recently, on hyperactivity. He has written three books for undergraduate social psychology classes.

Thomas D. Cook is professor of psychology at Northwestern University.

William D. Crano is professor of psychology and director of the Center for Evaluation and Assessment at Michigan State University.

David A. Kenny is associate professor of psychology at the University of Connecticut.

Louise H. Kidder is associate professor of psychology at Temple University.

Robert A. LeVine is Roy E. Larsen Professor of Education and Human Development at Harvard University.

Norman Miller is Mendel B. Silberberg Professor of Psychology at the University of Southern California.

Robert J. Richards is assistant professor in the Center for the Study of History of Science and Medicine, Committee on Conceptual Foundations of Science, and Committee on Biopsychology at the University of Chicago.

Paul C. Rosenblatt is professor of psychology in the Department of Family Social Science at the University of Minnesota.

Marshall H. Segall is professor of social psychology in the Maxwell School, Syracuse University.

Abner Shimony is professor of philosophy and physics at Boston University.

Peter Skagestad is assistant professor of philosophy at Williams College.

Stephen Toulmin is professor in the Committee on Social Thought at the University of Chicago.

William C. Wimsatt is associate professor in the College and in the Department of Philosophy, the Committee on Evolutionary Biology, and the Committee on Conceptual Foundations of Science at the University of Chicago.

Scientific Inquiry and the Social Sciences

A Volume in Honor of Donald T. Campbell

Marilynn B. Brewer
Barry E. Collins

1

Perspectives on Knowing:
Six Themes from
Donald T. Campbell

The contributions to this volume cross many of the traditional
boundaries among academic disciplines, a tribute to the breadth of
scholarship represented by Donald Campbell. Though diverse, the
various chapters share at least one common conceptual tie to the
philosophical position that there does exist—in both physical and
social domains—an objective reality apart from our experience of
it. Campbell has invited his audiences—from the most empirical
test constructor to the most philosophical epistemologist—to join
him in a thought experiment. Let us posit, he says, that there is an
objective reality that can be seen (known) by the perceiver. Let us
imagine that there is such a reality of independent objects and
relationships among objects and that reality exists separately from

the knower. Assume further that knowledge of this reality is neither direct nor infallible but is "edited" by the objective referent and reflected in the convergence of observations from multiple independent sources of knowing.

Campbell has labeled the ontological point of view represented in this thought experiment "hypothetical realism" or "critical realism." The conviction that there is an external reality is hypothetical because it is necessarily presumptive and philosophically underjustified. Campbell makes no claim of proof for this assumed reality; he says only that such an assumption makes sense out of the achievements of adaptation that characterize biological evolution, individual learning, and scientific knowledge alike.

The relationship between hypothetical realism and the scientific model is well illustrated in Peter Skagestad's summary (Chapter Four in the present volume) of the approach represented by philosopher Charles S. Peirce ([1871] 1958): "Since we have a historical experience of a general drift toward consensus within the scientific community, we may understand objective truth—and hence the notion of reality—by extrapolating from this experience the notion of a universal, unshakable consensus toward which inquiry is converging as a limit." This statement links the rules of replicability and consensual validation that characterize modern science to ontological realism, but it can also be viewed in light of a countervailing philosophical position which would hold that convergence to consensus reflects the presence of a common *theory* of reality rather than objective truth. In Campbell's home discipline of social psychology, the idea that shared biases can generate consensus on what is real in the absence of any objective basis has been the subject of classic empirical demonstration.

Muzafer Sherif (1935) made use of the autokinetic effect in visual perception to create a laboratory setting for the experimental investigation of the formation of norms (consensus) in social groups. One can easily produce the autokinetic effect by projecting a single small steady beam of light in an otherwise completely dark environment. In the absence of any visual reference for locating the light, it appears to move erratically in all directions and to change location if turned on and off. In the absence of any actual movement, the perception of movement *appears* to be objectively based. Sherif used judgments of the extent of this perceived movement (in inches) for his experimental studies of social consensus. Male university students made a long series of auto-kinetic movement judgments, either singly or in groups, under one of two experimental conditions. Some participants first made a series of judgments alone and then were placed into group settings where two or three individuals made their judgments (spoken aloud) in each other's presence. Other participants were placed in groups for their initial series of judgments and then made judgments alone in a subsequent session.

When initial judgments were made alone, individuals tended to develop a personal range of judgments around some internal reference point, but this point varied widely from subject to subject. The average judged movement for individuals ranged from as little as .5 inch to more than 10 inches, with a lack of consensus that reflected the absence of any objective reality. By contrast, when respondents made their judgments in the group setting, the two or three individuals rapidly converged toward a common average judgment—a not altogether surprising demonstration of social influence. Judged movement still varied from group to group, but the judgments of individuals within the same group became relatively homogeneous after a few trials. When students who had made their initial judgments in a group context were subsequently put in a situation where judgments were made alone, the effect of group consensus on individual judgments persisted. If the only data available had been the judgments made *alone* by individuals who had *previously* been in the same group, the results would show an apparent convergence of judgments obtained independently (that is, objectivity), whereas the "objectivity" was actually the residual product of prior shared experience.

The power of residual shared bias to affect perceptions of reality across time and across people was ingeniously demonstrated in a follow-up of Sherif's experiments undertaken by Jacobs and Campbell (1961). Under conditions where perceived autokinetic movement averaged around 3.8 inches, Jacobs and Campbell introduced an "arbitrary" group norm by instructing experimental accomplices (confederates) to give judgments averaging 15.5 inches. In the initial conditions of the experiment, three confederates were placed in a group with only one "naive" respondent (nonconfederate). In this social setting, the judgments of the one naive group member were strongly influenced by those of the confederates, averaging between 14 and 15 inches. After thirty trials, one confederate was removed from the group and replaced with another naive respondent, and this procedure was repeated until all the confederates had been replaced. Then the series of judgments were continued through eight more replacement cycles, with a new member added and the most experienced old member removed after every thirty judgments. Even after all the original respondents had been removed from the experiment, the influence of the confederate-generated norm was evident in the group consensus through four or five successive "generations" after their removal.

These social psychological experiments illustrate the nature of the challenge to hypothetical realism and set the agenda for scientific methodology. Theories held in common *do* affect our constructions of reality, and the problem is to disentangle shared perceptions from shared reality. Addressing this challenge has been the purpose behind almost all of Donald Campbell's contributions to social science theory and methodology. This basic theme has given rise to six conceptual subthemes that recur throughout his writings and are briefly summarized as follows:

1. *The natural selection model of knowing, knowledge accumulation, and moral tradition.* The "blind variation and selective retention" theme, which weaves through Campbell's work on epistemology (see, for example, Campbell, 1974) and his views on the experimenting society (see Campbell, 1969a), is the pivotal theme that unifies his psychological and philosophical contributions. The blind variation aspect stresses the idea that the initial generation of novel alternatives (theories, problem solutions, social institutions, and the like) contains no foresight as to their ultimate usefulness. The second aspect—selective retention—refers to the winnowing process whereby some innovations are maintained and others are not, as a function of their fit to some external selective mechanism or criterion. Campbell has been particularly creative in utilizing the evolution of perceptual systems as a biological analogy for the accumulation of knowledge at both the individual and group levels. Probably one of his most controversial applications of this model has been his proposal that cultural and moral traditions are the product of blind variation and selection processes in the course of social evolution (Campbell, 1975).

2. *Multiple perspectives and triangulation.* According to Campbell, any given view of reality reflects as much the perspective or methods of observation as it does the object being viewed. The assumption that there is an independent reality of knowable objects must contend with the fact that the "same" object appears in one way under some circumstances and in another way at different times or to different persons or from different vantage points. Any knowledge of that object, then, is in part determined by the nature of the object itself and in part by artifacts intrinsic to the perspective or instruments through which the object is viewed. Triangulation is the model that Campbell employs to convey the idea that multiple vantage points permit fixing on a common object of perception in a way that is impossible from a single point. This theme is central to his methodological papers, in his advocacy of multiple operationalism (Campbell, 1969b) and his recommendations for distinguishing between trait and method variance in psychological measurement (Campbell and Fiske, 1959). A related principle is his emphasis on heterogeneity of irrelevancies. In order to contribute effectively to a process of triangulation, different observations or measurements must contain different, non-overlapping perspectival biases or measurement artifacts. If irrelevant but systematic sources of error cannot be removed from any single observation or observational setting, then objectivity requires the use of multiple observers, multiple techniques of observation, and multiple methodologies.

3. *Entitativity and the unit of selection.* Campbell has long espoused a type of nonreductionism that recognizes the validity of studying systems at different levels of organization or different units of analysis. He has proposed that processes of natural selection operate directly, not only at the level of the cell or individual organism but also at higher levels of organization. While accepting the general reductionistic principle that achievements at higher levels require specific lower-level mechanisms and processes for their implementation, he also emphasizes that laws of the higher-level selective system determine in part the selection of lower-level events. Related to this position is his concern with the entitativity of social groups (Campbell, 1958) and the role of group selection in social evolution (Campbell, 1972, 1975). If selective processes are to operate at superorganismic levels of organization, then there

must be some organizational or structural principles that give such units boundedness or coherence, so that they can be perceived and acted on as a whole. In his own writings, Campbell has emphasized the principle of common fate as a major element in the definition of collective entities.

4. *Pattern matching and the contextual bases of knowledge.* Campbell's linkages to principles of Gestalt psychology are most evident in his treatment of the nature of perception and the dependence of judgments on contextual features. He has claimed, for instance, that our ability to recognize entities or make judgments of equivalence relies heavily on a process of matching *patterns* of stimuli (or stored representations of stimulus patterns), rather than a point-by-point matching of individual stimulus items. This conceptual theme is illustrated in Campbell's contributions to the methods of cross-cultural research, including his recommendations for assessing equivalence of instructions and translations of ideas across cultures and languages (Campbell, 1964; Werner and Campbell, 1970). At a different level of abstraction, Campbell also sees pattern matching as a model for the fit between scientific theory and data (Campbell, 1966) and for understanding the extent to which any advance in scientific knowing is grounded in an implicit trust in the great bulk of previous or commonsense knowledge.

5. *Assimilation, contrast, and adaptation level.* Related to the contextual bases of knowing are various phenomena of perceptual and social judgment that reflect the relativity of such judgments as a function of the perceiver's own position or experience. This relativity constitutes a particular form of the vantage-point problem, whereby the same object can be perceived as different at different times or by different persons at the same time. Most specifically, Campbell has been interested in the extent to which deviation from current or past experience enhances the salience and perceived intensity of a stimulus or experience, and he has applied these contrast effects to an analysis of a number of social phenomena, including the perception of person traits (Campbell and others, 1964), ethnocentric group perceptions (Campbell, 1967), and the relativity of hedonic satisfaction (Brickman and Campbell, 1971).

6. *Cue utilization and composite dispositions.* The weighting and combination of information derived from different sources

and across different modalities are closely related to the concept of triangulation. This theme in Campbell's work incorporates notions of different kinds of knowing or knowledge processes, including individual and social (vicarious) modes of knowing or acquiring behavioral dispositions (Campbell, 1959, 1974). In his treatment of social attitudes (Campbell, 1963) and of conformity (Campbell, 1961), Campbell illustrates that response dispositions can be viewed as the product of a composite of multiple sources of information. In explicating this particular subtheme, Campbell has attempted to reconcile explanatory concepts drawn from both S-R (behavioral) and Gestalt (cognitive) traditions in psychology.

Cutting across all six of these subthemes in his work is Donald Campbell's remarkable facility in moving across levels of generality, from the abstract to the concrete. For instance, the abstract concepts of triangulation and multiple operationalism are familiar to most social scientists in such concrete representations as the multitrait-multimethod matrix (Campbell and Fiske, 1959) and unobtrusive measures (Webb and others, 1966). The transition from general to specific, from abstract to concrete is characteristic of Campbell's work and his treatment of all six of the subthemes identified here. It also provides the basis for the organization of the present volume. Each of the subsections of the book contains contributions that derive from or expand on Campbell's ideas at a particular level of abstraction. The sections are introduced with specific illustrations from among Campbell's own papers that exemplify his treatment of issues at that level of generality and provide the background for the contributed papers within that section.

The first section deals with Campbellian constructs at their most abstract, with contributions that discuss issues surrounding the nature of knowing and the philosophical justification for science and empiricism. Although the authors of the chapters in this section do not address the subject matter of the social sciences directly, the relevance of epistemological issues to method and theory in social science has been central in Campbell's own work and is carried through in the remainder of the present volume. In the second section of the book, the implications for meta-methodological issues are illustrated, with papers that emphasize the need for convergence of multiple methodological approaches as essential to the social sciences. In the third section, the utilization of multiple methods to assess the validity of social psychological theories and their application is illustrated in the context of specific substantive issues. Following that section, Donald Campbell's own commentary provides an autobiographical perspective on many of the issues raised in this volume.

References

Brickman, P., and Campbell, D. T. "Hedonic Relativism and Planning the Good Society." In M. Appley (Ed.), *Adaptation-Level Theory: A Symposium*. New York: Academic Press, 1971.

Campbell, D. T. "Common Fate, Similarity, and Other Indices of the Status of Aggregates of Persons as Social Entities." *Behavioral Science*, 1958, *3*, 14-25.

Campbell, D. T. "Methodological Suggestions from a Comparative Psychology of Knowledge Processes." *Inquiry*, 1959, *2*, 152-182.

Campbell, D. T. "Conformity in Psychology's Theories of Acquired Behavioral Dispositions." In I. A. Berg and B. M. Bass (Eds.), *Conformity and Deviation*. New York: Harper & Row, 1961.

Campbell, D. T. "Social Attitudes and Other Acquired Behavioral Dispositions." In S. Koch (Ed.), *Psychology: A Study of a Science*. Vol. 6: *Investigations of Man as Socius*. New York: McGraw-Hill, 1963.

Campbell, D. T. "Distinguishing Differences of Perception from Failures of Communication in Cross-Cultural Studies." In F. S. C. Northrop and H. H. Livingston (Eds.), *Cross-Cultural Understanding: Epistemology in Anthropology*. New York: Harper & Row, 1964.

Campbell, D. T. "Pattern Matching as an Essential in Distal Knowing." In K. R. Hammond (Ed.), *The Psychology of Egon Brunswik*. New York: Holt, Rinehart and Winston, 1966.

Campbell, D. T. "Stereotypes and the Perception of Group Differences." *American Psychologist*, 1967, *22*, 817-829.

Campbell, D. T. "Reforms as Experiments." *American Psychologist*, 1969a, *24*, 409-429.

Campbell, D. T. "Definitional Versus Multiple Operationalism." *et al.*, 1969b, *2*, 14-17.

Campbell, D. T. "On the Genetics of Altruism and the Counter-hedonic Components in Human Culture." *Journal of Social Issues,* 1972, *28,* 21-37.

Campbell, D. T. "Evolutionary Epistemology." In P. A. Schilpp (Ed.), *The Philosophy of Karl Popper.* La Salle, Ill.: Open Court, 1974.

Campbell, D. T. "On the Conflicts Between Biological and Social Evolution and Between Psychology and Moral Tradition." *American Psychologist,* 1975, *30,* 1103-1126.

Campbell, D. T., and Fiske, D. W. "Convergent and Discriminant Validation by the Multitrait-Multimethod Matrix." *Psychological Bulletin,* 1959, *56,* 81-105.

Campbell, D. T., Miller, N., Lubetsky, J., and O'Connell, E. J. "Varieties of Projection in Trait Attribution." *Psychological Monographs,* 1964, *78* (entire issue 592).

Jacobs, R. C., and Campbell, D. T. "The Perpetuation of an Arbitrary Tradition Through Several Generations of a Laboratory Microculture." *Journal of Abnormal and Social Psychology,* 1961, *62,* 649-658.

Peirce, C. S. "Fraser's Edition of *The Works of George Berkeley,* Review." In A. Burks (Ed.), *Collected Papers of Charles Sanders Peirce.* Vol. 8. Cambridge, Mass.: Harvard University Press, 1958. (Originally published 1871.)

Sherif, M. "A Study of Some Social Factors in Perception." *Archives of Psychology,* 1935, No. 187.

Webb, E. J., Campbell, D. T., Schwartz, R. D., and Sechrest, L. B. *Unobtrusive Measures: Nonreactive Research in the Social Sciences.* Chicago: Rand McNally, 1966.

Werner, O., and Campbell, D. T. "Translating, Working Through Interpreters, and the Problem of Decentering." In R. Naroll and R. Cohen (Eds.), *A Handbook of Method in Cultural Anthropology.* New York: Natural History Press/Doubleday, 1970.

PART I

In spring of 1977, Donald Campbell delivered the William James Lectures at Harvard University under the general title "Descriptive Epistemology: Psychological, Sociological, and Evolutionary." Some of the contributors to this first section of the present volume attended those lectures, and others have discussed written drafts of the lectures with him. Their chapters reflect the stimulation of ideas and perspectives that the series inspired. A few quotations from those lectures will serve to explicate the six themes discussed previously in the general introduction and to introduce some of the philosophy and history of science issues discussed in the following chapters.

Models of
Knowing

Descriptive Epistemology

During the course of the James lectures, Campbell attempted to explicate these philosophical, historical, and methodological issues within the framework of a new disciplinary orientation that he labeled "descriptive epistemology":

> We are cousins to the amoeba, and have received no direct revelation not shared with it. . . . No nonpresumptive knowledge is possible, neither in ordinary knowing or in knowing about knowing. Our only

11

hope as competent knowers is if we be the heirs of a substantial set of well-winnowed presuppositions. . . . A descriptive epistemology that takes over from here will be more like science than philosophy. . . . It will be hypothetical, contingent, and presumptively descriptive of knowers and the world to be known. Within these limitations, it can still deal with the issue of truth: In what kind of world would what kind of procedures lead a knowing community to improve the validity of its model of the world? . . . For such endeavors, "theory of science" becomes technically more appropriate than the term "philosophy of science" . . . [and] under the general heading of theory of science not only history of science, but also such specialties as sociology of science and psychology of science.

I intend the term [descriptive epistemology] to describe a more modest, less definitive, enterprise than pure philosophical epistemology may aspire to. . . . Descriptive epistemology is going to be descriptive of how people go about it when they think they are acquiring knowledge, or how animals go about perceiving and learning when we think they are acquiring knowledge. At this level we are doing psychology, physiology, and sociology, without necessarily engaging epistemological issues. But I also want descriptive epistemology to include the theory of how these processes could produce truth or useful approximations to it. . . . Such theory of knowledge will have to be presumptive—will have to assume certain general truths about the world while probing and choosing among others. But within that limitation, it can and should, I believe, be hypothetically normative, with theory as to why science works, if and when it works, to produce valid knowledge; why it fails when it fails, and how one should go about it if one wants valid knowledge [Campbell, 1977, Lecture 1].

In the William James Lectures, Campbell provided a new coherence and direction for many of the conceptual themes developed in his earlier contributions to psychological theory and methodology. Descriptive epistemology incorporated Campbell's natural selection model of knowledge acquisition, which he developed most fully in his paper on "Evolutionary Epistemology" prepared for a volume in honor of Karl Popper (Campbell, 1974a) and which had appeared in earlier versions as the blind variation and selective retention model for perception (1956), creative thought (1960), and scientific discovery (1959a, 1974b).

Blind Variation and Selective Retention

As in these earlier papers, Campbell's William James Lectures advocated a blind variation and selective retention model for knowledge acquisition and accumulation as well as for biological evolution:

> For me, embodied knowledge falls within the general class of instances of "fit," of the fit of one system to another. . . . The fit between animal form and environmental opportunity is an important exemplar of fit. . . . All instances of fit are achieved, maintained, and improved by profoundly indirect processes, involving selection rather than prefitted variations [1977, Lecture 2].
> Learning and perception have evolved in a world in which they were useful. . . . Inevitably the equipment of perception and learning evolved to optimize the diagnosis of entifiable aspects of the envionment [1977, Lecture 3].

While thus reaffirming the theme of blind variations edited by external selectors, Campbell also acknowledged the role of other types of selection processes:

> My emphasis on selection-as-the-source-of-fit should not have been read as excluding other selective processes that select for other things than the fit of one system to environment or to an alien system. [The] emphasis on what I will now call *external* selection has, however, led to the neglect of several types of selection that also affect the shape of embodied knowledge. I am going to call one class of these *structural* selection. . . . [For instance,] before

a mutant on an old rule structure can be subject to external selection as a potential improved rule structure, it must first meet the prior selective requirement of being a functionable rule structure.

Vicarious selectors are another type of internal selector which share important features with both structural selectors and external selectors. They are, in fact, vicarious in being internal representations of external selectors. . . . Now it turns out that all of the organismic processes we would call learning involve vicarious selectors rather than or in addition to a direct encounter with reality itself. . . . The vicarious selector is only approximately accurate, and has fringes of inappropriateness which can produce illusions. . . . Even in ranges of optimal function, the built-in presumptions are approximate truths about past worlds and may no longer hold if the ecology has changed [1977, Lecture 2].

Context Dependence of Knowing

Campbell's William James Lectures also expanded on his earlier papers on the indirectness and the context dependence of knowing (1966, 1969a):

Our neurology of visualization and categorization, and our primitive ostensive linguistics, having been evolved and learned in adapting to an environment of highly entitative middle-sized objects, may be expected to be partially inappropriate when applied to entities not represented in that selective system, as in microphysics, cosmology, the conceptual spaces of higher mathematics, and perhaps even the social sciences. . . . I will use the concept of the doubt-trust ratio to introduce the theme of fallibilist holistic context dependence in conceptual change in science. . . . Science depends upon common sense even though at best it goes beyond it. Science in the end contradicts some items of common sense, but it only does so by trusting the great bulk of the rest of common-sense knowledge [1977, Lecture 4].

The James Lectures also expanded on Campbell's contributions to the sociology of science, including his analyses of the biases inherent in the social locus of scientific discovery and transmission (1959b), the social-organizational structure of scientific disciplines (1969b), and the phenomenon of multiple independent invention or discovery (Campbell and Tauscher, 1966):

> Not only is science conducted in the context of the elaborate social system of science, its products— "scientific knowledge" or "scientific truths"—are social products, incompletely specifiable in the beliefs of any one scientist or the writings in any one book. ... Before a scientific community can be a self-perpetuating social vehicle for an ever improving set of beliefs about the physical world, it must first meet the social structural requirements of being a self-perpetuating social system at all. ... A scientific community must recruit new members and reward old members well enough so that the young recruits can be attracted to a lifelong commitment to the field, and can justify the drudgery and the painful initiation rites. Journals must get published, purchased, and read. Members must keep loyal to the group and kept from defecting to other tribes. Jobs must be found for loyal followers. Social facilitators and leaders are needed to keep the group together, and must be rewarded for this role.
>
> But I want to assert that among these belief-preserving mutual admiration societies, all of which share this common human tribalism, science is also different, with different specific values, myths, rituals, and commandments, and that these different norms are related to what I presume to be science's superiority in the improving validity of the model of the physical world which it carries. ... In spite of the theory ladenness and noisiness of unedited experimental evidence, it does provide a major source of discipline in science. ... The experiment is meticulously designed to put questions to "Nature Itself" in such a way that neither the questioners, nor their colleagues, nor their superiors can affect the answer. ... In our iterative oscillation of theoretical emphases, in

our continual dialectic that never achieves a stable
synthesis, we are now ready for a post-post-positivist
theory of science which will integrate the epistemological relativism recently achieved with a new and
more complex understanding of the role of experimental evidence and predictive confirmation in science [1977, Lecture 5; also published as Campbell,
1979].

Overview of Chapters

These themes from Campbell's William James Lectures provide the backdrop for the contributions in the first section of the
present volume. The first two chapters in the section, by Stephen
Toulmin and Robert Richards, place the natural selection model in
historical context, comparing it with alternative philosophical
models for explaining the fit between human understanding and
objective reality. The following chapters, by Peter Skagestad and
Abner Shimony, focus on the philosophical arguments associated
with hypothetical realism; both suggest that Campbell is unneces-
sarily tentative in his defense of ontological realism and that the
position can be justified, in contrast to more phenomenological
orientations, on both logical and empirical grounds. Finally, William Wimsatt introduces the concept of robustness, which is
closely related to Campbell's concepts of triangulation and entitativity. Taken as a whole, these five chapters make clear the extent
to which our notions of scientific validity and objectivity are embedded in the particular epistemological framework that Donald
Campbell has applied to both methodological and substantive
issues.

References

Campbell, D. T. "Perception as Substitute Trial and Error." *Psy-
chological Review,* 1956, *63,* 330-342.
Campbell, D. T. "Methodological Suggestions from a Comparative
Psychology of Knowledge Processes." *Inquiry,* 1959a, *2,* 152-
182.
Campbell, D. T. "Systematic Error on the Part of Human Links in

Communication Systems." *Information and Control,* 1959b, *1,* 334-369.

Campbell, D. T. "Blind Variation and Selective Retention in Creative Thought as in Other Knowledge Processes." *Psychological Review,* 1960, *67,* 380-400.

Campbell, D. T. "Pattern Matching as an Essential in Distal Knowing." In K. R. Hammond (Ed.), *The Psychology of Egon Brunswik.* New York: Holt, Rinehart and Winston, 1966.

Campbell, D. T. "A Phenomenology of the Other One: Corrigible, Hypothetical, and Critical." In T. Mischel (Ed.), *Human Action: Conceptual and Empirical Issues.* New York: Academic Press, 1969a.

Campbell, D. T. "Ethnocentrism of Disciplines and the Fish-Scale Model of Omniscience." In M. Sherif and C. W. Sherif (Eds.), *Interdisciplinary Relationships in the Social Sciences.* Chicago: Aldine, 1969b.

Campbell, D. T. "Evolutionary Epistemology." In P. A. Schilpp (Ed.), *The Philosophy of Karl Popper.* La Salle, Ill.: Open Court, 1974a.

Campbell, D. T. "Unjustified Variation and Selective Retention in Scientific Discovery." In F. Ayala and T. Dobszhansky (Eds.), *Studies in the Philosophy of Biology.* London: Macmillan, 1974b.

Campbell, D. T. "Descriptive Epistemology: Psychological, Sociological, and Evolutionary." William James Lectures, Harvard University, 1977.

Campbell, D. T. "A Tribal Model of the Social System Vehicle Carrying Scientific Knowledge." *Knowledge,* 1979, *2,* 181-201.

Campbell, D. T., and Tauscher, H. "Schopenhauer (?), Sequin, Lubinoff, and Zehender as Anticipators of Emmert's Law: With Comments on the Uses of Eponymy." *Journal of the History of the Behavioral Sciences,* 1966, *2,* 58-63.

2

Stephen Toulmin

Evolution, Adaptation, and Human Understanding

What is Man, that he may know a Number?
And what is Number, that a Man may know it?

With these Pythagorean questions, the late Warren McCulloch posed his own central puzzle about the powers of the human intellect (McCulloch, 1965). In so doing, he showed unmistakably the genealogical connections linking contemporary natural science with the older traditions of Western philosophy. Many natural scientists today would like to "do without" philosophy and see their own "empirically based" inquiries as wholly divorced from earlier speculative debates. But this separation of science from philosophy can be more easily achieved in some fields than in others.

There is nothing very metaphysical about, for instance, the study of snails or hydrocarbons or about the design of electric motors; but in other fields of science—such as those concerned with fundamental matter theory, the origins of the physical universe, or the study of higher mental functioning—that divorce may never be final. In these fields, as Immanuel Kant argued two hundred years ago, we have a natural tendency to overreach the boundaries of our own rationality; so, however firmly we seek to expel philosophy by the front door, it comes in again at the back. In attempting to think *scientifically* about such topics as higher mental functioning, cognitive development, and the neurosciences, we find ourselves caught up—even today—in the continuation of a much more ancient *philosophical* debate, about "human understanding" and its significance for the larger world toward which our mental activities are directed.

In this respect, the sciences of control, communication, and information processing and the sciences of the brain and central nervous system stand in a line of epistemological inquiry that goes back through Hermann von Helmholtz and Immanuel Kant, John Locke, and René Descartes, to the philosophers of classical antiquity. And the same is true of the "evolutionary" arguments by which Donald Campbell has himself set out to reestablish the philosophical theory of human knowledge and discovery in its proper scientific context. At the very outset of philosophy, indeed, the pre-Socratic philosophers were already posing their own central problems in terms that Warren McCulloch consciously echoed:

What is Man, that he may understand the Nature of the World?
And what is the World, that Man may understand its Nature?

With this ancestry in mind, I shall attempt to do three things in this chapter. (1) I shall show how the issues in the current phase of this debate about human understanding are related to those that were the focus of concentration in earlier phases. (2) I shall seek to define the questions that call for reflection and analysis as a result. (3) I shall argue that the reflective analysis of these questions (to which Campbell has contributed so much) is of central importance, equally, for philosophy itself and for the sciences—especially for the human sciences.

Three Phases in the History of Epistemology

Throughout the entire development of epistemology, one central fact or phenomenon has been the starting point for discussion. That phenomenon appeared as wonderful to Albert Einstein in the twentieth century as it had to Pythagoras and Anaximander twenty-five hundred years before (Schilpp, 1949). It is the simple fact that the world of nature is intelligible to human beings at all—or, to speak more exactly, the fact that we humans find the world of nature intelligible to the notable extent we do. From the start, the challenge to philosophers and scientists alike has been to account for this fact; to explain, that is, how nature comes to be so consistently transparent to human efforts at mastering its workings, instead of proving totally inscrutable or revealing its character only spasmodically and unreliably. In some way or other, for some reason or other, human beings and their mental equipment are evidently "adapted to" that world, and the question accordingly arises: What is the deeper significance of this adaptation, and what is its fundamental source? Though this question has been attacked from several very different directions over the past twenty-five hundred years, the question itself has survived unchanged; and it needs to be answered with as much care and attention today as it did before Socrates and Plato put forward the first systematic theories of knowledge. If this entire development is now viewed as a single continuous argument, linking Thales and Anaximander to Warren McCulloch and Jean Piaget, it can conveniently be divided into three successive phases. During the first phase, the key terms were *order* and *cosmos*; during the second phase, *harmony* and *design*; and during the third, current phase—as Campbell and I would agree—they are *evolution* and *adaptation*.

To begin at the beginning: All the chief philosophical schools of classical antiquity, aside from that of Epicurus and his followers, believed in a basically unchanging cosmic order, which had either existed in its present form forever or else had come into existence at some definite moment in the past with all its characteristics already in their present form. This "order" represented an objective feature of the world, independent of human experience and interpretations; yet the human intellect seemingly possessed the insight required to grasp its principles. If that was so—that is, if the order of nature was not opaque to the human reason, it was because, in the ancients' view, human beings formed part of a single comprehensive *cosmos* along with physical and biological nature. One way or another, a single rationality or *logos* embraced the entire world, humanity and nature alike. Ultimately, the Stoics built up a picture of the world that integrated the order of human affairs (the *polis*) with the order of natural phenomena (the *cosmos*) to form a single unified system, which they called *cosmopolis* —the concept that survives today as a pale reflection of its former self in the idiomatic word *cosmopolitan.* Either way, the ancients found nothing particularly surprising about the intelligibility of nature: by their very nature, the human reason and the universal *logos* had a close affinity or analogy. Nature, society, and the human mind were all of them (so to say) alternative expressions of a single more comprehensive plan, or order (Sambursky, 1956, 1959).

After the theological debates of the Middle Ages, the secular argument about human understanding revived in the years around A.D. 1600; but this time the underlying assumptions of the discussion had basically changed. The self-styled "new philosophers" of the seventeenth century (Newton and Leibniz, quite as much as Locke and Descartes) saw human mentality and physical nature as belonging to distinct realms. To some extent, this separation was carried over from the theology of medieval Christendom; thus, the philosophical duality of mind and matter in seventeenth-century science perpetuated the preexisting theological distinction between soul and body. (As a result, those few philosophers—like Spinoza and Hobbes—who hoped for a more unified, monistic theory were criticized as irreligious.) To some extent, however, it reflected also some genuine differences that marked off the rules and principles of mental and intellectual life from the novel, mechanistic "laws of nature" by which seventeenth-century scientists were beginning to explain such natural phenomena as the movements of the planets around the sun. The rational operations of the human mind, seemingly, had nothing in common with the causal workings of the physical world.

In any event, the separation of mind from matter carried with it a parallel separation of humanity from nature, and this immediately gave the problems of epistemology a new urgency. If the human mind were entirely distinct from physical nature, how could the laws of the physical world be humanly intelligible at all? During the seventeenth and early eighteenth centuries, most of the plausible answers to that question shared the same general form and rested on the assumption that mind and matter alike—both the rational world of mental operations and the causal world of physical phenomena—were themselves dependent, in turn, on God the Creator. If a harmony was to be observed between the rational world of mind and the causal world of matter, that harmony was presumably a consequence of God's design. The Creator has fashioned the physical world, and its laws were an expression of His rationality; but He had also created the human mind, and in such a way that the human intellect—for all its uniqueness—was, thanks to Divine Providence, preadapted to make sense of the physical world. Matter and mind, *physis* and *logos,* might be fundamentally different in kind, but the beneficence and rationality of God's design ensured their harmonious cooperation.

So, when Galileo, Descartes, and Newton sketched the ground plan and laid out the foundations for natural science as we know it, their construction initially went on within a larger cosmological scaffolding. The world as they saw it—the world view within which their scientific theories were at first confined—was split into two separate realms. On the one hand, there was mind, the sphere of reason and freedom; on the other hand, there was matter, the sphere of causality and necessity; and these two realms were united together into a single harmonious whole, only by the deliberate design of the Creator.

We find ourselves in a very different situation today. As so often occurs with large building projects, the house of science quickly outgrew its original ground plan and burst through its scaffolding. The "new philosophers" generally took it for granted that the basic order and structure of nature had been unchanged since the creation, and also that its entire history was encompassed within a biblical time scale of some six thousand years. These two beliefs buttressed one another. Until a historical time scale millions, or even thousands of millions, of years in length could be taken seriously, there was no reason to call the stability of the natural order into question (Toulmin and Goodfield, 1965). Like the inhabitants of the Auvergne in central France, who lived for thousands of years among the cones and lava streams of long-dead volcanoes without ever seeing them for what they were, most people before A.D. 1800 were blind to surviving evidence of the true antiquity of the Earth, even when it lay plainly to hand. That part of the seventeenth-century scaffolding has been entirely discarded. Scientists now view the entirety of nature in historical terms, as having been subject to radical transformations over a time scale of several thousand millions of years, during which both the surface of our Earth and the living creatures inhabiting it have mostly been quite other than they are today.

Meanwhile, many other changes within natural science have helped to undermine the seventeenth-century barriers separating humanity from nature. The rise of brain physiology and comparative psychology during the last 100 or 150 years, together with the recognition of our evolutionary ancestry, has made any such absolute separation implausible; most recently, the demonstration that other animal species (notably, chimpanzees) can be taught to use human sign language—even crudely—has put in serious doubt the uniqueness of human mentality. As a result, the earlier mind-matter dichotomy—although it has by no means disappeared from our thinking—no longer has the central significance for the scientific world picture that it had three hundred years ago. Scientifically speaking, indeed, we are already halfway back toward a more "classical" view of the world, in which humanity and nature will be reintegrated into a single common order.

Given the place that humanity occupies within this new, historical vision of nature, the problem of human understanding is having to be rethought along with the rest of our scientific world view. Kant, Hegel, and Marx took some indispensable first steps, but the larger task (as we shall see) is still incomplete. Kant was a pivotal figure, but he still looked back quite as much as forward. He saw clearly enough that the seventeenth century's split vision of the world had created insoluble problems for philosophy, and he made gallant attempts to reformulate his agenda in terms that might escape those problems; but he lived too early to master the complexities of history and evolution, and he never took the depth and significance of historical change seriously enough. Indeed, Kant's own critical method for demonstrating the supposed necessity of our basic forms of thought and perception (the so-called "synthetic a priori") obliged him to treat those forms as cultural-historical universals, and so prevented him from entertaining the possibility that they might have had a significant historical development. To have done so would, in his eyes, have risked adulterating philosophy with "mere anthropology" and confounding the a priori with the empirical.

For his part, Hegel was ready to bring history into the picture; yet he still preserved the older attitude toward the division between human and natural affairs. To him, the word *history* meant the dialectical and progressive "unfolding" associated with the foreordained "life cycle" of human history, while nature went its quite separate way, according to its own cyclical, mechanical, and repetitive laws. Marx, in his turn, attempted to overcome Hegel's last division between humanity and nature; but his conception of cosmic history remained, in its own way, as "providentialist" as that of Hegel. Assuming that Friedrich Engels' *Dialectics of Nature* is a trustworthy guide to Marx's views also, the final outcome of historical change was—for Marx as much as for Hegel—the fulfillment of all previous history (Toulmin, 1972). For Marx as much as for any medieval theologian, that is to say, the entire course of time and history would reach their goal and justification at the Last Day. Although (by an ironical choice) Marx chose to dedicate *Kapital* to Charles Darwin, the Marxian unification of nature and history thus remained, in effect, one more in the long line of Judeo-Christian heresies. So we need not be surprised that in the 1960s the followers of Teilhard de Chardin found common cause with the French neo-Marxists. Both groups shared the same vision of "the last things" and saw the entire purpose of the universe as lying at the End of Time. Such a view can carry real weight only for those who have not yet fully completed the intellectual transition involved in passing from the second to the third phase in the historical development of epistemology.

Significance of Campbell's Evolutionary Epistemology

Suppose that we set earlier visions of cosmic order and divine design aside entirely and attempt to construct a full-scale historical world view around the idea of evolution. How, in that case, should we reformulate the central problems about human understanding?

What leading features must such an evolutionary epistemology possess, and what challenges will it pose both to philosophy and to natural science? Those are problems about which Donald Campbell has helped teach us to think clearly and constructively for the first time (Campbell, 1965, 1970, 1974).

Let us go back to the beginning. We face today the same old questions that philosophy has always faced:

> *What are Humans, that they can understand the World?*
> *And what is the World, that Humans can understand it?*

Or, alternatively:

> *How is it that human thought and practice are adapted*
> *to the tasks of dealing with the world? And what is*
> *the nature or measure of that adaptation?*

But today we can make a fresh demand on any answer to those questions before accepting it as satisfactory or convincing. Within a truly historical view of the world, the question "How is it that human thought and practice are well adapted?" will mean—quite simply—"How did, and do, human thought and practice *become* adapted?" Any harmony existing between the human mind and the world need no longer be seen as the direct product of divine creation, any more than the camouflage of the stick insect or the long neck of the giraffe is regarded that way. Instead, such harmony must represent the outcome of intelligible historical changes —in a general enough sense, the outcome of "evolution." So the fundamental phenomenon—the "adaptedness" of our modes of life and thought—has now to be accounted for in terms of its historical origins—the processes by which those modes *became* adapted. And the measures we place on this "adaptedness" must make sense in terms of the processes that are actually available for making our human procedures more "efficacious." It is pointless for philosophers to judge the products of our intellectual or practical procedures by standards that could not conceivably be achieved in any scientifically and historically intelligible manner.

If we are to reintegrate humanity and nature within our contemporary view of the world, we must reintegrate our ideas about the history and development of humanity with our ideas about the history and development of the natural world. This requirement creates problems for us on two levels, collective and individual. Collectively speaking, we must face questions about the social, cultural, and intellectual changes that are responsible for the historical evolution of our various modes of life and thought—our institutions, our concepts, and our other practical procedures. (These questions correspond to questions about *phylogeny* in evolutionary biology.) Individually speaking, we must face questions about the manner in which maturation and experience, socialization and enculturation shape the young child's capacities for rational thought and action—how the child comes to participate in his native society and culture. (These questions correspond to the questions about *ontogeny* in developmental biology.) And when Jean Piaget says that intelligence is "a biological adaptation," he challenges us all—himself included—to explain what sort of thing such an *adaptation* can be.

Adaptive Change in Evolution of Cultures

The first generation of historically minded philosophers (notably, Hegel) considered the differences between the processes of human history and the phenomena of natural change so profound that, in their view, nothing useful could be learned by comparing the two. But Hegel lived in the heyday of the mechanistic Newtonian physics. He was dead before Darwin provided his historical and ecological analysis of organic evolution, to say nothing of the subsequent revolutions in twentieth-century physics—the two scientific steps that, since 1850, have largely transformed our scientific picture of nature. So we start today from quite another point than Hegel did, and we can usefully learn from considering the likenesses and differences between the history of human activities and that of nature. Specifically, as Campbell has convincingly demonstrated, the basic patterns of *adaptive change* characteristic of Darwinian theory have clear parallels in human affairs.

The process of "populational adaptation"—the selective perpetuation of advantageous novel forms from among a pool of freely generated variants—was, indeed, already familiar within at least one human context *before* Darwin introduced it in evolutionary biology 120 years ago. The example I have in mind comes from historical linguistics. Suppose that we ask how the different Indo-European languages (for example, the various Romance languages, from Romanian to Portuguese) became differentiated during the years between the fifth and fifteenth centuries. The explanation of this process that carries conviction today closely resembles the accounts that we are given in evolutionary biology when we ask similar questions about the differentiation of organic species. In brief, novel words and variant speech forms are always coming into use in any language community; they are coined and employed sometimes consciously but often without any deliberate design. In each case, only a few of these variant forms will win an established or dominant position within the language concerned. Under certain special circumstances (such as isolation and local interaction), one or another entire way of speaking (originally, only one dialect among others) may achieve an autonomous existence on its own, so becoming differentiated from its forerunners and cognates and establishing itself as an independent language in its own right. In such circumstances, the basic question about successful adaptation becomes: "What was it about the specific situation within which this particular dialect appeared that explains why this form outlasted the other available forms within that specific situation?"

From an evolutionary point of view, the harmony between the human mind and the situations it has to deal with no longer requires us to presuppose some long-term destination or cosmic goal, toward which any "successful" historical change must be headed. Rather, the changes that become established at any particular place and time are those that best answer the specific demands appearing at that particular place and time. This approach requires us to view historical change in ecological rather than eschatological terms. The novelties that survive here and now (whether novel organic forms or linguistic usages, social institutions or scientific concepts)—the variants that become established as an outcome of what Campbell calls "editing" (that is, generalized "selection")—are those that make a specific contribution toward meeting demands that actually arise here and now (Campbell, 1974).

Dual Character of Evolutionary Change

If posed in this generalized form, the problem of *adaptation* falls into two parts. First, we have to understand the sources of *adaptability* present in any situation, whether these involve the ap-

pearance of novel variant forms of social greeting, political processes, religious liturgy, scientific explanation, or whatever. Second, we have to understand the measures and indices of *adaptedness,* the manner in which certain of these variant forms become
dominant, and the tests that they have to pass in the course of
doing so, whether their success is that appropriate to new legal
procedures or medical treatments, speech forms or refrigerator designs, scientific hypotheses, or whatever. The spread and acceptance
of new patterns of family structure, new styles of dress, new modes
of bureaucratic organization, or new methods for teaching natural
science are *dual* processes, quite as much as the origin of species
by variation *and* natural selection. All of these examples of spread
and acceptance involve both the tentative appearance of innovations within particular localities (communes, experimental schools,
trade unions, scientific societies, or whatever) and also their subsequent spread and acceptance as providing improved solutions to
outstanding problems (see Thompson, 1917).

At an abstract, mathematical extreme, this evolutionary approach to the problems of historical change shades over into a generalized form of economic analysis. (The etymological connections
between the terms *ecology* and *economics* are no accident.) Yet
we must also take care to ask ourselves whether there may not also
be *qualitative* differences between the "evolutionary" processes
characteristic of different spheres. In each case, that is, we must
ask how specific "demands" arise within particular historical situations, what kinds of situations create "opportunities" and "needs,"
how the creative fertility of natural phenomena or human inventiveness turns up novel "variants" in any kind of situation, how
certain variants prove that they meet these demands (how well
"adapted" they are to the demands of this particular situation),
and how they subsequently become "dominant" in the sphere of
activities concerned, unless prevented from doing so by rigid con-
servatisms built into the situation.

Consider, for example, the issues that have preoccupied
philosophers of science over the last fifty or sixty years. For much
of that period, the philosopher's goal was to account for the
"good" and the "bad" of science—to explain what makes novel
scientific concepts and hypotheses worthwhile and even justifies
us in incorporating them into the current "body of knowledge"—

in terms drawn from formal logic. Surely (it was thought) inductive logicians could develop some "confirmation theory" or other formal algorithm for *measuring* the virtues and defects of new scientific ideas (see Hempel, 1959; Suppe, 1974). During the last fifteen years, it has become increasingly clear that this was only a pipe dream. The virtues of scientific theories are not *formal* but *functional,* and we cannot avoid looking directly at the actual historical processes by which novel concepts and theories in fact win a place in science (see Kisiel, 1974).

This process, too, turns out to be a dual process of variation and selective perpetuation, just as much as the processes of change that affect organic populations and linguistic forms. Novel concepts first enter a science tentatively, so forming a pool or "population" of competing variants. Subsequently, they are selected among as meeting the specific theoretical demands of the particular scientific situation involved, and so as being "better or worse adapted" to the current problem situation than their competitors. The evolution over time of scientific ideas—the *phylogeny* of science, so to say—is thus an "evolutionary" process, just as much as the evolution over time of organic species, languages, and other "historical entities" (Toulmin, 1972, ch. 5).

Evolution of Social and Political Forms

The same pattern of analysis is also relevant to the theory of social and political change. For a long time, social and political philosophers found it hard to account for the "adaptedness" of human institutions except by interpreting them as the outcome of *foresight.* On the one hand, the existence of successful institutions might be a tribute to the ingenuity of human beings themselves, and the foresight in question was then human foresight. So understood, the adaptedness of social and political forms was presumably the result of human beings' calculating the prospective consequences of various alternatives and consciously choosing the "best-adapted" option. But human foresight and calculation certainly do not account at all convincingly for any but small-scale, short-term institutional changes. The larger-scale, longer-term history of social and political institutions was beyond human calculation or explanation of how collective social and political life originated in the first place. From A.D. 1700 on, therefore, political and social philosophers (Vico is the classic example) looked elsewhere for the "foresight" required to account for these larger-scale features of the *polis*. The adaptedness of human institutions was still seen as the outcome of foresight, but that foresight was now the infinite, unconditioned foresight of Divine Providence. (The Latin term *providence,* after all, means "fore-sightedness.") The operation of Divine Providence, in this view, ensured that human beings would eventually succeed in developing effective institutions, if only unwittingly, because the World had been fashioned in such a way that this development of effective institutions was guaranteed in advance.

For twentieth-century social scientists, of course, such appeals to Divine Providence appear hardly satisfactory, and social theorists have looked around for alternative kinds of "adaptive processes" to explain the structure and efficacy of human institutions. The theoretical models that have won most attention from sociologists hitherto are analogies with physiology. That is true both of Émile Durkheim, who was directly influenced by the great French physiologist Claude Bernard (Hirst, 1975), and of the American sociologist Talcott Parsons, who was also influenced by Claude Bernard's ideas (Parsons, 1971), though more indirectly through L. J. Henderson. Both men have analyzed the working of society as involving the operation of "social systems," which maintain "healthy" or "normal" social equilibria in the same way that the digestive system maintains "healthy" or "normal" physiological equilibria in the human body.

Still, such physiological analogies between society and the organism are scarcely very satisfactory. In the late twentieth century, indeed, they are even more questionable than they were when Plato first made systematic use of them in the *Republic,* more than two thousand years ago. For, as Claude Bernard himself knew very well, explanations in terms of physiological systems may be able to account for the maintenance of equilibria in the working of the body—for example, the constant body temperatures of warm-blooded animals—but they cannot explain either the initial development of these physiological systems themselves or the subsequent appearance of successful new forms and/or features. Inevitably, a theory of "homeostasis" is a theory of *stasis,* not of *change*; and a theory of social or political homeostasis, such as Durkheim and Parsons put forward, is, equally inevitably, a *conservative* theory, which can explain only how the institutions of a society can defend themselves *in their present form*.

Once again, however, there is an alternative way ahead. Within general biology, physiology can no longer stand on its own; its necessary historical complement is a theory of organic evolution. The existence of successful, well-adapted homeostatic mechanisms is no longer self-explanatory. Instead, it is just one more phenomenon to be accounted for, in evolutionary terms, as the outcome of successive adaptations to former conditions of life. In the social realm, likewise, a theory of social and political "structures" calls equally for a theory of social and political evolution, as its natural complement. We need to ask ourselves how institutional (or other social) variants make their appearance in the first place; how they show their merits as solutions to outstanding social problems; and how they succeed—given favorable conditions —in spreading and establishing themselves more widely. That is, we must focus our attention less on the internal organization of self-maintaining institutions or procedures and more on the historical processes by which innovations demonstrate their "adaptedness" to the current demands and some come through the political "editing" process with flying colors.

Varieties of Adaptation

The "adaptedness" of human thought and practice, accordingly, no longer presents itself to us as rooted in fixed cosmic order (as in antiquity) or as a product of God's providential design (as in the philosophy of the Renaissance), but rather as the historical outcome of evolution and ecology. Those of us who argue in these terms are not placing naive reliance on analogies with "organic evolution" in biology, in the way that was done, abortively, by such earlier philosophers as Ernst Mach and John Dewey. Rather, we are arguing for a more comprehensive conception of "historical development" as an *evolutionary* process, in which efficacious (or "well-adapted") forms of life and modes of thought are selectively perpetuated from among the whole population of variants in circulation within any given milieu.

Such a comprehensive conception of evolutionary change, however, must take into account both the full variety of respects in which different forms demonstrate their "adaptedness" and also the full variety of processes and mechanisms capable of generating such "adaptation." When the so-called sociobiologists argue that the adaptedness of human social behavior (for example, our capacity for altruism) is the outcome of genetic factors, they adopt too narrow a view both of what constitutes "adaptedness" in human social life and of the processes by which the adaptedness of social behavior is maintained. They insist, first, on using the term *adaptation* in its narrowest biological sense, which is defined entirely in terms of the survival, or differential fertility, of populations. Our social behavior has these adaptive features (they imply) because all the rival populations of early hominids that lacked them are by now extinct. And they insist, also, on attributing the transmission of these features to our "genes" (see Wilson, 1975; Midgley, 1978).

On both accounts, the sociobiologists' position is oversimplified. We know so little about the manner in which human genes find eventual expression in adult behavior (let alone adult social behavior) that we are in no position to *insist on* anything. Imagining that a propensity to altruism can itself be encoded in the DNA of our cell nuclei is entirely premature; plenty of other factors (including social and cultural factors, such as language) deserve to be taken just as seriously as the genes themselves. Furthermore, it is not the risk of extinction alone against which "well-adapted" social and cultural forms defend us. Adaptation is important to human societies in many other respects besides sheer survival. Certainly, a human population that was incapable of surviving at all would be gravely "ill adapted" to its conditions of life. Once that extreme threat is lifted, however, the "adaptedness" of our modes of thought and practice can, surely, be measured quite as legitimately by the manner in which they protect us against want and need and ignorance of many other kinds.

The evolution of human culture and society has, in fact, generated individual activities and collective enterprises of many kinds: arts and sciences, modes of family organization and social institutions, orchestras and theaters, industries and universities. In a broad sense, each of these activities and enterprises has had its own evolutionary history, within which the dual process of diversification and selective perpetuation has continuously helped to re-fine and "adapt" our thought and conduct to our conditions of life. (We get rid of our own ineffective modes of action by selec-tion or superannuation, instead of killing off entire rival popula-tions.) This mental "adaptation" is never perfect, any more than the physical "adaptation" of bodily structures is ever perfect. Still, in most respects, known processes and mechanisms are capable of generating and maintaining an "adaptation" that is good enough for existing practical purposes. (In highly conservative societies, of course, these adaptive processes may be given no opportunity to operate, but that is another matter.) At any rate, the actual degree of harmony between mind and the world—between human modes of life and thought and the world that is their object—need no longer be seen as a metaphysical mystery. It is testimony, simply, to the fact that "adaptive processes" have been permitted to oper-ate often enough, and for long enough, to bring us to the ade-quately but incompletely "adapted" position in which we find ourselves today.

Two qualifications are needed in conclusion. First, the *goals* of adaptation vary substantially from one context to another and from one enterprise to another. Different arts and sciences, differ-ent cultural enterprises and modes of social organization are adap-tive in different respects. In the continuing competition between rival scientific ideas, the merits that are selected for are, manifestly, quite other than those that are selected for in the refinement of techniques in the fine arts or the development of more effective procedures of public administration (Toulmin, 1972, ch. 6). In the course of human adaptation, all of these different "evolutionary processes" go along in parallel, and on occasion they may well rub up against, and even conflict with, one another. Some of the most active debates in professional ethics today, indeed, arise at such points of conflict (Toulmin, Rieke, and Janik, 1979, ch. 17).

Second, the *processes and mechanisms* of human adaptation that operate within all these activities and enterprises are of several different kinds. (1) To some limited extent, of course, such adap-tation can claim to be the outcome of conscious thought and fore-sight; we can to some limited extent deliberate, calculate, and act in a consciously and actively "adaptive" manner, thereby making our procedures more "apt." But this is only one way, and probably not the commonest way, in which "adaptation" comes about. (2) In addition, the generation of innovations and the selective perpetuation of favored variants go on continuously, without either the need for or even the possibility of conscious foresight and calculation. As a result, human adaptation often takes place by the operation of "populational" processes of a typical "evolutionary" kind. (3) Once advantageous new possibilities have been identified, they must be given some practical embodiment if they are to make themselves effective and survive for longer than a single generation. The favored innovations must be somehow "represented" or "encoded" in a form that makes them available for transmission to subsequent generations, as new institutions, new concepts, or other kinds of self-sustaining procedures. Underlying these modes of transmission, in turn, may be factors of several kinds, ranging from the macromolecular structure of the genes, by way of the subtleties and complexities of human language, to the simple and easily mastered customs and learning procedures by which children enroll themselves in a culture.

These varied mechanisms and processes should not, of course, be regarded as rival ways of achieving the same result. Rather, they are complementary; each has its own proper part to play in the larger and more complex processes of human adaptation and cultural evolution. About that overall process no brief or simple account can be given. This is an area of inquiry where the chief source of risk is our own temptation to oversimplify issues. The organic theory of the state, Leibniz's "preestablished harmony," the exaggerations of sociobiology and behaviorism, Piaget's attempts to transform *autorégulation* into a theory of learning, Chomsky's "innate" language capacity—all illustrate once again our tendency to "overreach the boundaries of our own rationality" and to be charmed by simple, tidy intellectual solutions to problems that in fact defy simplification. Our only protection is a firm resolve to be vigilant in respecting the *minimum* complexity that is needed in our theories if they are to do justice to an *intrinsically* complex set of interlocking historical processes.

Finally, in talking about the "processes" and "mechanisms" involved in human evolution, we are by no means offering a *deterministic* account of social and cultural history: on the contrary. Though political philosophers and social scientists may often have exaggerated the role of conscious human foresight and calculation in social and political life, human choice and reasoning do, of course, have a very significant part to play, both there and in many of the other processes of adaptation. In fact, the whole story of human adaptation, as presented from the standpoint of an evolutionary epistemology, is a story not about the effects of mechanistic causes but rather about human actions that are performed "for reasons"—all those human selections and preferences, priorities and choices that determine which procedural innovations shall survive and be perpetuated. Still, the realm of reasons and the realm of causes can no longer, for us, be *separate* realms, like the separate Cartesian "substances" of mind and matter. Rather, human life intertwines reasons and causes in as complex a manner as it does all other aspects of the "mental" and the "material," the "human" and the "natural."

If we are to heal the wounds created by the Cartesian split, and reintegrate humanity with nature, it will follow as a result that human actions, too, are performed *within* the world of natural processes; and the older philosophical barriers separating rationality from causality will have to be dismantled along with all the others. A world of nature into which humanity has been reintegrated will no longer be an impersonal, mechanistic world. Rather, it will be a world within which the human reason itself is a causally efficacious agency, within which—as the ancients recognized but the philosophers of the seventeenth century denied—we have the elbowroom that we require to exercise the autonomy that is a chief mark of our humanity.

References

Campbell, D. T. "Variation and Selective Retention in Socio-cultural Evolution." In H. R. Barringer, G. I. Blanksten, and R. W. Mack (Eds.), *Social Change in Developing Areas*. Cambridge, Mass.: Schenkman, 1965.

Campbell, D. T. "Natural Selection as an Epistemological Model." In R. Naroll and R. Cohen (Eds.), *A Handbook of Method in Cultural Anthropology*. New York: Natural History Press/Doubleday, 1970.

Campbell, D. T. "Evolutionary Epistemology." In P. A. Schilpp

(Ed.), *The Philosophy of Karl Popper*. La Salle, Ill.: Open Court, 1974.

Hempel, C. G. *Aspects of Scientific Explanation*. New York: Free Press, 1959.

Hirst, P. Q. *Durkheim, Bernard and Epistemology*. London: Routledge and Kegan Paul, 1975.

Kisiel, T. "New Philosophies of Science in the U.S.A." *Zeitschrift für allgemeine Wissenschaftstheorie [Journal of Universal Scientific Theory]*, 1974, *5*, 138-191.

McCulloch, W. S. *Embodiments of Mind*. Cambridge, Mass.: M.I.T. Press, 1965.

Midgley, M. *Beast and Man*. Ithaca, N.Y.: Cornell University Press, 1978.

Parsons, T. "On Building Social Systems Theory: A Personal History." *Daedalus*, 1971, *99* (4), 828-831.

Sambursky, S. *The Physical World of the Greek*. London: Routledge and Kegan Paul, 1956.

Sambursky, S. *Physics of the Stoics*. London: Routledge and Kegan Paul, 1959.

Schilpp, P. A. (Ed.). *Albert Einstein: Philosopher-Scientist*. La Salle, Ill.: Open Court, 1949.

Suppe, F. *The Structure of Scientific Theories*. Urbana: University of Illinois Press, 1974.

Thompson, D. W. *On Growth and Form*. Cambridge, England: Cambridge University Press, 1917.

Toulmin, S. *Human Understanding*. Princeton, N.J.: Princeton University Press, 1972.

Toulmin, S., and Goodfield, J. *The Discovery of Time*. New York: Harper & Row, 1965.

Toulmin, S., Rieke, R., and Janik, A. S. *An Introduction to Reasoning*. New York: Macmillan, 1979.

Wilson, E. O. *Sociobiology*. Cambridge, Mass.: Harvard University Press, 1975.

3   *Robert J. Richards*

Natural Selection and Other Models in the Historiography of Science

The writing of science history may itself be regarded as a scientific enterprise, involving evidence, hypotheses, theories, and models. I wish to investigate seven historiographic models and their variants. While these undoubtedly do not exhaust the store available to imaginative historians of science, they nonetheless represent, I believe, those that have played the significant roles in the development of the discipline, either as models that have long functioned in historical writing or as models more recently proposed in meta-historical works.

Note: Segments of this chapter were presented at the annual meeting of the Society for the Social Studies of Science (Indiana University, November 1978) and at the monthly gathering of the Chicago Group for the History

The several models discussed in the first part of this chapter are idealizations of major assumptions that have guided the construction of histories of science since the Renaissance. They thus embody directive ideas concerning the character of science, its advance, and the nature of scientific knowing. Since the models so abstracted are idealizations, they do not always precisely reflect the structures of particular written histories. Yet they can serve to elucidate those controlling assumptions that have shaped our understanding of science and its history.

The second part of this chapter will attend to that class of models that appears the most powerful for capturing the actual movement of science: evolutionary models. I will briefly examine two instances of this class, the models of Popper and Toulmin; consider their deficiencies; and then develop a natural selection variant, which, I believe, escapes their liabilities. The resources of this variant will be tested against Lakatos's model of scientific research programs, perhaps its strongest competitor. In advocating a natural selection model of scientific evolution, I will make ample use of Donald Campbell's hypotheses concerning the mechanisms of creative thinking in science.

Five Models in the Historiography of Science

I have a fourfold purpose in offering the following dissections of historiographic models: first, to show that historians of science have typically constructed their narratives in light of distinct sets of assumptions; second, to display the major forms these assumptions have taken during the development of the discipline; third, to explore the advantages and disadvantages of the models; and, finally, to provide comparative standards by which to judge the virtues of the natural selection model proposed in the second part of the paper.

of the Social Sciences (University of Chicago, April 1980). I owe thanks to Len Berk, Richard Blackwell, John Cornell, Lindley Darden, Sophie Haroutunian, Gary Kahn, Marcy Lawton, Jacques Quen, and William Wimsatt for reading earlier versions of this paper and offering many helpful suggestions. Allen Debus introduced me to the study of the history of the historiography of science, and his advice has always been a guide. This study was supported by NIH Grant PHS 5 S07 RR-07029-12.

The Static Model. Many historians and scientists of the late Renaissance and early Enlightenment shared R. Bostocke's conviction, as expressed in *The Difference Betwene the Auncient Phisicke and the Latter Phisicke* ([1585] 1962), that God infused certain men (such as Adam or Moses) with scientific knowledge, which was passed on to successive generations intact. Even Newton employed a static model and contended that his *Principia* was a recovery of wisdom known to the ancients (McGuire and Rattansi, 1966).

Use of a static model in history of science accorded with the Renaissance presumption that ancient thought embodied the highest standards of knowledge and style. But another consideration also promoted the acceptance of the model. This may be found in Olaus Borrichius's *De Ortu et Progressu Chemiae Dissertatio* ([1668] 1702), a standard textbook of the history of chemistry during the late seventeenth and early eighteenth centuries. In accord with the tradition, Borrichius credited Tubalcain, a descendant of Cain and a figure he identified with Vulcan, as having received from God this divine knowledge of chemistry. The Cartesian argument he used to fortify his baroque sentiments displays an important justification for use of the static model. He reasoned that "the priests of Tubalcain would have been unable to discover, shape, and form the metals of iron and copper except that their *ratio* was prior known; that the natures of these minerals might be investigated and that they might be cooked, purged, and segregated could not occur except that knowledge of this were divinely inborn. Once this knowledge is had, however, these techniques follow for any skillful people" ([1668] 1702, p. 1). Borrichius, tinctured with the Cartesian spirit, was assured that chemical knowledge and science in general must be innate, at least in their fundamentals; for unilluminated natural induction could never of itself lead to such scientific achievements as his age witnessed. And if the essential features of a science had this kind of origin, then from its first discoverer it could only be passed on (or rediscovered again by succeeding generations). This model of the origin and course of science can be detected in transmogrified form in Thomas Kuhn's Gestalt model (described below), which assumes that in a moment of insight the transformed vision of an inspired genius may establish the framework and fundamental premises of a science, the details of which may be left to the normal plodding of his disciples.

The Growth Model. After the late Renaissance, historians of science increasingly discarded the static model, replacing it with one still in use today. By the eighteenth century, the growth model clearly prevailed, as Freind's *The History of Physick from the Time of Galen to the Beginning of the Sixteenth Century* (1725) and Watson's essay "On the Rise and Progress of Chemistry" (1793) testify. Indeed, Freind's history may be read as a sustained argument against the Renaissance tendency to overprize the ancients and to assume that the essential concepts and principles of science lay with them, only to be ornamented by succeeding generations. Freind proposed to show that the knowledge of medicine did not begin and end with Hippocrates and Galen. Instead, as a careful study of the writings of subsequent physicians demonstrated, "Physick was still making progress 'till the Year 600" (1725, vol. 1, p. 298). (He charted the gradual advance of the science since the beginning of the Medieval period in the second volume of his history.) As a consequence of the particular model he had chosen, that of gradual, cumulative growth, Freind could recommend reading in the history of medicine as "the surest way to fit a man for the Practice of this Art" (1725, vol. 1, p. 9). This is a piece of advice annulled by historians advocating other models, as we will see.

Watson's essay highlights an assumption of the growth model that was to have particular importance in later controversies; namely, that science in its conceptual development is relatively isolated from other human occupations, even from the technology that fostered it. Watson was assured of this independence, since he understood science to have a rational integrity not found in the less "liberal and philosophical" pursuits (Watson, 1793, p. 30). This presumption also bound together the various parts of that monument to the growth model, the *Encyclopédie* of Diderot and D'Alembert. In the *Discours Préliminaire* to the *Encyclopédie,* D'Alembert (1758) projected the prescriptions of the growth model back even to prerecorded thought, suggesting that primitive sensory awareness might gradually have established the foundational principles of scientific advance.

In the nineteenth century, geologists came to recognize the power of the uniformitarian principle that directed historians of the earth to interpret apparent cataclysmic upheaval in terms of slow cumulative change. William Whewell, although persuaded of catastrophism in geology, yet applied the uniformitarian perspective to his *History of the Inductive Sciences* (1837). Thus he discomfits those historiographers advancing the strong psychosocial model, which supposes harmony in a scientist's cognitive style. Whewall rejected the idea that discontinuous intellectual upheaval marked the development of the various sciences: "On the contrary, they consist in a long-continued advance; a series of changes; a repeated progress from one principle to another, different and often apparently contradictory" (1837, vol. 1, p. 9). Did the progress of science occur by contradictory ideas replacing one another, there would not, of course, be organic growth, but revolutionary saltation. That is why Whewell urged his reader to remember that the contradictions were only apparent: "The principles which constituted the triumph of the preceding stages of the science may appear to be subverted and ejected by the later discoveries, but in fact they are (so far as they were true) taken up in the subsequent doctrines and included in them. They continue to be an essential part of the science. The earlier truths are not expelled but absorbed, not contradicted but extended; and the history of each science, which may thus appear like a succession of revolutions, is, in reality, a series of developments" (1837, vol. 1, p. 10).

The central assumptions embodied in the growth model are compendiously present in the work of George Sarton, the doyen of historians of science in the middle of this century. His several observations on the nature of science therefore afford a convenient summary of the implications of the model. The primary note of the model is its affirmation of the unalterable and clearly discernible progress of science toward the fullness of truth, progress that can only be momentarily delayed by retarding forces: "The history of science . . . is an account of definite progress, the only progress clearly and unmistakably discernible in human evolution. Of course, this does not mean that scientific progress is never interrupted; there are moments of stagnation and even regression here or there; but the general sweep across the times and across the countries is progressive and measurable" (Sarton, 1962b, p. 21).

The steady advance of science, accomplished by the rationally exact methods of quantification and experimentation, and "its astounding consistency (in spite of occasional, partial, temporary contradictions due to our ignorance) prove at one and the same time the unity of knowledge and the unity of nature" (1962b, p. 15; see also 1962a, p. 31). Since the unity and continuity of knowledge, which are grounded in the unity and intelligibility of nature, are not, in Sarton's estimation, enjoyed in other human pursuits, these latter are unable conceptually to affect the course of science. Moreover, the clear evidence of history gives no support to attempts at the sociologizing of scientific knowledge. The internal progress of science, as its history reveals, has a force beyond the vicissitudes of men's passions and the subtle pressures of social life. To be sure, science does not grow in a social vacuum: men need food, they are called to war; money for equipment is required. But, as Sarton avows, the man of science remains ultimately untouched in his theoretical endeavor by the ideologies or conditions of society: "Nobody can completely control his spirit; he may be helped or inhibited, but his scientific ideas are not determined by social factors" (1962b, p. 13). Insofar as the history of science is independent of the cultural life of the larger community, it can serve as a standard of truth and error in those other domains: "The history of science describes man's exploration of the universe, his discovery of existing relations in time and space, his defense of whatever truth has been attained, his fight against errors and superstitions. Hence, it is full of lessons which one could not expect from political history, wherein human passions have introduced too much arbitrariness" (1962b, p. 21).

The Revolutionary Model. A brief examination of the history of the term *revolution* suggests that its application to scientific thought is not necessarily derived from analogies with political overthrow. The *Oxford English Dictionary* indicates that its use to describe dramatic changes in thought antedates by a considerable period its use to designate political upheavals. By the late eighteenth century, the term was widely employed to denote important transformations in the course of science. When Kant referred to particular "revolutionary" events in the history of science, he employed the word in the manner of contemporary historians of science: to describe a profound shift in thinking, after which there is relatively smooth scientific progress to the present time. For Kant, as well as for most recent historians using the model, revolution in a science is a one-time affair. In the preface to the second edition of the *Kritik der reinen Vernunft* (*Critique of Pure Reason*), Kant pictured the intellectual revolution undergone by the mathematical and physical sciences, before which we had no science proper and after which we had unimpeded advance into the modern period. Mathematics had to grope during the Egyptian era, but with the Greeks came the revolution that set it on its present course. Natural science had to wait a bit longer for its revolution, as Kant explained: "It took natural science much longer before it entered on to the road of science; for it is only about a century and a half since the proposal of the ingenious Bacon of Verulam partly fostered its discovery and, since some were already on its trail, partly gave encouragement. But this can be explained only as a suddenly occurring revolution in the mode of thought (*eine schnell vorgegangene Revolution der Denkart*)" (Kant, [1787] 1956, p. 23 [B xii]).

While the use of the term *revolution* to describe radical changes in thought is older than its use in the specifically political context, the political analogy is often implied and does seem justified. Political revolutionaries have particular enemies with whom they wage their ideological and bloody battles; the scientific revolutionaries of the sixteenth and seventeenth centuries also had their foes: Aristotle, Ptolemy, Galen, the Scholastics. Political revolutionaries have the aim of overturning an undesirable system and replacing it with one that will perdure and serve as a base for further progress; their scientific counterparts had the same goal. Significant political revolutions are not usually spontaneous; they have their doctrinal basis formed in the work of men who may be long dead before the revolution. Historiographers of scientific revolutions also acknowledge necessary foundations: the groundwork of modern physical science laid, for example, by the Merton school of mathematical physics or the Paduan Aristotelians of the early Renaissance. The historical significance of a political revolution lies more in the fruit of the new ideas and systems that the revolution inaugurates—fruit that may take time in maturing. Those writing of the scientific revolution fomented by Copernicus, Kepler, Galileo, Harvey, and Descartes construe the ideas of these scientists as establishing the foundations for thoroughly modern science, even though their specific conceptions may no longer be acceptable.

The most influential historian to employ the revolutionary model was Alexandre Koyré, whose views set out its essential features. To the history of science Koyré brought the philosopher's eye for metaphysical assumptions and the intellectual historian's concern for doctrinal context. For him the scientific revolution of the sixteenth and seventeenth centuries was an outward expression and consequence of a more fundamental turn of mind, of a "spiritual revolution" having two basic features. There was, first of all, the Platonically motivated dismissal of the qualitative space of Aristotle and the Scholastics and its replacement with abstract geometrical space; Galileo's contribution to the scientific revolution was precisely his insistence on mathematical reasoning rather than on sense experience as the foundation for scientific success (Koyré, 1968). But this alteration in thought about the universe was only a phase, though the crucial one, of a more pervasive revolution, one that brought about "the destruction of the Cosmos, that is, the disappearance, from philosophically and scientifically valid concepts, of the conception of the world as a finite, closed, and hierarchically ordered whole . . . and its replacement by an indefinite and even infinite universe which is bound together by the identity of its fundamental components on the same level of being. This, in turn, implies the discarding by scientific thought of all the considerations based upon value concepts, such as perfection, harmony, meaning, and aim, and finally the utter devalorization of being, the divorce of the world of value and the world of facts" (Koyré, 1957, p. 2).

Insofar as the revolution had banished from explanatory rule such concepts as "perfection, harmony, meaning, and aim," conceptual historians of science under the banner of Koyré have felt justified in dismissing from serious consideration Neoplatonic mysteries and Paracelsian occultism, which were contemporary with what has become known as the new science. Perhaps paradoxically, Koyré himself was quite willing to consider the influence of such spirits on the new science, though he rejected the notion that the influence was specifically scientific (1957, p. 54). (The role of occult influences on the development of science is highly controverted. The dispute may be followed in the discussions by Yates, 1964, 1968; Rattansi, 1973; and Hesse, 1973. The various parties are brought together in Stuewer, 1970.)

Models in historiography, as well as in science, provide more than merely a heuristic for investigation. They focus attention, exclude possibilities, and reveal connections that are not manifest. Whether as covert assumptions or consciously accepted devices, models intervene (inevitably, I believe) between the historian and his subject. Yet the sensitive historian is not often led far astray by the odd magnifications his model might produce; his distorted perspective can be corrected by the hard feel of facts that he continues to accumulate. Moreover, the use of a model and the application of its embedded hypotheses require an artful intelligence, of the kind that individualizes the shape of the crafted product. Thus, the historians who generally employ the conceptual version of the revolutionary model may offer different perspectives on the same issues. Crombie (1961, vol. 2, p. 125), Hall (1966, pp. 370-371), and Gillispie (1960, pp. 8-16), for example, in some contrast to Koyré, locate the revolution in scientific thought in the application of mathematics to mechanics and the resultant construction of formal mathematical systems for the construal of nature. Hall (1966, pp. 217-243) believes that the instruments and techniques developed by craftsmen have provided a stimulus and auxiliary to the new sciences; while Koyré virtually ignores the crafts, since the science of Galileo and Descartes "is made not by engineers or craftsmen, but by men who seldom built or made anything more real than a theory" (1968, p. 17). Hall (1966, p. 370) regards pre-seventeenth-century investigations of nature as essentially discontinuous with science after that period. Crombie (1953, 1961), who devotes considerable attention to the Medieval development of the foundations of modern science, believes that "a more accurate view of seventeenth-century science is to regard it as the second phase of an intellectual movement in the West that began when philosophers of the thirteenth century read and digested in Latin translation the great scientific authors of classical Greece and Islam" (1961, vol. 2, p. 110). Gillispie (1960, pp. 8-16), too, acknowledges the debt of Renaissance science to Greek mathematical rationalism.

Yet those who generally employ the revolutionary model agree—and this constitutes the essential feature of the model—that a revolution in thought, a decisive overthrow of distinctly ancient modes of conception, is necessary to set a discipline on the smooth course of modern science. Hall clearly highlights the core of the model. For him the Medieval period did have its quasi-science; and though that enterprise set the stage for the appearance of modern science, yet the mathematical methods of the latter were radically different from the methods of its predecessor: "Rational science, then, by whose methods alone the phenomena of nature may be rightly understood, and by whose application alone they may be controlled, is the creation of the seventeenth and eighteenth centuries" (Hall, 1966, p. xii). It is the method of rational science that guarantees its further progress—without fear of taking fundamentally wrong paths. Whatever revisions in science have come since the revolution are revisions in content only, not in structure (1966, p. xiii).

The Gestalt Model. In recent years ideas from particular currents within the social and psychological sciences have joined those springing from conceptual studies in the history of science, especially those studies whose epistemological channels run to neo-Kantianism. From this confluence has emerged a model that, for want of a better designation, may be called a Gestalt model. Among those most influential in employing this model in the history of science and in dealing with metahistorical problems are Norwood Russell Hanson and Thomas Kuhn.

Both Hanson (1970) and Kuhn (1970b) explicitly use devices drawn from Gestalt psychology and the psychology of perception. The Necker cube, the goblet-faces display, pictures of creatures looking alternately like birds or antelope, and similar puzzles illustrate for them the ways in which context, past experience, and familiar assumptions control our perceptual and conceptual experience of things. In the scientific domain, as construed by Hanson, it is the well-entrenched theory that determines the perception of facts: "Physical theories provide patterns within which data appear intelligible. They constitute a 'conceptual Gestalt.' A theory is not pieced together from observed phenomena; it is rather what makes it possible to observe phenomena as being of a certain sort, and as related to other phenomena" (Hanson, 1970, p. 90). Likewise in Kuhn's judgment: "Assimilating a new sort of fact demands more than additive adjustment of theory, and until that adjustment is completed—until the scientist has learned to see nature in a different way—the new fact is not quite a scientific fact at all" (Kuhn, 1970b, p. 53).

If facts and their organizing theories are mutually implicative and constitute a perceptual-conceptual whole—a paradigm, to use Kuhn's by now worn term—and if "the switch of Gestalt . . . is a useful elementary prototype for what occurs in full-scale paradigm shift" (Kuhn, 1970b, p. 85), then the model of scientific advance through the gradual increment of new facts and ideas proves inadequate for the historian's needs: "The transition from a paradigm in crisis to a new one from which a new tradition of normal science can emerge is far from a cumulative process, one achieved by an articulation or extension of the old paradigm. Rather, it is a reconstruction that changes some of the field's most elementary theoretical generalizations as well as many of its paradigm methods and applications" (Kuhn, 1970b, pp. 84-85).

The Gestalt model makes principally two demands: first, that the historian should attempt sympathetically to assimilate and reconstruct the context of scientific discourse of a given period, and in this way to determine the social, psychological, and historical influences that controlled the ways scientists patterned their theoretical concepts and perceived through them facts constituting the domain of scientific inquiry; and second, that the historian should regard the history of science not as an internal and smooth flow of observations and theoretical generalizations across the ages but as the sudden shift of different world views, linked only by the extrinsic contingencies of time and place.

The Gestalt model as employed by Hanson and Kuhn bears similarities to the revolutionary model—and, of course, Kuhn's express aim is to describe the structure of revolutions in science. But the differences between the revolutionary model as commonly used and the Gestalt model are marked. Those employing a revolutionary model discover in the course of a particular science a signal awakening of thought, an overturning of what the model characterizes as a decidedly archaic mode of thinking, and the establishment of a lasting foundation for future progress by, as Hall (1966, pp. xiii-xiv) puts it, "accretion." Since the logic of the revolutionary model hinges on the dichotomy between ancient and modern methods of scientific thought, historians of this persuasion usually assume that revolution in a science is a one-time affair. The Gestaltists, however, emphasize multiple "scientific revolutions," no one of which secures a position that is any more scientific or more stable than others that have occurred. The revolutionists believe that revolutions happen for good *reasons,* reasons that sustain the future growth of a science. The Gestaltists, as is consonant with the source of their model, tend to stress psychological and sociological factors in scientific change. In their view, scientific change is rarely the result of good reasons; indeed, reasons have weight only against a background of commonly accepted theory. The revolutionists view science as a search for truth about the world. The Gestaltists argue that there is no truth about the world; truth is a function of the coherence of the theoretical arrangement which holds at any one time; there are no independent, theory-free standards against which a hypothesis might be measured to assess its truth (Hanson, 1970, p. 15). The revolutionists are apt to regard postrevolutionary science as better than or more true than prerevolutionary science. The Gestaltists believe that the perceptual-conceptual paradigm adopted by a given community of scientists is incommensurable with those assumed by their predecessors: in the Gestalt switch of the goblet-faces display, the goblet is no better or truer than the faces (Kuhn, 1970b, pp. 170-171).

The Gestalt model encourages the historian to interpret scientific ideas as parts of a larger complex of meanings; it emphasizes the mutual determination of these elements. The hermeneuticist of the scientific Gestalt begins with a mode of experience or paradigmatic idea (for example, Harvey's functional teleology, Descartes's distinction between mind and body, Newton's universal gravity) and moves laterally, interpreting one symbol in the pattern in terms of the others, ultimately including socially and culturally entwined meanings. Another recent model, however, suggests that the interpretive relation is vertical and unidirectional, that scientific patterns of thought merely reflect deeper and more covert social or psychological structures.

The Social Psychological Model. From the time of the ancient period, scientists have frequently justified their theories by appeal to more general doctrines—metaphysical, religious, or social—to which those theories have been related. Samuel Clarke defended Newton's science, since it would "confirm, establish, and vindicate against all objections those great and fundamental truths of natural religion" (Clarke, [1717] 1956, p. 6). But it was only at the beginning of our century, after transformations in the social and psychological sciences (by Marxism, Durkheimian social anthropology, Freudianism, and other developments), that historians seriously attempted to organize their narratives under the assumption that scientific programs might be fueled by social interests and psychological needs. What united both socially oriented and psychologically disposed historians was the conviction that apparently extrinsic conceptual structures, whether embedded in social relationships or psychological complexes, might covertly determine the generation, formulation, and acceptance of scientific ideas. Moreover, though Freudians would insist on the primacy of sedimented attitudes, they usually admitted that these originated in certain real or imagined social situations. Correspondingly, Marxist historians have recognized that the effects of class stratification are mediated by subtle patterns of individual belief. Because of these common features, social and psychological models may be considered as forming one class of historiographic models.

Social psychological models can be divided into those prescribing weak determination and those prescribing strong determination of scientific development. The weak version of the model is the central organizing device of J. D. Bernal's four-volume *Science in History*. The model guided Bernal in mapping an enlarged field of investigation. "Science," he proposed, "may be taken as an institution; as a method; as a cumulative tradition of knowledge; as a major factor in the maintenance and development of production; and as one of the most powerful influences molding beliefs and attitudes to the universe and man" (Bernal, 1971, vol. 1, p. 31). Such a generous conception of science reciprocally compelled him to trace the social and psychological patterns in the terrain of science. For example, he initially explained Darwin's hypothesis of natural selection as a reformulation of Malthusian economics in other terms—that is, as a biological construction of a "theory built to justify capitalist exploitation" (Bernal, 1971, vol. 2, p. 644). For Bernal the sources of scientific thought, the institutions of science, its methods, the economic forces that drive it, and its impact on society were all fit subjects for social-psychological analysis. Yet science as a "cumulative tradition of knowledge" was not.

Bernal could not bring himself to extend his admittedly Marxist vision to the heart of science. He confessed that the cumulative nature of science distinguished it from such other human pursuits as law, religion, and art. Though science, like these other enterprises, grows in a field of social relations and class interests, its claims, unlike theirs, can be checked directly "by reference to verifiable and repeatable observations in the material world" (1971, vol. 1, pp. 43-44). The weak model thus protects the internal logical and justificatory structure of science from the hands of the sociologist and the psychologist. That is, it does so when the science goes right.

When it goes wrong, the historian has a sure sign that extrinsic social or psychological factors have intruded. For instance, Erik Nordenskiöld, in his influential *History of Biology*, felt constrained to invoke the weak model in his account of the unwarranted (as he believed) acceptance of Darwinian theory by scientists in the latter half of the nineteenth century: "From the beginning Darwin's theory was an obvious ally to liberalism; it was at once a means of elevating the doctrine of free competition, which had been one of the most vital cornerstones of the movement of progress, to the rank of natural law, and similarly the leading principle of liberalism, progress, was confirmed by the new theory. . . . It was no wonder, then, that the liberal-minded were enthusiastic; Darwinism must be true, nothing else was possible" (Nordenskiöld, [1924] 1935, p. 477).

It is ironic that one powerful tradition in the sociology of science, led by Robert Merton (1973) and Joseph Ben-David (1971), endorses the weak model as the only one appropriate for treating the cognitive content of science. Ben-David, for example, admits that socially conditioned biases and ideology "might have played some role in the blind alleys entered by science." In those darkened corners, sociology can be illuminating. But the main roads of science are "determined by the conceptual state of science and by individual creativity—and these follow their own laws, accepting neither command nor bribe" (Ben-David, 1971, pp. 11-12). Sociologists in this tradition confine their empirical analyses to questions of institutional organization, the spread of scientific knowledge, social controls on the focus of scientific interest, and public attitudes toward science. They regard the cognitive content of science as the reserve of those intellectual historians whose concerns are principally logical and methodological.

But even with the support of the dominant tradition in the sociology of science, can the weak model be justified? Those employing it usually fail to supply a convincing reason why social or psychological analyses might explain error in science, but not truth. The persuasiveness of this model is further diminished when one considers that, in a strict sense, most past science is "erroneous," at least by contemporary standards. Hence, if the logicist assumptions of the weak model are consistently heeded, the content of virtually all past science ought to be amenable to social and psychological interpretation. Nor should contemporary science be exempt, since there is no reason to suspect that it has achieved final truth.

The logic of the preceding line of reasoning appears to have persuaded, implicitly at least, those employing a strong version of the model. Margaret Jacob (1976), in *The Newtonians and the English Revolution,* interpreted the seventeenth-century mechanical philosophers as insisting on the passivity of matter (that is, particles do not move themselves) and divine ordering of natural law *in order* to express their latitudinarian social and religious ideology. The sociologist of science David Bloor detected this use of the strong version of the model in Jacob's work and defended it. On the basis of Jacob's account and his own investigations, Bloor concluded that the seventeenth-century scientist Robert Boyle and his colleagues and opponents "were arranging the fundamental laws and classifications of their natural knowledge in a way that artfully aligned them with their social goals" (Bloor, in press). The lesson Bloor drew from this drama of Restoration science was that quite generally in the history of science, "the classification of things reproduces the classification of men" (Bloor, in press).

The strong version of the model, then, asserts that the structure of scientific knowledge is determined not by nature but by social patterns or psychological complexes. The model stipulates that logic and appeal to natural facts are surface, that what really matters in comprehending the work of scientists are dominance struggles with the father—as in the case of Mitzman's (1971) reconstruction of Weber's social science—or the social practices of a society—as in the case of Bloor's (1976, pp. 95-116) account of Greek mathematics.

The strong version of the social psychological model, despite initial implausibility, does focus the historian's sight on a cardinal feature of scientific development: that science depends on norms—norms suggesting what is appropriate both to investigate and to accept. Norms, however, are dictated not by nature but by the decisions of men. The logic of scientific argument cannot coerce, except that men feel moved to abide by its rules and adopt its premises. Ultimately, the acceptance of meta-rules and first premises appears to be a function of social enculturation, of psychological training and conditioning. For, as Aristotle pointed out, only the fool tries to demonstrate the principles on which all his arguments are based.

Nonetheless, the strong version of the social psychological model seems too strong. It is liable to a "tu quoque" response. Why, after all, should we be convinced by the account of a historian who uses the strong version, if that account itself merely reflects his inferiority complex or his Calvinistic upbringing? The destruction of scientific rationality also undermines the plausibility of historical argument. To stanch the destructive relativism of both the social psychological model and the Gestalt model, while preserving the edge of their insights, is one of the chief tasks for which evolutionary models have been constructed.

Evolutionary Models of Scientific Development

The use of evolutionary theory in explanations of cultural phenomena can easily be traced to the mid-nineteenth century. John Lubbock, Walter Bagehot, Lewis Henry Morgan, Edward Tylor, Herbert Spencer, and a host of others applied evolutionary concepts to societal institutions in an effort to account for the descent from primitive thought (Burrow, 1966; Stocking, 1968). More recently, the specialized use of evolutionary notions has, aping its biological counterpart, proceeded from the macroconsideration of culture to the microconsideration of the development of ideas, particularly scientific ideas. Gerald Holton (1973, pp. 392-395), for instance, makes detailed and explicit use of the evolution analogy in his *Thematic Origins of Modern Science*; and in a reconsideration of his theory of paradigms, Kuhn (1970a, p. 264) has suggested that the appropriate approach to science history is evolutionary. But others have understood evolutionary theory to offer more than a vague analogy; its formal structure provides, so it is argued, the very explanation of scientific growth. Not only are ideas conceived but, like Darwin's finches, they also evolve. Karl Popper, Stephen Toulmin, and Donald Campbell are leading proponents of this sort of strict epistemological Darwinism. In what follows, I will briefly examine the proposals of Popper and Toulmin, show the deficiencies of their models, and then elaborate a natural selection version that is a refinement of Campbell's conception. The analysis of the natural selection model will include an evaluative comparison with Lakatos's model of scientific research programs and will conclude with a consideration of its historiographic advantages over other models.

The Models of Popper and Toulmin. In *The Logic of Scientific Discovery,* Popper (1934) describes the scientific community's selection of theories not as a process by which a given theory is *justified* by the evidence but as one by which a theory survives because its competitors are less fit. Thus, he argues that the preference for one theory over another "is certainly not due to anything like an experimental justification of the statements composing the theory; it is not due to a logical reduction of the theory to experience. We choose the theory which best holds its own in competition with other theories; the one which, by natural selection, proves itself the fittest to survive" ([1934] 1968b, p. 108).

In Popper's estimation, our scientific and pedestrian quests for knowledge always begin not with pure observation but with a problem that has arisen because some expectation has not been met. In confronting the problem, the cognizer makes unrestrained conjectures about possible solutions, much as nature makes chance attempts at solving particular survival problems (Popper, 1972, p. 145). These conjectures are then tested against empirical evidence and rational criticism. The rational progress of science, therefore, consists in replacing unfit theories with those that have solved more problems. These latter, according to Popper, should imply more empirical statements that have been confirmed than their predecessors (1968a, pp. 215-250). This condition enables us to describe successor theories as closer to the truth and, consequently, more progressive. I will not expand further on Popper's conception, since Lakatos has already done this with concision, fashioning from it a model of scientific research programs, which will be examined below.

The evolutionary model permits Popper to avoid the presumption that theories are demonstrated by experience; it also allows him to dismiss the view that theories and creative ideas arise from any sort of logical induction from observation. Thus, the older and newer problems of induction are skirted. Popper believes that the model directs one to interpret scientific discovery as fundamentally an accidental occurrence, a chance mutation of ideas. He consequently fails to emphasize that the intellectual environment not only selects ideas but restricts the kinds of ideas that may be initially entertained by a scientist. Attention to the environment of scientific ideas is, however, precisely what Toulmin requires for an adequate account of scientific growth.

Toulmin's (1961, 1972) thesis is that scientific disciplines are like evolving biological populations; that is, like species. Each discipline has certain methods, general aims, and explanatory ideals that provide its coherence over time, its specific identity, while its more rapidly changing content is constituted of loosely related conceptions and theories, "each with its own separate history, structure, and implications" (1972, p. 130). To comprehend the evolution of a science so structured requires that one attend to the cultural environment promoting the introduction of new ideas, as well as to the selection processes by which some few of these ideas are perpetuated.

The content of a discipline, according to Toulmin's scheme, adapts to two different (though merging) environmental circumstances: the intellectual problems the discipline confronts and the social situations of its practitioners. Novel ideas emerge as scientists attempt rationally to resolve the conceptual difficulties with which their science deals; but often those new sports will also be influenced by institutional demands and social interests. Therefore, in explaining the appearance of innovative ideas within an evolving science, one must consider both reasons and causes. After such variations are generated, however, one must turn to the processes, rational and socially causal, by which the variations are selected and preserved.

The processes that shape the growth of a discipline—the selection processes—also occur within particular intellectual and social settings. The intellectual milieu consists of the immediate problems and entrenched concepts of the science and its neighbors. Within this environment, rational appraisal by the scientific com-

munity tests the mettle of new ideas. The survivors are incorporated into the advancing discipline. The social and professional
conditions of the discipline also work to cull ideas, sanctioning
some and eliminating others. Both of these selection processes—
selection against intellectual standards and against social demands
—may act in complementary fashion or oppositionally. But both
must be heeded, in Toulmin's judgment, if one is to understand
the actual history of a science. The historian, however, will look
only to intellectual conditions when he or she requires a rational
account of the development of a particular science. When he or
she wishes to discover, for instance, the causes accelerating or retarding scientific growth, he will turn to the social and professional
institutions of that science (Toulmin, 1972, pp. 307-313).

It is the mark of the recent past in the historiography of science that it is the rational continuity of science that appears to demand explanation. Toulmin has proposed his evolutionary model
to meet this need. In his view, the continuity of disciplines, like
the continuity of biological species, involves transmission of previously selected traits to new generations. In science this process
is, according to Toulmin, one of enculturation: junior members of
a discipline serve an apprenticeship in which they learned by tu-
tored doing, by exercising certain "intellectual techniques, procedures, skills, and methods of representation, which are employed
in 'giving explanations' of events and phenomena within the scope
of the science concerned" (1972, p. 159). What principally gets
inherited, he believes, is not a disembodied set of mental concepts
but particular constellations of explanatory procedures, techniques, and practices that embody the explicating representations
and methodological goals of the science. Through the active participation in an ongoing scientific community, the novice inherits
two kinds of instantiated concepts. The first comprises the specific
substantive ideas and theories, the special explanations and techniques that solve recognized problems at any one period in the
evolution of a discipline. The second kind of inheritance remains
continuous over much longer periods and only slowly changes. It
consists of the explanatory ideals, the general aims, and the ultimate goals that distinguish the disciplines from one another. It is
within this more general inherited tradition that large-scale conceptual changes in substantive theory occur "by the accumulation

of smaller modifications, each of which has been selectively per-petuated in some local and immediate problem situation" (1972, p. 130). But such changes should not suggest, as they do for those adopting the Gestalt model, that there are not good reasons for the shifts. The basic structure against which reasons can be mea-sured is the continuity of explanatory aims and ideals which a dis-cipline manifests through long periods of its history.

Toulmin further attempts to ensure that his model will allow rational criteria to operate in science by adjusting it with the pos-tulate of *coupled evolution*. The neo-Darwinian theory of organic evolution requires that variability within a species be independent, "decoupled" in Toulmin's terms, of natural selection. According to his interpretation of the modern synthesis, there is no preselec-tion or direction given classes of variations. But coupled evolution, which he regards as another species of the larger genus of evolu-tionary process, postulates that variation and selection "may in-volve related sets of factors, so that the novel variants entering the relevant pool are already preselected for characteristics bearing di-rectly on the requirements for selective perpetuation" (1972, p. 337).

What Toulmin has suggested by his postulate of coupled evo-lution is not, however, a Darwinian sort of mechanism, in which generation of variation is blind or random, but a Lamarckian one, in which conscious acts preshape the material in anticipation of the exigencies of survival. Accordingly, the cardinal feature of the Darwinian perspective, competitive struggle against environmental demands, is largely obviated. Natural selection has no pivotal role in Toulmin's scheme.

But Toulmin need not have abandoned the device of natural selection so quickly. For the neo-Darwinian synthesis does recom-mend clearly acceptable senses in which individuals within a species might be described as preselected or preadapted to an altered envi-ronment: when, for example, heterozygote superiority retains alleles that would be fit in different circumstances; or when link-age holds in a population certain alleles, which would enhance adaptation to changed surroundings; or when alleles at certain loci have fixed rates of mutation. Such mechanisms for storing varia-tion act as constraints on selection, making specific kinds of adap-tive responses to a given situation more likely. It is, of course, true that the variations stored and the methods of their preservation are presumed to be products of generations of previous selection. In any case, classes of variations characteristic of elephants are not likely to occur in the species *Rattus rattus*. The genetic background of a species will restrict, and in that sense preselect, the kinds of variations that are immediately possible. In a moment I will indicate what this feature of biological evolution suggests for understanding conceptual evolution.

The Natural Selection Model. Popper's version of the evolutionary model of scientific growth emphasizes that theories succeed one another something like species: that theory is selected that solves more problems than its competitors. Toulmin's version complements Popper's by focusing on the cultural environment in terms of which new ideas appear and are incorporated into an evolving discipline. But Toulmin relinquishes a formal Darwinian device in an attempt to capture the way problem solutions originally emerge. As was just proposed, abandonment of a natural selection mechanism is not necessary in order to model the birth of new scientific theories. In this section, I want to build upon the Popperian and Toulminian variants and thereby refine a natural selection model for historiographic use. I will do this in two stages: first, by further specifying exactly what it is that evolves in scientific change; and second, by adding a psychosocial theory of idea production and selection—basically a development of Campbell's views.

According to Toulmin's model, the species-like entity that evolves is the intellectual discipline. But this, I think, is the wrong analogue. Intellectual disciplines are, after all, composed of heterogeneous ideas, methods, and techniques, while a species is a population of interbreeding individuals that bear genetic and phenotypic resemblance. Disciplines, moreover, are organized formally into subdisciplines and overlapping and competing specialties (Campbell, 1969; 1977b, pp. 9-10) and are interlaced informally with invisible networks (Crane, 1972). Disciplines seem more like evolving ecological niches, consisting of symbiotic, parasitic, and competing species. The proper analogue of species is, I believe, the conceptual system, which may be a system of theoretical concepts, methodological prescriptions, or general aims. The gene pool constituting such a species is, as it were, the theory's individual ideas, which are united into genotypes or genomic individuals by the bonds of logical compatibility and implication and the ties of empirical relevance. These connecting principles may themselves, of course, be functions of higher-order regulatory ideas. Biological genotypes vary by reason of their components, the genes, and the specific linkage relations organizing them; and these genotypes display different phenotypes accordingly, both as they have slightly different components and componential relationships and as they react to altered environments. Analogously, the cognitive representation of a scientific theory—its phenotypic expression in terms of the model here proposed—will vary from scientist to scientist by reason of the slightly different ideas constituting it, their relations, and the changing intellectual and social environment that supports it. So, for instance, Darwin and Wallace both advanced specifically the same evolutionary theory, though the components of their respective representations were not exactly the same and the intellectual problems to which they applied their views and for which they sought resolutions also differed in some respects. Yet we still want to say that Darwin and Wallace developed the "same"—specifically the same—theory of evolution by natural selection. Constructing the model in this way also allows us to appreciate that, like the boundaries between species, the boundaries separating theories may be indefinite and shifting.

If a historiographic model of scientific development proposes that conceptual systems, like biological species, evolve against a problem environment, that model, to tighten the Darwinian analogy, should include a mechanism accounting for adaptive change in scientific thought. During the last quarter century, Donald Campbell (1956, 1959, 1960, 1965, 1974a, 1974b, 1977a, 1977b) has worked to construct a psychological theory of idea production and selection that meets this demand. His mechanism of "blind variation and selective retention" not only illuminates a fundamental feature of creative thinking in science (and in other cognitive pursuits as well) but, as an unintended consequence, also explains why some ideas seem to come (as Toulmin believes) preadapted to their intellectual tasks. Let me first sketch the essential aspects of Campbell's natural selection mechanism and then add some refinements, similar to ones I have recently suggested (Richards, 1977) and Campbell (1977a) has endorsed.

In the Darwinian scheme, species become adapted to solve the problems of their environment through chance variations and selective perpetuation. Campbell believes that the creative thinker exhibits counterpart cognitive mechanisms, which blindly generate possible solutions to intellectual problems, select the best-adapted thought trials, and reproduce consequently acquired knowledge on the appropriate occasions. A distinctive postulate of this model is that cognitive variations are produced blindly, which is to say that initial thought trials are *not justified* by induction from the environment, or by previous trials, or by "the eventual fit or structured order that is to be explained" (1974b, p. 150). The production of thought variations by the scientist—or the creative thinker in any realm—is therefore precisely analogous to chance mutations and recombinations in organic evolution.

The bones of Campbell's conception can be fleshed out in ways that make it appropriate for the historiographic model I have in mind. The following additional postulates serve this function.

1. The generation and selection of scientific ideas, both as the hypotheses that guide a scientist's work and as the relatively sedimented doctrine of the scientific community, should be understood as the result of a feedback mechanism. Such a mechanism, which we may consider only formally without worrying about its physiological realization, will generate ideas in a biased rather than in a purely random fashion. For without some restraints on generation, a scientist might produce an infinity of ideas with virtually no probability of hitting on a solution to even the simplest problem. But, of course, even mutations and recombinations of genes do not occur completely at random. The constraints on idea production are determined by the vagaries of education and intellectual connections, social milieu, psychological dispositions, previously settled theory, and recently selected ideas and schemes—all of which form, in Campbell's terms, a "nested hierarchy" (1974a, p. 419). This postulate therefore suggests that, though ideas may come serendipitously, their generation is not unregulated but can be comprehended by the historian. Thus, for example, when Darwin began musing on the nature of a mechanism to explain species change, he did so in a conceptual environment partly formed of ideas stimulated by his *Beagle* voyage and acquired from his grandfather, from Lamarck, and from a host of authors whom he read

between 1836 and 1838 (see Richards, 1979). These ideas not only determined the various problems against which successful hypotheses were selected, but they also initially fixed the restraints on the generation of trial solutions. It is within a certain (albeit vaguely defined and shifting) conceptual space that chance variations are displayed. And it is because of such constraints that even a scientist's rejected hypotheses can make sense to the historian.

2. To think scientifically is to direct the mind to the solution of problems posed by the intellectual environment. Novel ideas are not produced, however, when perceptual or theoretical situations are settled. As Popper (and Dewey before him) has contended, for thinking to occur, there must be a troubled, unsettled cognitional matrix; the perceived environment must be changing. Conversely, alterations in the intellectual situation that are not perceived or that are ignored must lead to the arrest of *scientific* thought and the eventual extinction of a *scientific* system.

3. The initial perception of the problem situation cannot be a function of generation and selection—lest an infinite regress ensue. A primary appreciation of the problematic is necessary in order to adjust both the constraints on trial generation and the criteria for selection. After a primitive determination of the problem, however, adjustment of constraints may occur as the perceived structure of the problem is transformed in light of advances already achieved.

4. Ideas and ultimately well-articulated theories are originally generated and selected within the conceptual domain of the individual scientist. Only after an idea system has been introduced to the scientific community (or communities, since scientists usually belong to several interlocking social networks) does public scrutiny result. The broader conceptual environment established by the community may present somewhat different problem situations and standards of competitive survival. To the extent, however, that the problem environment of the individual and that of the community coincide, individually selected ideas or theories will be fit for life in the community. If the historian neglects to consider the processes of idea generation and evaluation at the individual level, then it will appear that scientific ideas come mysteriously preadapted to their (public) environment.

5. Finally, if this model is to be used in construing the ac-

quisition of knowledge in *science,* then one must suppose that selection components operate in accord with certain essential criteria: logical consistency, semantic coherence, standards of verifiability and falsifiability, and observational relevance. These criteria may function only implicitly, but they form a necessary subset of criteria governing the development of scientific thought throughout its history. Without such norms, we would not be dealing with the selection of *scientific* ideas. They thus aid the historian of science in distinguishing his subject from other cognitive occupations. It should be stressed, however, that these selection criteria are themselves the result of previous generation and continuous selection, processes which have brought us from protoscience to science—just as mammals have been brought up from the reptiles. The complete set of selection criteria define what in a given historical period constitutes the standard of scientific acceptability. The above-specified criteria are only elements of this more comprehensive norm.

    Natural Selection Model Versus Scientific Research Programs. Since the natural selection model (NSM for short) is a model, it implicitly represents a theory about science, its structure, growth, and rationality. The model and its imbedded theory portray scientific conceptual systems as quasi-organisms that compete for survival; and it proposes that the system that best solves the problems of its cultural environment will survive, gradually displacing its competitors. In order to assess the model's viability at this early stage of its development, we might compare it with another recently advanced model, which appears to offer it the keenest competition— Lakatos's model of scientific research programs (SRP). Lakatos (1970, 1971) designed SRP to serve both as a standard for appraising the scientific and rational status of contemporary conceptual systems and as a historiographic device for constructing explanatory accounts of science's growth. Because of this explicit intention and the model's rigorous formulation, SRP furnishes an exceptional standard by which to evaluate NSM.

    Lakatos formulates his model expressly for the purpose of interpreting the history of science as rationally progressive. He contrasts his conception with the Kuhnian model, which he regards, correctly I believe, as forbidding judgments of general scientific progress across problem shifts and as supplying no criteria for distinguishing scientific rationality from doctrinaire opinion (Lakatos, [1970] 1978, pp. 8-10). Yet, like Kuhn, Lakatos chooses a larger unit of analysis than the solitary idea or theory, since he recognizes the historical and epistemological fact that ideas cannot be evaluated in isolation from auxiliary concepts, which specify normal conditions, relevant evidence, and theoretical pertinence. He takes this larger conceptual scheme, the SRP, as the entity to be judged as progressing (or degenerating), as competitive with other programs, and as the basis for estimating the rationality of a particular scientific enterprise.

As Lakatos characterizes its structure, SRP has a *hard core* of central principles and a *belt of surrounding auxiliary hypotheses* that continue to change during the life of a program ([1970] 1978, pp. 48-52). Newton's program, for example, had a stable center consisting of his three laws of dynamics and a principle of attraction; it also had a belt of hypotheses composed of assumptions about the gravitational center of large bodies, the viscosity of different resisting media, the paths of planets, the distance of the fixed stars, and a host of other boundary conditions. If a program is to be pursued, the hard core embodying its defining ideas must be protected, especially in the early stages of growth, from noxious facts and the harmful competition of rival programs. The program's *negative heuristic,* then, bids falsification attempts be deflected to the auxiliary hypotheses. It is the protective girdle of hypotheses that is challenged by the facts and adjusted to escape the force of contrary evidence. The *positive heuristic* of the program complements the negative imperative by proposing means of advancing the empirical content of the program through development of the auxiliary hypotheses. It discharges this function principally by suggesting replacement hypotheses when evidence or internal logic require and by setting the plan which the program will stubbornly follow in the face of anomalies and the claims of rival programs.

Lakatos offers his model as a refinement of Popper's. It nevertheless differs from Popper's selection model on an important point. Popper sometimes suggests that theories can be falsified directly, by infection from toxic facts, and that such falsified theories are (or scientific honor demands that they should be) rendered immediately extinct (Popper, [1934] 1968b, pp. 86-87).

Lakatos, in contrast, recognizes that theories may accumulate anomalies but that scientists properly adjust their auxiliary hypotheses to avoid them, even, if possible, to turn them into dramatic corroborations of the program. Darwin, for example, was initially stumped by the seemingly inexplicable adaptations of neuter insects—they left no progeny to inherit favorable variations. But when after several years he finally developed his mechanism of kin selection, what originally threatened to falsify his theory became the strongest evidence for it (Richards, in press). Such manipulations, however, can be abused.

To prevent ad hoc alterations from turning rational science into empirically immune pseudoscience, Lakatos stipulates that such adjustments should be capacious enough to extend the empirical content of the theory beyond the confuting cases, so that such extension yields predictions of new facts: "A given fact is explained scientifically only if a new fact is also explained with it" ([1970] 1978, p. 34). Predictive extension is for Lakatos the mark of a scientifically authentic research program. If a program accounts for the empirical content of rivals but generates further predictions, then the program is *theoretically progressive* and thus *scientific*. If the predictive excess is corroborated, then the program is also *empirically progressive*; otherwise, it is *degenerating*. But if a program confronts incompatible facts that its constituent theories cannot neutralize, while theories of a rival program can give them account, and if that rival also generates further predictions, some of which are corroborated, then the original program is *falsified*. Programs that persist in the face of a rival's success can only be judged *pseudoscientific* (Lakatos, [1970] 1978, pp. 32-35).

Lakatos has constructed his model as a device for appraising research programs, both recent and historically remote ones. He believes that appropriate standards are required if the historian is to do his job properly. SRP allows the historian to distinguish the internal history of science, which expresses the rational growth of objective knowledge, from "empirical (sociopsychological) 'external history' " (Lakatos, [1971] 1978, p. 102). Since the principal meaning for "science" is "accomplished objective knowledge," the internal history of science captures all that is essential to it. SRP allows the historian to select out of the morass of historical clutter precisely his subject, the internal logical development of theories. After all, "most theories of the growth of knowledge are theories of the growth of disembodied knowledge: whether an experiment is crucial or not, whether a hypothesis is highly probable in the light of the available evidence or not, whether a problem shift is progressive or not, is not dependent in the slightest on the scientists' beliefs, personalities, or authority. These subjective factors are of no interest for any internal history" ([1971] 1978, p. 118). Indeed, with the help of SRP, the historian of science should construct an ideal history, the normative fabula that the logic of a given research program demands. Lakatos offers Niels Bohr's program as illustrative: "Bohr, in 1913, may not have even thought of the possibility of electron spin. He had more than enough on his hands without the spin. Nevertheless, the historian, describing with hindsight the Bohrian program, should include electron spin in it, since electron spin fits naturally in the original outline of the program. Bohr might have referred to it in 1913. Why Bohr did not do so is an interesting problem which deserves to be indicated in a footnote" ([1971] 1978, p. 119).

Despite his comic exaggeration, Lakatos does not intend that the historian should literally write a bilevel history, one narrative in the text and another in the footnotes. But his model does require a history structured with a logically distinct internal core, which is to control the significance assigned to external social and psychological events. This core, in his view, should express not only the research program that some scientist actually established but also an enlarged program that contains features compatible with those of the original.

In Lakatos's application, SRP exudes a peculiar Platonic odor, which most historians, I think, would find offensive. For the model appears to demand not a historian but a Laplacean demon who could extract from a program all that was logically or compatibly contained therein. Insofar as SRP is used by a human historian, it seems to urge him to read history backward, to find in earlier, inchoate concepts the results of more recent research. In Lakatos's hands, SRP would obscure the vision of the historian wishing to detect the emergence of scientific ideas from previously developed ideas, community expectations, and personal aims.

Because an instrument is badly used does not mean, of

course, that it is defective. SRP conceivably could be given a historically justified employment. But the model itself is, I think, radically deficient for historical work. A comparison with NSM will convince us, I believe, of the advantages of the latter. In the following analysis, I will use episodes in the development of Darwin's conceptual system as the test base. Lakatos's selective use of examples from the history of physics has prejudiced his case.

1. As Lakatos structures it, a research program is essentially immutable; its hard core remains stable and defines the period of the program's existence. For Darwin's program, certainly the mechanism of natural selection must be regarded as a core principle. Yet when he initially attacked the problem of species change, he developed and used several other mechanisms before hitting upon natural selection. And even after formulating that key principle, he continued to modify both its logic and its scope of application. Nor did he abandon those original mechanisms of species change; they were integrated into his evolving conceptual system. Despite changes and reorganization of core principles, Darwin's conceptual system retains a historical identity.

NSM allows for this kind of alteration. Evolving conceptual systems may undergo fundamental modifications, changes more basic than simple adjustment of peripheral principles. A system will be regarded as forsaken only when historical continuity has been broken and the problem situation vacated. Barring this, NSM encourages the expectation that introduction of fundamental ideas will alter a developing system's more remote principles and that changes in these latter—the adaptations by which a conceptual system more immediately meets the requirements of its environment —will in turn affect the central principles. Expectation of reciprocity, not unilateral alteration, is the methodological rule.

2. SRP evaluates a program as nondegenerating only if it continues to make novel predictions that are empirically confirmed. What success rate must be maintained in order to keep a favorable evaluation is, however, unspecified. More seriously, if this criterion of progress were actually operative in the mid-nineteenth century, it would have counseled the immediate rejection of Darwin's conceptual system; for his theory made no real predictions (certainly not comparable to Lakatos's favorite—Einstein's forecast of starlight bending near the sun). Darwin simply did not

use his system as a predictive instrument; and any quasi-predictions derivable from it would have found substantiation only in those facts already interpreted through the eyes of a convinced evolutionist. Darwin fairly estimated that the advantage of his theory was that it made sense of a medley of facts. Its cogency lay in uniting what before seemed disparate.

NSM interprets a conceptual system as progressive much in the way we want to regard biological systems as progressive—if they continue to solve the problems of their environment. For conceptual systems we must judge such progress by using the measuring standards provided by the intellectual environment. For example, astrology continues to exist within a rather specialized cultural niche; but its central environment is not that of contemporary science. The requirements for the survival of scientific ideas have changed over the course of ages, making new demands that astrology could not meet. The conceptual system, which at one time existed within the same intellectual milieu as ancient astronomy, has now migrated to a more logically tolerant climate.

3. SRP assumes that competitive programs vie head to head, each claiming the same explanatory ground, with one inching out the other by a few more predictions. This is hardly ever the case. Rival theories usually have preferred evidentiary bases, which at best only partly overlap. Thus, Darwin's conceptual system could explain the evolution of neuter insects, but Spencer's could more easily account for coadaptive evolution of organs. Appraisal in this situation is impossible if it is guided by SRP alone; and any attempts at the employment of SRP would undoubtedly fall prey to naive presentism—the justification by hindsight of what we presently know to be the successful system (see Stocking, 1968, pp. 2-12, for a diagnosis of this ill).

NSM recognizes, by comparison, that competing conceptual systems may occupy partly coincident but not identical problem spaces and that in such cases the preferred evidentiary ground of each—that which offers the strongest support for their particular claims—will likely differ. For the historian using NSM, this is ordinarily expected. But it does not mean that comparative evaluations are precluded. On common ground arbitration is possible. But when there is no overlap, NSM directs the historian to assay the central intellectual environment of the scientific community to determine what problems it regarded as significant at the time. That Darwin could explain the wonderful instincts of animals, especially the behavior of neuter insects, was perceived as particularly dramatic, since the complex instincts of animals constituted the province many prominent natural theologians had reserved as the most supportive of their account. NSM, in this respect, does not evaluate all explanations according to the same scale. It weighs the significance of a particular explanation or prediction by the importance the scientific community placed on it at the time. SRP permits no such discriminative evaluation.

4. SRP should require that every emerging program be judged as falsified, since immature programs cannot usually compete with rivals in empirical coverage or predictive success. SRP fails to recognize that the existence of a conceptual system depends on the character of several intellectual and cultural environments. The continued viability of a conceptual system is, first of all, a function of the set of problems the individual scientist has determined for himself, which may only partly coincide in scope and significance with the problem set of the larger scientific community. Emerging conceptual systems, then, are rationally pursued when they solve those immediate problems. As the scientist publicizes his efforts, he thereby introduces his system into a different environment. There it may compete with rivals and, depending on the terrain, will survive or perish (or mutate or hybridize or migrate).

5. SRP extirpates conceptual systems from their historical situation. The only relation that can exist between systems is that of fundamental opposition (for if their cores were logically similar, they would constitute the same program). There is no sense that conceptual systems may evolve into different systems, or branch off from a parent system, or merge with close relatives to form a hybrid system, or exist as part of the intellectual environment of other systems. These vital historical relationships are obscured by an essentialistic model of the kind SRP represents, but they are highlighted by NSM.

6. Since SRP is designed only to provide a standard of appraisal, it cannot direct the historian in his attempt to capture ideas as they are born, or to explore the immediate environment which shapes their content. Appraisal is a procedure of justifica-

tion—as Lakatos construes it, "public" justification—which for the historian is only part of the story: the historian wants also to record the birth of new ideas and chronicle their growth. NSM, by contrast, functions both for appraisal and for guiding the historical reconstruction of the environment of discovery. It suggests appraisal of a conceptual system from three perspectives: that of the problems of the individual scientist; that of the problems of the scientific community (or communities, since the scientist may be a member of more than one); and that of the problems of subsequent communities. Insofar as a system continues to solve the problems that the individual scientist recognizes as important, it is *rational* to pursue development of the system. (This supposes, of course, that among the particular standards an individual scientist sets for a successful solution are those few essential criteria of consistency, empirical pertinence, and the like.) If the problems he resolves are also those recognized by his scientific community or subsequent communities, we can describe his system as *scientific*. And if his system adapts more effectively than its rivals to newly uncovered problems in his community or those of subsequent communities, it is to that extent *progressive*. These evaluations can be made with some confidence, of course, only in retrospect: they are distinctively historical as opposed to merely philosophical judgments.

In regard to the context of discovery, NSM urges a reconstitution of the scientist's own beliefs, which form the most intimate environment out of which his ideas are generated and against which they are selected. It is that environment that shapes those central features of a system which perdure as the system adjusts to the demands of the wider community. Close scrutiny of the private space out of which theories arise indicates to the historian, therefore, the logical shape of scientific ideas. To understand the core of a conceptual system, then, often necessitates precisely the kind of investigation of an individual's beliefs that SRP does not support and Lakatos's own employment of it actually denies.

7. SRP stipulates that conceptual systems be judged as resolving problems only if they meet certain contemporary criteria of scientific acceptability (for example, dramatically confirmed predictions): SRP is insistently presentistic. Though there are, I believe, some few "eternal principles" of rational discourse in em-

pirical science (for example, logical consistency, observational pertinence, procedures of verification and falsification), yet it is clear that different ages also invoked special standards (for example, compatibility with theological belief, inductive support, axiomatic formulation), which must be considered if we are to appreciate the rational objective of and conceptual forces molding scientific ideas (Laudan, 1977, pp. 128-133). SRP ethnocentrically requires that all criteria of scientific reason conform to our own. NSM, on the other hand, establishes several standards of evaluation: those appropriate to the scientist's own conception of his problems; those operating in the scientific community of his period; and those utilized by subsequent communities, including our own. In this way, scientific appraisal is contextualized, and thereby made into a truly historical evaluation.

8. Finally, SRP directs the historian to separate the scientific domain from the social-psychological domain. It treats scientists of past ages as if they were prototypes of the neo-Popperian philosopher. NSM, however, recognizes that within the private conceptual milieu of the individual scientist such fixed boundaries cannot be observed. The relations governing his thought constructions may be logical or psychological, sanctioned by the scientific community of the time or derived from social and religious concerns—or any one of these, depending on the standard of comparison. The scientist likely could not make these distinctions, nor could the scientific community. They are allowed to the historian only in respect of the norms he has chosen. NSM, unlike SRP, directs the historian to make such discriminations against several criterial environments: the private milieu, the larger social and scientific communities, and succeeding communities. The muses of a Popperian third world have no final say here.

Conclusion: The Natural Selection Model as a Historiographic Model

Every model bears to its primary analogue both similarities and dissimilarities. I have stressed the analogies of NSM to biological selection and evolution. Of course, one may, as a purely logical exercise, construct a theory of scientific cognition and development which simulates the theory of biological evolution. But even if an exact fit were possible between the formal structures of the two domains, only the metaphysical mind would find this frightfully compelling. What the historian of science wants to know is how a model of this kind is to be of use. This I have tried to specify in the preceding comparative analysis. I will conclude by indicating more generally what I see as the principal historiographic advantages of NSM.

1. NSM is an articulated model having definite implications, though—as with any model—craft is required for its application. That models are typically used by historians I have attempted to show in the first section of this paper. That they must be used is perhaps less obvious but, I think, epistemologically demonstrable. Hence, if models are necessary, one which is well worked out and which forces explicit and conscious application will have greater value for the historian.

2. NSM, like its biological counterpart, is flexible enough to serve as a higher-order model for more specialized theories of scientific advance. Darwin's scheme could subsume particular theories of, say, ontogenesis, while at the same time becoming more securely anchored to its empirical base by ties with these lower-level theories. Just so, NSM receives instantiation by its relation to well-founded epistemological, psychological, or social theories of particular domains. For example, Darden and Maul's (1977) conception of the role of interfield theories (theories formed from parent theories in closely connected scientific fields) in the development of a science can be interpreted, in terms of NSM, as a more particularized characterization of the general phenomenon of hybridization in the evolution of theoretical systems. Several such well-founded lower-level theories can be systematized under NSM and thereby provide a unified but highly resolved vision for the historian.

3. NSM preserves the traditional distinction between the process of discovery—that is, when ideas are generated and generational criteria continually adjusted—and the process of justification—that is, when ideas are selected, both in the problem situation of the individual scientist and in that of the scientific community. The model recognizes that at times the environment in which discovery occurs will differ extensively from that of justification, but that both must be scrutinized by the historian. Yet it also indi-

cates that often the environments will largely coincide, that the considerations promoting discovery will be similar to those serving as justificatory norms. That is to say, in the actual movement of science, the same set of criterial ideas and situations might be both generational and selective, just as in the biological sphere epistatic relations and ultimately a given environment may control the kinds of allelelic alternatives available, and consequently select a set of alleles for perpetuation once they are generated.

4. NSM guides the careful survey of central environments for the generation and selection of ideas—that is, those environments constituted by the specific problems of the individual scientist and his community; but it also directs the exploration of intersecting and neighboring niches formed by other kinds of cultural concerns. That is, the model encourages the historian to attend not only to the logic and particular content of scientific theory development but also to its psychology, sociology, economics, and politics. NSM thus protects the historian against the self-refuting reduction of the strong version of the social psychological model by focusing his interpretive efforts on central environments. At the same time, NSM avoids the insularity of the traditional growth and revolutionary models, and opposes head on the newer isolationism of Lakatos's SRP by building bridges to recent studies in the psychological, sociological, and wider intellectual aspects of scientific activity.

5. NSM suggests, in conformity to the data of science's history, that a science advances neither by reason of a fixed set of universal standards, as implied by the growth and revolutionary models, nor through irrational macro-saltations of radically distinct paradigms, as required by the Gestalt model. It does lead the historian to recognize that standards of scientific acceptability themselves evolve, while yet retaining some stable features; to regard the raw material of scientific evolution, its ideas, as discrete but genetically related to prior conceptual states; to hold the usual source of variability to be recombinations of ideas rather than novel mutations; and, generally, to assume that conceptual systems will change more slowly in some climates, more rapidly in others, but never in radically discontinuous fashion.

6. NSM makes intelligible the nonprogressive character of some conceptual systems in the history of science. The growth and

revolutionary models must regard such systems as actually non-scientific, since they did not lead to modern science. The Gestalt model and the strong version of the social psychological model must consider them as logically and scientifically indistinguishable from other competing systems. But NSM is able to construe them either as systems which failed to respond adequately to a changing intellectual environment, though interstitially perpetuated for a period, or as systems which continued to develop, but in an environment different from that of the main stem of the scientific community. Thus, NSM does not compel the historian to ignore such systems as not being authentically scientific, or to treat them as if they were conceptually equivalent to the more direct forebears of contemporary science.

7. NSM allows the historian to achieve both a diachronic and a synchronic perspective on his subject. That is, at each stage in the development of a science, the historian will be prompted to isolate the selection criteria for scientific ideas, so that the value of a given conceptual system can be judged by relevant standards, and so that one is not constrained to describe something as pseudo-science, or crank science, or mysticism simply because it does not conform to all contemporary norms of scientific acceptability. From a diachronic point of view, the historian can discriminate patterns of early science in a way that its practitioners themselves could not, distinguishing it from religion, superstition, myth, and other human pursuits. Thus, the synchronic approach permits one to understand science on its own terms, while the diachronic approach aids in determining the progenitors of more recent science.

In sum, NSM has decided advantages over the other historiographic models we have examined: it renders normative what sensitive historians do instinctively.

References

Ben-David, J. *The Scientist's Role in Society.* Englewood Cliffs, N.J.: Prentice-Hall, 1971.

Bernal, J. *Science in History.* (4 vols.) (3rd ed.) Cambridge, Mass.: M.I.T. Press, 1971.

Bloor, D. *Knowledge and Social Imagery.* London: Routledge and Kegan Paul, 1976.

Bloor, D. "Durkheim and Mauss Revisited." In J. Law (Ed.), *The Language of Sociology*. Keele, England: University of Keele, in press.

Borrichius, O. *De Ortu et Progressu Chemiae Dissertatio* [*The Rise and Progress of Chemical Dissertations*]. In J. Manget (Ed.), *Bibliotheca Chemica Curiosa* [*Library of Chemical Curiosities*]. Geneva: Chouet, 1702. (Originally published 1668.)

Bostocke, R. *The Difference Betwene the Auncient Phisicke and the Latter Phisicke*. Selections in A. Debus, "An Elizabethan History of Medical Chemistry." *Annals of Science*, 1962, *18*, 1-29. (Originally published 1585.)

Burrow, J. *Evolution and Society*. Cambridge, Mass.: Cambridge University Press, 1966.

Campbell, D. T. "Perception as Substitute Trial and Error." *Psychological Review*, 1956, *63*, 330-342.

Campbell, D. T. "Methodological Suggestions from a Comparative Psychology of Knowledge Processes." *Inquiry*, 1959, *2*, 152-182.

Campbell, D. T. "Blind Variation and Selective Retention in Creative Thought as in Other Knowledge Processes." *Psychological Review*, 1960, *67*, 380-400.

Campbell, D. T. "Variation and Selective Retention in Socio-Cultural Evolution." In H. R. Barringer, G. I. Blanksten, and R. W. Mack (Eds.), *Social Change in Developing Areas*. Cambridge, Mass.: Schenkman, 1965.

Campbell, D. T. "Ethnocentrism of Disciplines and the Fish-Scale Model of Omniscience." In M. Sherif and C. W. Sherif (Eds.), *Interdisciplinary Relationships in the Social Sciences*. Chicago: Aldine, 1969.

Campbell, D. T. "Evolutionary Epistemology." In P. A. Schilpp (Ed.), *The Philosophy of Karl Popper*. La Salle, Ill.: Open Court, 1974a.

Campbell, D. T. "Unjustified Variation and Selective Retention in Scientific Discovery." In F. Ayala and T. Dobzhansky (Eds.), *Studies in the Philosophy of Biology*. London: Macmillan, 1974b.

Campbell, D. T. "Discussion Comment on 'The Natural Selection Model of Conceptual Evolution.'" *Philosophy of Science*, 1977a, *44*, 502-507.

Campbell, D. T. "Descriptive Epistemology: Psychological, Socio-logical, and Evolutionary." William James Lectures, Harvard University, 1977b.

Clarke, S. *The Leibniz-Clarke Correspondence.* Manchester, England: Manchester University Press, 1956. (Originally published 1717.)

Crane, D. *Invisible Colleges.* Chicago: University of Chicago Press, 1972.

Crombie, A. *Robert Grosseteste and the Origins of Experimental Science, 1100-1700.* New York: Clarendon Press, 1953.

Crombie, A. *Medieval and Early Modern Science.* (2 vols.) (2nd ed.) Cambridge, Mass.: Harvard University Press, 1961.

D'Alembert, J. *Discours Préliminaire* [*Preliminary Discourse*]. In D. Diderot and J. D'Alembert (Eds.), *Encyclopédie ou Diction-naire Raisonne des Sciences, des Arts et des Métiers* [*Encyclo-pedia or Rational Dictionary of the Sciences, Arts, and Crafts*]. Vol. 1. (2nd ed.) Paris: Lucquois, 1758.

Darden, L., and Maul, N. "Interfield Theories." *Philosophy of Science,* 1977, *44,* 43-64.

Freind, J. *The History of Physick from the Time of Galen to the Beginning of the Sixteenth Century.* (2 vols.) London: Walthe, 1725.

Gillispie, C. *The Edge of Objectivity.* Princeton, N.J.: Princeton University Press, 1960.

Hall, A. *From Galileo to Newton.* New York: Harper & Row, 1963.

Hall, A. *The Scientific Revolution, 1500-1800.* Boston: Beacon Press, 1966.

Hanson, N. *Patterns of Discovery.* Cambridge, England: Cambridge University Press, 1970.

Hesse, M. "Reasons and Evaluation in the History of Science." In M. Teich and R. Young (Eds.), *Changing Perspectives in the History of Science.* London: Heinemann Educational Books, 1973.

Holton, G. *Thematic Origins of Scientific Thought: Kepler to Einstein.* Cambridge, Mass.: Harvard University Press, 1973.

Jacob, M. *The Newtonians and the English Revolution.* Ithaca, N.Y.: Cornell University Press, 1976.

Kant, I. *Kritik der reinen Vernunft* [*Critique of Pure Reason*]. (2nd ed.) In W. Weischedel (Ed.), *Immanuel Kant Werke in*

sechs Bänden [*The Works of Immanuel Kant in Six Volumes*].
Vol. 2. Wiesbaden: Insel, 1956. (Originally published 1787.)

Koyré, A. *From the Closed World to the Infinite Universe.* Baltimore: Johns Hopkins University Press, 1957.

Koyré, A. *Metaphysics and Measurement: Essays in Scientific
Revolution.* Cambridge, Mass.: Harvard University Press, 1968.

Koyré, A. *Newtonian Studies.* Chicago: University of Chicago
Press, 1969.

Kuhn, T. "Reflections on My Critics." In I. Lakatos and A. Mus-
grave (Eds.), *Criticism and the Growth of Knowledge.* Cambridge, England: Cambridge University Press, 1970a.

Kuhn, T. *The Structure of Scientific Revolutions.* (2nd ed.) Chicago: University of Chicago Press, 1970b.

Lakatos, I. "Falsification and the Methodology of Scientific Re-
search Programmes." In J. Worrall and G. Currie (Eds.), *The
Methodology of Scientific Research Programmes: Philosophical
Papers of Imre Lakatos.* Vol. 1. Cambridge, England: Cambridge
University Press, 1978. (Originally published 1970.)

Lakatos, I. "History of Science and Its Rational Reconstructions."
In J. Worrall and G. Currie (Eds.), *The Methodology of Scientific Research Programmes: Philosophical Papers of Imre Lakatos.* Vol. 1. Cambridge, England: Cambridge University Press,
1978. (Originally published 1971.)

Laudan, L. *Progress and Its Problems: Towards a Theory of
Scientific Growth.* Berkeley: University of California Press,
1977.

McGuire, J., and Rattansi, P. "Newton and the 'Pipes of Pan.'"
Notes and Records of the Royal Society of London, 1966,
21, 108-143.

Merton, R. *The Sociology of Science.* Chicago: University of Chicago Press, 1973.

Mitzman, A. *The Iron Cage: An Historical Interpretation of Max
Weber.* New York: Grosset & Dunlap, 1971.

Nordenskiöld, E. *The History of Biology.* (2nd English ed.) New
York: Tudor, 1935. (Originally published in Swedish 1920-
1924.)

Popper, K. *Conjectures and Refutations: The Growth of Scientific
Knowledge.* (2nd ed.) New York: Harper & Row, 1968a.

Popper, K. *The Logic of Scientific Discovery.* (2nd English ed.)

New York: Harper & Row, 1968b. (Originally published in German 1934.)

Popper, K. *Objective Knowledge.* Oxford, England: Oxford University Press, 1972.

Rattansi, P. "Some Evaluations of Reason in Sixteenth- and Seventeenth-Century Natural Philosophy." In M. Teich and R. Young (Eds.), *Changing Perspectives in the History of Science.* London: Heinemann Educational Books, 1973.

Richards, R. "The Natural Selection Model of Conceptual Evolution." *Philosophy of Science,* 1977, *44,* 494-501.

Richards, R. "Influence of Sensationalist Tradition on Early Theories of the Evolution of Behavior." *Journal of the History of Ideas,* 1979, *40,* 85-105.

Richards, R. "Instinct and Intelligence in British Natural Theology: Some Contributions to Darwin's Theory of the Evolution of Behavior." *Journal of the History of Biology,* in press.

Sarton, G. *The History of Science and the New Humanism.* Bloomington: Indiana University Press, 1962a. (Originally published 1937.)

Sarton, G. *Sarton on the History of Science: Essays by George Sarton.* Cambridge, Mass.: Harvard University Press, 1962b.

Stocking, G. *Race, Culture, and Evolution.* New York: Free Press, 1968.

Stuewer, R. (Ed.). *Minnesota Studies in the Philosophy of Science.* Vol. 5: *Historical and Philosophical Perspectives on Science.* Minneapolis: University of Minnesota Press, 1970.

Toulmin, S. *Foresight and Understanding: An Enquiry into the Aims of Science.* New York: Harper & Row, 1961.

Toulmin, S. *Human Understanding.* Oxford: Oxford University Press, 1972.

Watson, R. "On the Rise and Progress of Chemistry." In *Chemical Essays.* Vol. 1. (6th ed.) London: Evans, 1793.

Whewell, W. *History of the Inductive Sciences.* (3 vols.) London: Parker, 1837.

Yates, F. *Giordano Bruno and the Hermetic Tradition.* Chicago: University of Chicago Press, 1964.

Yates, F. "The Hermetic Tradition in Renaissance Science." In C. Singleton (Ed.), *Science and History in the Renaissance.* Baltimore: Johns Hopkins University Press, 1968.

4

Peter Skagestad

Hypothetical Realism

This chapter will address itself to one of Donald Campbell's most interesting contributions to contemporary philosophy; namely the wedding of epistemological naturalism to ontological realism. By "epistemological naturalism" is here meant the view that our knowledge *of* the world is itself a fact *in* the world and that the theory of knowledge may therefore be regarded as a part of empirical science. By "ontological realism" is meant the view that the world of which we have knowledge exists independently of our

Note: For comments and criticism the author wishes to thank William B. Ober, at Hackensack Hospital, Nils Roll-Hansen, at the Norwegian Research Council, and Abner Shimony at Boston University. None of these, of course, is in any way responsible for the contents of this chapter.

knowledge of it. I shall begin by supplying some background material on the naturalist tradition in epistemology and proceed to outline what I take to be Campbell's position. Next, I shall indicate
some unresolved questions arising from Campbell's formulation of
realism; and, finally, I shall suggest a further elaboration of realism, which I hope will turn out to be an acceptable supplement to
the ideas developed by Campbell.

Historical Introduction

An erudite student of the history of ideas (as well as of numerous other disciplines), Campbell has long been conscious of
being heir to an intellectual tradition, to which he has repeatedly
paid generous tribute (see especially Campbell, 1960, 1974a). This
tradition, which I shall call the tradition of epistemological naturalism, is an interdisciplinary one, including a long roster of distinguished students of the biology, psychology, sociology, and history
of man's individual and collective cognitive activities. It is also an
international tradition, including the Austrians Mach and Boltz-
mann, the German Simmel, the Frenchmen Poincaré and Sourieau,
the Scotsman Bain, and the Americans James and Baldwin. While
learning from this tradition and acknowledging his indebtedness to
it, Campbell has, however, departed from its mainstream on one
crucial philosophical issue: the issue of the *objectivity* of truth,
understood as a relation of *correspondence* to a real, external
world. This notion has been frequently thought to be out of place
in a naturalistic account of man as a knower. Once we have described how human beings experience knowledge processes, so it
has been argued, we have thereby said all there is to be said, and
we add nothing by postulating an objective, external reality to
which man's experience is supposed to "correspond." Truth (so
the argument continues) can only be a relation *among* experiences;
as such, it is something relative to human interests, needs, and de-
sires; in effect, it consists in the most practical, convenient, or economical adaptation of ideas to experience and to each other. The
notion of an objective truth, involving the notion of a "real"
world beyond experience, has in this perspective been thought to
be a purely idle notion.

William James, with his happy gift for hitting on striking literary similes, once ridiculed the notion of objective truth by quoting, from Ernst Mach, these lines by the poet Lessing (James, 1908, p. 220):

> Sagt Hänschen schlau zu Vetter Fritz,
> "Wie kommt es, Vetter Fritzen,
> Dass grad' die Reichsten in der Welt,
> Das meiste Geld besitzen?"

Or, in a somewhat prosaic translation of my own:

> Little Hans says slyly to Cousin Fritz,
> "Why is it, Cousin Fritzen,
> That precisely the wealthiest in the world
> Own the most money?"

James goes on to assert a similarity among a number of words ending in -th; these words may *seem* to denote qualities that cause certain processes to arise, whereas they really denote *only* those processes themselves. "Wealth," for instance, is not a *cause* of having much money; it *is* just having much money. Analogously, James argues, health, strength, and truth consist only in those processes from which we infer their presence. Specifically, the truth of an idea consists in the process by which it works satisfactorily to adjust our ideas to each other and to experience in a maximally simple, secure, and labor-saving manner (James, 1908, p. 58). Later, James (1909) tried to show that his account of truth was not really in conflict with realism. It is debatable whether he succeeded in this (see, for instance, Lovejoy, 1963); be that as it may, the text of his earlier book stands as one of the clearest and most forthright statements of the antirealism which characterized the classic era of naturalistic epistemology. This antirealism appears under a variety of names—most notably, pragmatism, instrumentalism, conventionalism, phenomenalism, and idealism. Elsewhere (Skagestad, 1979), I have argued for a restricted use of the word *pragmatism*; here, however, I shall follow Campbell and use the word in its ordinary sense.

As stated by James, this naturalistic antirealism seems almost self-evidently true. Nor is it purely a philosopher's quirk. It has seeped down to the behavioral sciences, where it has long been influential under the name of epistemological relativism. (Campbell employs this term for purposes of his own and has coined the term *ontological nihilism* for the doctrine here in question; see Campbell, 1977, p. 22. I cannot follow him in this nomenclature, since I do not expect anyone to confess to being an ontological nihilist.) A moderate version of epistemological relativism holds that the question of the absolute truth or validity of beliefs cannot be meaningfully raised in the behavioral sciences. Once we have ascertained the economic, sociological, or psychological *causes* of the origin of a belief, as well as the *functions* through which it has been maintained, there is no further question to be asked about truth or validity; we can at most ask how well the belief fulfills its particular function within its social and cultural context. By largely focusing in their substantive work on those irrational causes and functions which tend to distort our perception of reality, epistemological relativists have frequently—though often, no doubt, unintentionally—implied a more extreme thesis, approaching skepticism. When Erikson and Fromm describe the role of anxiety and identity crises in the formation of religious beliefs, and when Kuhn describes the role of indoctrination, peer pressure, and Gestalt switches in the formation of scientific beliefs, the reader is led to wonder whether human belief, being subject to this kind of causation, is the sort of thing that ever *can* be true.

Campbell's Hypothetical Realism

In recent years a handful of thinkers have recognized the problem of reconciling realism with epistemological naturalism and have tackled it from different perspectives; among them are Hilary Putnam, Abner Shimony, Karl Popper, and Donald Campbell. I do not mean that these thinkers necessarily call themselves naturalists; the term is here and throughout used in a generic sense to denote a broad and heterogeneous tradition. Both the problem and his own desiderata for a solution were forthrightly stated by Campbell in his contribution to a volume on Popper:

> For both Popper and the present writer the *goal of objectivity* in science is a noble one, and dearly to be cherished. It is in true worship of this goal that we remind ourselves that our current views of

reality are partial and imperfect. We recoil at a view of science which recommends we give up the search for ultimate truth and settle for practical computational recipes making no pretense at truly describing a real world. Thus our sentiment is to reject pragmatism, utilitarian nominalism, utilitarian subjectivism, utilitarian conventionalism, or instrumentalism, in favor of a critical hypothetical realism. Yet our evolutionary epistemology, with its basis in natural selection for survival relevance, may seem to commit us to pragmatism or utilitarianism [Campbell, 1974a, p. 447].

Campbell's answer to the dilemma is that the "truth" which is commonly ascribed to scientific theories is *often* nothing more than a pragmatically justified fit between theory and data, although the exact nature of such fit is usually more complex than most instrumentalists have made it out to be. Thus, nonrealist epistemologies provide beneficial antidotes to the naively realistic belief in the literal and exact truth of existing scientific theories. Scientific theories are and remain conjectural; they are blind variations, originally unjustified and later selectively retained for their presumptive truth value. The accumulation of knowledge is a profoundly wasteful process of trial and error; we err most of the time in the original invention of hypotheses, and we frequently err in our judgment of which hypotheses to retain. Finally, and perhaps most importantly, we have learned from the skeptical tradition in analytic epistemology, best exemplified by Berkeley and Hume, that we never can have any assurance of the absolute truth of any belief, no matter how well tested and confirmed. We are "cousins to the amoeba," and we can know nothing for certain. Insofar as pragmatism and instrumentalism help bring home this lesson, they perform the valuable office of undermining epistemological complacency and dogmatism.

On the other hand, the recognition that we can never know the truth for certain in no way implies that there *is* no objective truth. Following Popper, Campbell conceives of truth as the goal toward which science is striving; we can never claim to have reached the goal, but we can always try to get nearer to it. On the way we tentatively accept beliefs on the basis of their pragmatic or empirical adequacy, but we know that each one of our beliefs may subsequently have to be rejected as false; so the truth of a belief cannot consist in that adequacy from which we tentatively infer its truth. A belief which has been pragmatically adequate for generations may still turn out to have been false. Hence, as Campbell expressed it in his Williams James Lectures in 1977, his position is "to accept the *correspondence meaning of truth and goal of science,* and to acknowledge *coherence* as the major but still fallible *symptom of truth*" (1977, p. 20). The hypothesis of correspondence, it is true, is always underdetermined by the evidence of coherence; but *all* hypotheses are similarly underdetermined by the available evidence for them.

The notion of objective truth, then, is *compatible* with epistemological naturalism. In his 1974 essay, however, Campbell goes further and at least strongly suggests a more intimate kinship between naturalism and realism. The strategy of naturalistic epistemology (which Campbell refers to variously as "descriptive" and "evolutionary epistemology") is to study human knowledge processes the way we study animal knowledge processes. The paradigm case is, of course, the trial-and-error processes of the rat in the maze. The rat knows only its own trials (that is, theories) and the sensory data by which those trials are checked and corrected. But the experimenter has no qualms about including the maze in his description of the learning process, even though the rat knows the maze only through its sensory data. The hypothetical realist merely transfers this triadic model of theory-data-environment onto the study of man, hypothesizing that here, too, there is an external environment responsible for producing the sensory data: "Having thus made the real-world assumption in [the animal] part of his evolutionary epistemology, he is not adding an unneeded assumption when he assumes the same predicament for man and science as knowers" (Campbell, 1974a, p. 449).

What has now happened to the vacuity which James charged against the concept of objective truth? Campbell's answer is that, if he were doing the epistemology of his own personal knowledge, it would indeed be gratuitous and vacuous to assume an independently existing real world, to which that knowledge corresponds. But Campbell is not really addressing the question "How can *I* know?" That question, he holds, was shown by Hume to be unanswerable; I can have no guarantee that my beliefs correspond to anything real. Still, in a weak sense of "knowing," I have tentative and fallible knowledge of other knowers *and* of their environments. In an "epistemology of the other one," there is nothing vacuous in making use of my knowledge both of the knower and of the environment which he inhabits—the "real world." I have here the same kind of knowledge of the real world as I have of the knower and his beliefs, and I am therefore fully entitled to compare these and to describe the beliefs as corresponding or not corresponding to the world. Such a comparison becomes vacuous in cases where I know the real world only via the knower's beliefs, as is the case when this epistemology is extended to modern physics: "It is true that in extending this 'epistemology of the other one' to knowledge of modern physics, no separate information on the world-to-be-known is available with which to compare current physical theory. But this practical limitation does not necessitate abandoning an ontology one is already employing" (Campbell, 1974a, pp. 449-450). As Campbell has made clear elsewhere (1974b, esp. pp. 139-140), we can still, without tautology, remain hypothetical realists with respect to modern physics. That is, we can still—presumptively and without compelling evidence—hypothesize that there is a real world with which physical theory is attaining a gradually increasing fit, although we lack the data for actually carrying out the description of how *current* physics is accomplishing this.

Such, in its briefest outline, is Campbell's reconciliation of hypothetical realism with epistemological naturalism. It would have been pleasant, at this point, to go on to show his concrete applications of these ideas in the field of social psychology. But a mere repetition of Campbell's ideas would be a poor tribute to a great mind who relishes a critical discussion above all else. I shall therefore devote the remainder of this paper to pursuing certain questions of realism which I believe Campbell has left unanswered, or has not answered satisfactorily. These questions are: What exactly does hypothetical realism say? What does it explain? And how can it be tested? In groping toward answers to these questions, I find myself in an intellectual debt to the same tradition to which Campbell pays homage. More specifically, though, I am indebted to one particular strain within this tradition, a strain which goes back at least to Charles S. Peirce and which is represented today by, among others, Abner Shimony and, if I understand him correctly, Hilary Putnam.

Content of Realism

My reservations about Campbell's evolutionary model are on record elsewhere (Skagestad, 1978) and need not be repeated here. I believe Campbell will probably agree to a formulation of hypothetical realism along the following lines: There is a real, theory-independent world, which exists independently of what we know about it and even independently of our existence as knowers. Science, as a historical entity, has progressed and is progressing by way of a gradual approximation toward a true description of this real world. Absolute truth, in the sense of an exact correspondence or fit between scientific theory and the world, is something which science cannot at any time claim to possess; it is a goal toward which science can only hope to converge in the indefinite long run. Note that no attempt is here made at giving an analytic definition of the word *science*. By *science* I am referring to a concrete tradition, which arose at some particular (albeit unspecifiable) point in time and which may some day die out. Like "the Protestant ethic" or "the spirit of capitalism," science is what Max Weber ([1920] 1958, pp. 47-48) called a "historical individual." Its ontological status is thus not unlike that of a biological species, a parallel of which Campbell (1974a, 1974b) has made much use.

If this is what realism says, then it seems to me that Campbell is conceding too much to the opponents of realism when he appears to grant that there is no evidence of fit between contemporary physical theory and the real world. Thus, he begins his article "Unjustified Variation and Selective Retention in Scientific Discovery" (Campbell, 1974b, p. 139) by asking, among other questions: "Do you marvel at the achievements of modern science, at the fit between scientific theories and the aspects of the world they purport to describe?" He then proceeds to consider an objection raised by many philosophers: "How can we claim a fit between the theories of science and the real world, when we know that real world only through the theories of science?" This question, however, Campbell declines to answer: "I myself have no quarrel with such philosophers. I must concede that I do not have compelling evidence of fit in any logically entailing sense, neither by deductive nor by inductive logic (were such to exist). In seeing the fit, and the puzzle, I do so only on the basis of presumptions which go beyond my capacity to verify or compellingly to demonstrate to another person" (1974b, p. 140). This, I believe, is a good enough retort to the total skeptics, who refuse to believe anything without logically compelling evidence. I, too, have no quarrel with such philosophers. As a realist, however, I do feel an obligation to answer those phenomenalists, instrumentalists, pragmatists, and conventionalists who have shared Campbell's (and my own) marvel at the achievements of science, but who have felt that the assertion of a fit between science and a real world revealed to us only through science is vacuous, and who have concluded that scientific progress must therefore consist in something else than increase in such fit. These philosophers are not necessarily asking for compelling evidence; but they are surely justified in asking for *some* evidence which may persuade them that realism is at least not tautological.

Campbell's use of the analogy with the rat in the maze will not, I fear, do for an answer. A phenomenalist, for example, may construe the rat, its data, and the maze in terms of sense data which exist only relatively to man's sensory organs, with no reference to a real world. This, he may say, is the proper ontology for studying rats in mazes, and this is the ontology which we may legitimately transfer to the study of man. If, in the latter case, we hypothesize a real world, then we are adding an extra assumption which will be vacuous unless it can be shown to have independent empirical consequences.

There is a different answer which I believe Campbell may make. The realist hypothesis, as sketched at the beginning of this section, does *not* assert an exact fit between current physical theory and the real world. Quite the contrary: it is part of the realist hypothesis that the fit is not exact and that there is room for indefinite future improvement in the fit. So the real world, on this hypothesis, is a different world from the one revealed to us by current physics, and the claim that current physical theory *approximately* fits the real world is at least not logically vacuous—although it may prove to be empirically vacuous. In order to secure the empirical content of hypothetical realism, we may follow up a suggestion which Campbell made in 1959, and which I believe he might have made more use of in his later advocacy of realism. In answering an objection from Arne Naess against "maze epistemology," Campbell acknowledged the limitations of such an epistemology for the study of the behavior of contemporary physicists; but he then added: "This limit is perhaps avoidable by the study of the problem-solving behavior of scientists at the earlier stages of a now well-developed science" (Campbell, 1959, p. 157). Once we admit the indirect strategies of argument by analogy and extrapolation, we can bring empirical considerations to bear on the truth or falsity of realism and thereby satisfy ourselves that it has indeed an empirical content.

Recently this same idea has been elaborated by Abner Shimony in a reply to the relativism of Thomas Kuhn. Against Kuhn's rejection of the idea of a "match" between theory and reality, Shimony proposes the "Hypothesis of Verisimilitude," which states that *"for well-established theories in a science in which high critical standards have been achieved, such a match exists to some good degree of approximation"* (Shimony, 1976, p. 574). Admitting the vagueness of this formulation, Shimony goes on to propose that the notion of a good degree of approximation to a match can be given a substantive content if one draws on analogies from the history of science, because "the relation between any actual theory . . . and the ideal theory which exactly matches the truth is an extrapolation from the relation holding between two actual theories" (1976, p. 576). This idea is reminiscent of certain ideas promulgated by Peirce, whose influence Shimony has repeatedly acknowledged. Peirce's approach was that, since we have a historical experience of a general drift toward consensus within the scientific community, we may understand objective truth—and hence the notion of reality—by extrapolating from this experience the notion of a universal, unshakable consensus toward which inquiry is converging as a limit, and which we would reach if inquiry were to be pursued for an indefinite length of time (Peirce [1871], 1958a, p. 12). Clearly, there is no point in time at which we can claim to know what that final consensus will be; yet it is characterized as the outcome of a process described in empirical terms, in terms of our historical experience of scientific inquiry. In this way, by means of extrapolation, we can give empirical—even operational—content to the concept of a reality, although the knowledge of that reality may elude us indefinitely. This formulation enables us to answer, not indeed the total skeptics, but at least our instrumentalist and phenomenalist opponents within the tradition of epistemological naturalism.

This is so far a somewhat barren statement of hypothetical realism. It is, I hope, clear that realism is neither logically nor empirically empty; it may still be thought to be an idle theory which does no real work within a naturalistic epistemology. To show that it is a fruitful and interesting theory, we shall have to show that it explains otherwise puzzling facts and that we can rationally judge of its truth or falsity by devising relevant empirical tests.

Explanatory Power of Realism

I am not sure whether Campbell thinks of the realist hypothesis as being truly explanatory. In one article at least, he describes the fit between scientific theories and reality as itself a puzzling fact which requires explanation (Campbell, 1974b). Now, an explanation may often itself stand in need of a higher-order explanation; for instance, an empirical law explains particular facts and is itself explained by an overarching theory. So, though Campbell seems to think of realism as something that needs to be explained, I believe he will agree that realism is not an observed or assumed fact, which believers in scientific progress are entitled to take for granted. Realism is a *theory*; and, as Peirce frequently insisted, "nothing can justify a theory except its explaining observed facts" (Peirce [1897], 1931, p. 170).

What, then, does realism explain? Peirce suggested, among others, two puzzling facts which are explained by the realist hypothesis. One is the fact of error elimination, or, on the phenomenal level, the fact of exchanging one belief for another. There would be no point in changing beliefs only in order to go on changing beliefs indefinitely. This activity of changing one's beliefs can be rendered rational only as an attempt to reach a belief which will not thereafter need to be changed. Thus, the real—understood as the object of a stable belief not subject to further correction—

serves here as an explanatory notion (Peirce [1868], 1935, p. 311). A second fact for which Peirce proposed realism as an explanation is the puzzling fact, already alluded to, of a drift toward consensus in science; that is, the fact that independent investigators, starting from different assumptions and different observations, tend ultimately to arrive at the same conclusion. This fact, according to Peirce is explained by the hypothesis that their different inquiries are directed toward one and the same reality: "This hypothesis I say removes the strangeness of the fact that observations however different yield one identical result. It removes the strangeness of this fact by putting it in a form and under an aspect in which it resembles other facts with which we are familiar" (Peirce [1873], 1958b, p. 335).

It is necessary here to enter certain caveats, of which Peirce was certainly aware. In the first place, although in the first quote the notion of the real is invoked to account for the distinction between stable and unstable belief, it does not follow that we can simply infer the truth of a belief from its stability to date. Stability of belief may be the effect of limited experience or of sheer stubbornness, and a false belief—such as the belief that heavier objects fall faster than lighter ones—may persist for ages. A true belief is one which will never be overthrown, and we have no guarantee that any one of our beliefs enjoys this status. Still, when a belief—such as the belief that heavy objects fall when unsupported —persists unchallenged through major scientific and conceptual revolutions, we may explain this stability either by saying that the belief is true or by saying that it is sheer coincidence, and the latter is no explanation at all. It is, of course, *conceivable* that all beliefs once thought to be true will turn out to be false; in that case, there will be no empirical phenomenon for realism to explain.

In the second place, when Peirce invokes realism to explain agreement among independent investigators, it is important to note that we can rarely be assured, in any individual instance, of the genuine independence of the investigators. We are all too familiar with the situation where two investigators, trained in the same research tradition, "independently" arrive at the same tentative conclusion and are both convinced of its truth merely by discovering their mutual agreement. In a recent article in *Encounter,* the psychiatrist Michael Shepherd quotes an amusing example from

Hugh Trevor-Roper. Having studied the records of Hitler's child-hood, the psychoanalyst Walter Langer started relating the symp-toms to a female colleague, who quickly interrupted him and pro-duced a diagnosis. "And," Langer later reported, "to my utter amazement, she was right!" (Shepherd, 1979, p. 37). Trevor-Roper was not amazed but proceeded to comment that he had no difficulty imagining two demonologists in seventeenth-century Württemberg, exchanging gossip about a certain suspicious old lady and simultaneously reaching the same conclusion: "and 'by God!' exclaims one to the other, slapping his thigh and pouring out another gurgling Pokal, 'you're right!' " (Shepherd, 1979, p. 37). Such agreements on specific cases among people sharing the same theory require no special explanation. More puzzling are indepen-dent simultaneous discoveries of novel scientific theories. Even here, the independence may be spurious. A well-known example is the simultaneous discovery of natural selection by Darwin and Wallace. They had similar background and training, they had simi-lar experience as tropical field naturalists, and they both testified to having been influenced by reading Malthus's *Essay on the Prin-ciple of Population*. Even more strikingly, on the one issue where they differed, that of the evolution of man, the difference is ac-counted for by biographical differences. Darwin was the product of a culture which assumed the racial superiority of Englishmen; for him it was easy to see a gradual evolution from primates, via savages, to civilized men. Unlike Darwin, Wallace had an intimate knowledge of people living under primitive conditions; he had lived for years among the Malays and had found them as intelli-gent as, and morally rather superior to, his own countrymen. Hence, for Wallace there was between apes and men a chasm which could not be bridged by the piecemeal and gradual action of natural selection. (See Eiseley, 1961, ch. 11.)

One could go on like this; and if, in each case of agreement reached in science, there turned out to be a hidden hand of com-mon influences guiding the investigators, there would, once again, be nothing for the realist hypothesis to explain. But nothing of this sort has been established, and the pervasiveness of increased consensus in science and the ubiquity of independent discovery re-main unexplained data, calling for an explanation. In a sum, what realism explains are the facts that well-established beliefs in science tend to remain stable and that investigators independently addressing the same question tend ultimately to come up with the same answer. These facts, as we have seen, could conceivably be different; but, as far as we fallibly know from our partial and limited historical experience, they are not. We can easily imagine worlds in which realism would have no explanatory power, but realism does explain what we believe to be facts in our world. Moreover, to the best of my knowledge, these facts have not been doubted by the instrumentalist or phenomenalist opponents of realism—a point which, by the way, is well brought out by Shimony (1976, pp. 574-575).

Testability of Realism

Explanatory power is intimately yet ambivalently related to testability; this relationship, it appears to me, reflects a certain internal ambivalence in the concept of explanatory power itself. It has often been thought that explanatory power and testability are directly correlated: the more a hypothesis explains, the more highly testable it is, and conversely. So, for instance, a theory such as that of universal gravitation, which explains both terrestrial and celestial motion, can be tested both by laboratory experiments and by astronomical observations. Likewise, a theory which explains both electricity and magnetism can be tested by a wider range of facts than one which explains only magnetism. At the same time, it has been recognized that a theory becomes untestable after it has become too powerful. Thus, divine providence, sufficient reason, economy of nature, the class struggle, and childhood repression, for example, are hypotheses which explain everything that can conceivably happen and which cannot, therefore, be refuted by any conceivable event; as a result, we cannot devise empirical tests by which to judge of their truth or falsity. In these cases, we would also be inclined to dismiss the proposed hypotheses as unsatisfactory explanations—without thereby impugning their explanatory *power*.

All this seems to indicate that the goodness of an explanation is indeed linked with its testability, but the linkage is not a simple correlation between degree of testability and degree of explanatory power. An explanation loses its explanatory value once it becomes too powerful. We might say that the goodness of an explanation consists in that happy medium in which the explanation explains neither too much nor too little, but just enough to admit of being tested. I have given some examples of explanations which explain *too much* to be testable; at the opposite extreme are those which explain too little to be testable. To this latter category belong a great many explanations of particular and unrepeatable events. Historical explanations may often be of this kind, as may medical diagnoses, especially post mortem diagnoses. If, for instance, a patient dies during surgery and no observable cause of death can be assigned, his death might be explained as due to an idiosyncratic reaction to anesthesia. Since this reaction is idiosyncratic, there is no known evidence of it for the pathologist to look for during autopsy; and since the patient is dead, the diagnosis cannot be tested by what it tells us about his probable future reaction to anesthesia. Yet such reactions do occur, and they can cause a patient's death. If, in a particular case, the hypothesis of an idiosyncratic reaction were true, it would explain the patient's death. Once all other explanations have been ruled out, this one remains as the only one compatible with the evidence. Yet no pathologist would think of it as a good or satisfying explanation; it differs from an admission of ignorance only by its implication that all known causes have been investigated and ruled out.

In order to be testable, then, realism must not explain everything which might happen, but it must explain something more than that initial body of data for which it has been invoked as an explanation and for which its chief rival, phenomenalism, also claims to provide an explanation. One thing we must frankly admit is that there can be no crucial experiment whereby we can effect a neat Popperian falsification of one view or the other. We have already seen that what realism purports to explain are general classes of facts, and not individual events. Therefore, no single experimental result can overthrow realism. Any realist would concede that there are exceptions to both the stability of scientific beliefs and the drift toward convergence in science. To be forced to abandon realism, he would have to be persuaded that there is not *in general* any trend of stability or convergence, and these negatives can be established only by the accumulation of inductive evidence. It is to be expected that whatever else realism explains will be classes of facts of the same general nature. So realism cannot be experimentally disproved. The same holds for its rivals, such as phenomenalism, instrumentalism, and pragmatism. It does not follow that we cannot rationally choose one or another of these alternatives on empirical grounds; if, on some issue of contention, the inductive evidence were to accumulate consistently against realism and in favor of phenomenalism, there would come a point at which realism would no longer be rationally defensible. This type of testability is expounded with especial clarity by Hilary Putnam, in his *Meaning and the Moral Sciences*. Putnam takes realism to be the hypothesis that theoretical terms in science refer to real objects; and this hypothesis, he holds, explains the contribution which science makes to the success of our total conduct. Noting that certain theoretical terms—such as *phlogiston* and *ether*—have been found in the past not to refer to real objects, Putnam suggests that the history of science may serve as a testing ground for inductively overthrowing realism through the cumulative confirmation of its denial: "Eventually the following meta-induction becomes overwhelmingly compelling: *just as no term used in the science of more than fifty* (or whatever) *years ago referred, so it will turn out that no term used now* (except maybe observation terms, if there are such) *refers*" (Putnam, 1978, p. 25). The possibility of confirming this meta-induction ensures the testability of realism through inductive disconfirmation, which may in the end attain the cumulative strength of refutation.

Putnam has, I think, convincingly shown how a theory of the scope of realism may be testable; what remains is to devise a test which focuses specifically on the difference between realism and its known rivals. For this purpose, as Shimony (1976, pp. 572-573) has emphasized, we need to lay down certain ground rules for discussion. As has been demonstrated by Berkeley, Mach, and James, among others, a phenomenalist may always formulate his theory in such a manner that there can be no rational considerations by which to decide between it and realism. In such a case, Shimony (1976, p. 573) insists, the phenomenalist must be regarded as having withdrawn from the arena of rational discussion by reducing the disputed issue to a pseudoproblem. The realist-phenomenalist issue is a real one only to the extent that the phenomenalist will admit a formulation of his theory which permits empirical evidence or other rational considerations to be brought to bear on the choice between the two theories. Until phenomenalists themselves produce such a formulation, they are rationally obliged to accept the formulation offered by their realist opponents.

One testable difference between realism and phenomenalism may be formulated as follows. If Peirce and Shimony are right, both the stability and the drift toward convergence in the development of science can be accommodated by phenomenalists after the fact, but they cannot be predicted by phenomenalists, since they are accounted for only as coincidences. What this suggests is that the testing ground for the realist hypothesis lies in the intersection between the history of philosophy and the history of science—specifically, in the numerous episodes where scientists have been guided (or misguided) by philosophical considerations. In those episodes, on the hypothesis that scientific progress consists in gradually unveiling the truth about the real world, we should expect realistically inspired scientists to take the lead over phenomenalistically inspired ones in promoting novel theories and accepting novel discoveries. The reason for this is the following. It is easy for a phenomenalist to account for all the facts acknowledged by science at any one time, and it is not impossible for him to account for all the scientific change that has already taken place. Mach (1893), for instance, excelled on both counts. If scientific change typically takes place the way phenomenalists say it does— to wit, by adjusting theories to observed phenomena in an increasingly convenient and economical manner—then there is no reason why phenomenalists should not also initiate scientific change. But if, as the realist believes, we make progress in science by introducing new ideas about how things are, and only afterward discover phenomena by which to test these ideas, then it would be quite irrational for phenomenalists to take the lead. To do that, they would have to abandon the already achieved adjustment between theory and phenomena and commit themselves to a less well-adjusted theory, in the hope that, as the conceptual, mathematical, and experimental apparatus of the new theory was developed, a still better adjustment would eventually result. And we may rationally hope this only if we believe that our theory refers to real objects, which may later manifest themselves through phenomena that we have not yet discovered. On the realist view of scientific progress, therefore, it follows that realists have rational grounds for commiting themselves early to new theories, while phenomenalists are rationally obliged to adopt a wait-and-see attitude. On the assumption that scientists are on the whole rational, at least in their scientific pursuits, we can then put realism to an empirical test by looking at what has actually happened in the history of science.

Having thus specified how realism may be treated, I have completed the agenda I proposed for this chapter; I do not propose on this occasion to go on and actually carry out the test. I cannot, however, refrain from at least listing a few cases that do seem to confirm the realist hypothesis: Galileo versus Bellarmino on the reality of the heliocentric system, Pasteur versus Bernard on the existence of microorganisms in ferments, Planck versus Mach on the existence of atoms, and Bateson versus Pearson on the existence of genes. Of these, probably the most clear-cut case is that of Planck versus Mach, where Mach told Planck fairly explicitly that he himself had no need of atoms in his physical theory, *but* he would have no difficulty in making room for atoms should they turn out to exist (Mach, 1919, pp. 10-11). In a way, posterity has been right in according Mach the victory in the philosophical dispute; Planck was knocking his head against a wall by citing the evidence of contemporary physics against a version of phenomenalism which had been rendered immune to that kind of counterevidence. But what is significant for the kind of test I have devised is the historical fact that Mach could score a philosophical victory at the same time as he was being left behind by the progress of the science of his day. To repeat, though, there can be no crucial experiment, and nothing can be made to depend on a single case. The phenomenalist, no doubt, will be able to cite different cases or give different interpretations of the ones I have mentioned. It may be noted, for instance, that Pasteur himself fell behind in his dispute with Berthelot over the possibility of biochemical reduction (see Roll-Hansen, 1974, pp. 72-76) and that Bateson was scarcely more of a realist than Pearson (see Coleman, 1970). A fair test must consider the evidence cited on both sides and judge on what side the balance comes down. It might be a draw, in which case we should have to think up some other test or else be forced to regard the issue of realism versus phenomenalism as indeed a pseudoproblem.

In this chapter I have suggested tentative answers to the three questions of what realism says, what it explains, and how it may be tested. In so doing, I have found it necessary to go beyond Campbell's work and to draw on the writings of C. S. Peirce, Abner Shimony, and Hilary Putnam, while adding some (hopefully) novel suggestions of my own. In closing, I wish to emphasize that these questions not only arise logically out of Campbell's work but have arisen in my own mind largely as a result of Campbell's inspiring and stimulating influence—an influence exercised chiefly through lectures and conversations during his stay at Harvard in the spring term of 1977. Whether or not Campbell will agree with my answers, I hope at least to have given a small stimulus to a further pursuit of the questions and to the continuation of that fruitful exchange of opinion which is the lifeblood and essence of philosophy, as well as of science.

References

Campbell, D. T. "Methodological Suggestions from a Comparative Psychology of Knowledge Processes." *Inquiry,* 1959, *2,* 152-182.

Campbell, D. T. "Blind Variation and Selective Retention in Creative Thought as in Other Knowledge Processes." *Psychological Review,* 1960, *67,* 380-400.

Campbell, D. T. "Evolutionary Epistemology." In P. A. Schilpp (Ed.), *The Philosophy of Karl Popper.* La Salle, Ill.: Open Court, 1974a.

Campbell, D. T. "Unjustified Variation and Selective Retention in Scientific Discovery." In F. Ayala and T. Dobzhansky (Eds.), *Studies in the Philosophy of Biology.* London: Macmillan, 1974b.

Campbell, D. T. "Descriptive Epistemology: Psychological, Sociological, and Evolutionary." William James Lectures, Harvard University, 1977.

Coleman, W. "Bateson and Chromosomes: Conservative Thought in Science." *Centaurus,* 1970, *15,* 228-314.

Eiseley, L. *Darwin's Century.* New York: Doubleday, 1961.

James, W. *Pragmatism: A New Name for Some Old Ways of Thinking.* New York: Longman, 1908.

James, W. *The Meaning of Truth: A Sequel to "Pragmatism."* New York: Longman, 1909.

Lovejoy, A. O. *The Thirteen Pragmatisms.* Baltimore: Johns Hopkins University Press, 1963.

Mach, E. *The Science of Mechanics.* La Salle, Ill.: Open Court, 1893.

Mach, E. *Die Leitgedanken meiner naturwissenschaftlichen Erkenntnislehre und ihre Aufnahme durch die Zeitgenossen* [*The Keynote of My Scientific Theory Model and Its Assimilation of the Modern*]. Leipzig: Barth, 1919.

Peirce, C. S. "Notes on Scientific Philosophy." In C. Hartshorne and P. Weiss (Eds.), *Collected Papers of Charles Sanders Peirce.* Vol. 1. Cambridge, Mass.: Harvard University Press, 1931. (Original manuscript 1897.)

Peirce, C. S. "Some Consequences of Four Incapacities." In C. Hartshorne and P. Weiss (Eds.), *Collected Papers of Charles Sanders Peirce.* Vol. 5. Cambridge, Mass.: Harvard University Press, 1935. (Originally published 1868.)

Peirce, C. S. "Fraser's Edition of *The Works of George Berkeley.*" (Review.) In A. Burks (Ed.), *Collected Papers of Charles Sanders Peirce.* Vol. 8. Cambridge, Mass.: Harvard University Press, 1958a. (Originally published 1871.)

Peirce, C. S. "The Logic of 1873." In A. Burks (Ed.), *Collected Papers of Charles Sanders Peirce.* Vol. 7. Cambridge, Mass.: Harvard University Press, 1958b. (Original manuscript 1873.)

Putnam, H. *Meaning and the Moral Sciences.* London: Routledge and Kegan Paul, 1978.

Roll-Hansen, N. *Forskningens frihet og nödvendighet* [*The Freedom and Necessity of Scientific Research*]. Oslo: Gyldendal, 1974.

Shepherd, M. "The Psychohistorians." *Encounter,* 1979, *52* (3), 35-42.

Shimony, A. "Comments on Two Epistemological Theses of Thomas Kuhn." In R. S. Cohen, P. K. Feyerabend, and M. W. Wartofsky (Eds.), *Essays in Memory of Imre Lakatos.* Dordrecht, Netherlands: Reidel, 1976.

Skagestad, P. "Taking Evolution Seriously: Critical Comments on

D. T. Campbell's Evolutionary Epistemology." *The Monist*, 1978, *61* (4).

Skagestad, P. "Peirce and Pearson: Pragmatism Versus Instrumentalism." Paper presented at the Boston Colloquium for the Philosophy of Science, Boston University, February 6, 1979.

Weber, M. *The Protestant Ethic and the Spirit of Capitalism*. New York: Scribner's, 1958. (Originally published in German 1920.)

5

Abner Shimony

Integral Epistemology

This chapter is both an appreciation of the epistemological contributions of Donald Campbell and a statement of an epistemological program which is different from his in several respects.

In a lecture to the Boston Colloquium for the Philosophy of Science in 1977, he said:

> What I am doing is "descriptive, contingent, synthetic epistemology." ... I make a sharp distinction between the task and permissible tools of descriptive epistemology on the one hand and tradition-

Note: The research for this chapter was supported in part by the National Science Foundation.

al, pure, analytic, logical epistemology on the other. Descriptive epistemology is a part of science rather than philosophy, as that distinction used to be drawn by philosophers. It is science of science, scientific theory of knowledge, were those terms not too pretentious for the present state of the art.

While I want descriptive epistemology to deal with normative issues, with validity, truth, justification of knowledge—that is, to be epistemology—descriptive epistemology can only do so at the cost of presumptions about the nature of the world and thus beg the traditional epistemologist's question [Campbell, 1977a, p. 1].

Campbell's resolute restriction of his investigations to descriptive epistemology is both his great strength and his weakness. It is his strength because it frees him, at one stroke, from the slow-paced type of inquiry that dominates the literature of analytic epistemology; for example, "When I see a tomato there is much that I can doubt" (Price, 1932, p. 33). His investigations lead out of the study into the open air. There is a wonderful sweep in his survey of the stages of cognitive development (Campbell, 1974, pp. 422-434). In the perspective that Campbell offers, nature is in no way subservient to humans; however, because of the sequence of adaptations to nature which occurred in the human phylogeny, we are able to achieve something approaching objective knowledge.

The weakness in Campbell's program is that the traditional problems of analytic epistemology continue to be haunting, especially when answers to some of them are postulated as preconditions for his own investigations. He postulates an ontology of independent real objects: that there are entities in the universe which do not depend for their existence upon their being perceived or known by human beings. He accepts without argumentation the correspondence theory of truth: that a sentence is true if and only if the state of affairs which it expresses is the case, so that the truth of a sentence does not depend upon its being believed, or upon the utility of believing it, or upon the existence of evidence supporting it. He postulates a causal theory of perception: that one can understand the content of perceptual experience only by taking into account the causal relations which link the perceiver to objects existing independently of the perceiver's faculties. If Campbell's descriptive epistemology entirely abstained from normative questions, then there would be nothing wrong in principle with such postulation. It would be analogous to putting one's trust in the mathematicians and assuming the correctness of useful mathematical theorems without checking the proofs oneself. Since, however, Campbell wishes to deal with normative questions, especially with the justification of knowledge, he cannot avoid the problem of justifying his postulates.

This chapter will propose an *integral epistemology,* in which certain methods of descriptive epistemology (which Campbell espouses) and certain methods of analytic epistemology (from which he abstains) are combined for the purpose of rationally assessing claims to human knowledge. It is anticipated that the results of scientific investigation about human beings will shed light on the reliability of human cognition, and reciprocally that adequate justification can be given for the presuppositions of scientific investigations. The proposed integral epistemology is unequivocally naturalistic, following Campbell not only in his general thesis that a necessary condition for understanding human cognition is to see man's place in nature but also in his insistence that detailed attention to the sciences is indispensable for solving epistemological problems. The proposed approach differs from his in its envisagement of a dialectical structure of epistemology and in its resort to methodological, decision-theoretical, and semantic analysis.

A disclaimer should be made at the outset with regard to novelty, not just because of the usual obligation to acknowledge intellectual indebtedness but because an integral epistemology is a synthesis by its conception. A naturalistic view of human knowledge is at least as old as Aristotle's *De Anima,* though it has been greatly expanded by applications of the theory of evolution; and the thesis that epistemology has a dialectical structure goes back, of course, to Plato. Sustained attempts to incorporate naturalistic epistemology into a dialectical framework are, however, uncommon. Among classical philosophers, Peirce seems to come closest to the integral epistemology which I envisage, and the contemporary philosophers who approach it most closely—notably, Quine (1969) and Rescher (1977)—acknowledge their affinity to him. Rescher explicitly characterizes his epistemology as dialectical, but

Quine (1974, p. 137) does not seem to like the word, in spite of his espousal of a philosophical methodology which seems to me unequivocally dialectical. A dialectical naturalism is also outlined in Shimony (1970). In any case, much work remains to be done, since epistemology is an architectural discipline, and its components from various sources must be combined into a harmonious edifice. The work is not likely to be completed in the foreseeable future, if only because the components supplied by biology and psychology can be expected to continue their dramatic development.

Some Dialectical Theses

A dialectical method offers a reasonable resolution to a difficulty which is intrinsic to the enterprise of epistemology. If epistemology is a comprehensive study of human cognition, then the question of the validity of its own procedure falls within its domain. How, then, does the enterprise rationally begin? One radical type of answer to this difficulty is to identify incontrovertible initial propositions, guaranteed, for example, by clarity and distinctness, by self-certification, or by intuition; a radical answer at the opposite extreme is skepticism. In the work of a number of philosophers who call themselves "dialectical," one finds several characteristic features which permit them to navigate with control between these two extremes. Tentative suppositions are accepted at the beginning of inquiry, but they are subject to criticism and may be revised and refined on examination. Fundamental principles are, then, the end rather than the beginning of inquiry. The terms in initial suppositions are often vague, clarification of meaning being inseparable from the assessment of propositions. Most important, the dialectic is open, with no foregone conclusions and no suppositions that are so entrenched that they cannot be critically evaluated. In particular, the openness applies to the dialectical method itself. Inquiry may show that the dialectical method is in need of supplementation or that it is only a temporary expedient on the way to a deeper method.

Because it is characteristic of dialectical methods to examine the initial suppositions of inquiry and to reflect on the dialectical process itself, there is a widespread opinion that these methods entail a commitment to a coherence theory of truth—the theory that the truth of a sentence consists in its properly fitting into a system of assertions. The opinion is correct concerning some historically important versions of dialectic—notably, Hegel's—but it is not correct as a generalization. The integral epistemology envisaged in this paper agrees with Campbell's recommendation "to accept the *correspondence meaning of truth and goal of science,* and to acknowledge *coherence* as the major but still fallible *symptom of truth*" (1977a, p. 20).

In the subsequent discussion of connections between description and justification, four procedural or organizational theses will be deployed. For the most part, these theses are implicit in the general features of dialectical methods which have already been noted.

1. *Commonsense judgments about ordinary matters of fact have a prima facie credibility, which should not be discounted without clear positive reasons.* The distinctions between truth and falsity, between reliability and unreliability, between appearance and reality are habitually made in ordinary discourse, which consists largely of expressions of the kinds of judgments mentioned in thesis 1. Perhaps anticipations of these distinctions can even be found in stages of individual development before ordinary discourse is mastered, but one can hardly doubt that a secure achievement of a commonsense view of the world is a precondition to any kind of sophisticated examinaton of these distinctions. Consequently, a sweeping skepticism about the reliability of commonsense judgments undermines the rough-hewn contrast between, for instance, true and false judgments or real and apparent properties of things, with the danger that the dialectical process of refining these epistemological distinctions will be aborted at an early stage. It must be emphasized that the presumption of truth for the specified class of judgments does not imply a full commitment to them.

2. *The road to inquiry should not be blocked.* This thesis is Peirce's "supreme maxim of philosophizing" ([1897] 1931, p. 135). The thesis is implicit in the openness of the dialectic, which was noted previously, but it is a reminder that openness is not preserved without some effort. In a later section (headed "Evolutionary Considerations"), this thesis will be applied in the course of evaluating criteria of meaningfulness.

3. *Epistemology and knowledge of the world (a vague phrase, intended to refer both to the sciences and to metaphysics) should mesh and complement each other.* This thesis really has the status of a desideratum, for there is no a priori guarantee that we are capable of "closing the circle," in the sense of giving an objective account of the knowing subject as an entity in the world in a manner that adequately explains its cognitive capacities. Nevertheless, thesis 3 is reasonable, and its failure to hold in a particular philosophical system is prima facie evidence that something is amiss, unless convincing reasons are given why it should fail (as Kant, of course, tries to do in the *Critique of Pure Reason*).

4. *A vindicatory argument—that is, an argument to the effect that a certain method, M, will yield good approximations to the truth in a certain domain if any method will, so that nothing is lost and possibly something is gained by the use of M—is an acceptable form of epistemological justification.* This thesis has a decision-theoretic character. It is similar to recommending a strategy that performs as well as or better than all other strategies in a game. The thesis has been used by Hans Reichenbach, Herbert Feigl, Wesley Salmon, and others in discussions of the justification of inductive inference. It provides, incidentally, a good example of a somewhat vague concept that becomes clarified as a result of dialectical examination. In common sense there is a vague concept of justification, the explication of which has been a major philosophical problem. The retreat, under the critical scrutiny of Hume and others, from attempts to provide a justification in a strong sense for inductive inference has constituted a remarkable extended dialogue, and the resort to vindication has been a way of salvaging something from the epistemological optimism of an earlier epoch. An important warning is that, despite its reasonableness as stated, thesis 4 is difficult to apply in actual epistemological argumentation, because one seldom can be confident that the condition "nothing is lost and possibly something is gained" is satisfied.

Although these four theses are obviously in need of further discussion, it may be more illuminating to aim at clarification in the course of applying them to concrete epistemological problems than to analyze them further in an abstract way. The following three sections will examine the epistemological relevance of three different bodies of factual information about human cognition, together with some nondescriptive considerations. The section headed "Empirical Psychology and the Reliability of Perceptual Judgments" will discuss the reliability of perceptual judgments in the light of psychological studies of perception and cognition. The section on "Evolutionary Considerations" will consider the implications of evolutionary biology for the scope of human knowledge. And the section headed "Inductive Inferences and Vindicatory Arguments" will outline a program for the justification of inductive inference, involving among other things some reliance on the history of science (which is entitled to be recognized as part of descriptive epistemology). It must be emphasized in advance that answers to epistemological problems cannot be anticipated simply as corollaries of scientific results. An appropriate deployment of factual information is impossible without analysis, if only because of the considerable differences in the questions typically posed by scientists and by epistemologists. This warning is not a revocation of claims for the power of a naturalistic point of view but is, rather, a reassertion of the conception of an integral epistemology, in which factual information is indispensable but not self-sufficient.

Empirical Psychology and the Reliability
of Perceptual Judgments

An index of the significance of empirical psychology for epistemological purposes is the breadth of the overview that it provides of some of the traditional debates of analytic epistemology. The strong points on opposite sides of debates are often seen to fall into place; and, even more impressively, the naturalistic point of view is often completely at ease with conclusions which analytic epistemologists seem to have reached reluctantly, with a sense of abnegation and retrenchment. I shall illustrate these claims by examining the epistemological problem of the reliability of perceptual judgments. (Another illustration, concerning the extent to which observation is theory laden, is given in Shimony, 1978.)

Within empiricism there has been a major debate (Swartz, 1965) between sense data theorists (such as Price and C. I. Lewis) and their critics (such as Firth and Quinton). The issue can be posed more generally than it usually is: Are there intermediate steps between the physical stimulation of the sense organs and the final perceptual judgment which justify or support the perceiver's final judgment? The candidates for the intermediate steps are not restricted to sense data. Few, if any, authors dissent from the proposition that there are unconscious causal links, at least partly neural in character, in the total process of perception, but there is disagreement about the significance of these links for claims of knowledge.

Among the arguments presented by sense data theorists, there is one that generically favors epistemic mediation, without specifying its character. The argument is that a perceptual judgment, of the sort that can be expressed in a statement about physical objects, has a content that could not be summarized in any finite body of observational data; the judgment implies, for example, what would be seen if the identified object were viewed under normal conditions from any of an infinite number of vantage points. But a judgment of such strength must be an extrapolation from the actual observational base. That sense data are the entities that mediate between physical stimuli and final perceptual judgments can be demonstrated, according to sense data theorists, in an indirect way by consideration of illusions. Since an imitation often is classified perceptually as a real object of a certain kind, something common must be presented to the observer in the two cases. A reasonable candidate for the common element is *appearance,* which is interpreted as a network of sense data.

This reasoning has frequently been criticized. Quinton (1965), for example, argues that the identification of the locution "This appears to be ϕ" with "There is a ϕ sense datum" is mistaken. He claims that the primary usage of "appears" is not to signal a definite claim to a meager bit of knowledge about the state of the subject but, rather, to signal a tentative belief in a proposition about ordinary objects. He does, however, admit that there is a secondary sense of "appears," which is applicable when persons turn their attention away from things and toward their own awareness—as, for example, in painting impressionistically or taking audiological tests. This possibility of switching to what he calls a "phenomenological frame of mind," however uncommon its occurrence may be, seems at first to support the claim that sense data are revealed directly by introspection. The revelation occurs when one performs upon oneself the operation called "perceptual reduction," in which one attempts to eliminate all elements of experience that are not unequivocally "given"—in particular, all conceptual contributions (Firth, 1965, p. 235).

The critics of sense data theories insist, however, that the phenomenological frame of mind is incompatible with the normal object-perceiving state and that characteristics discerned in the former by introspection cannot legitimately be imputed to the latter (Firth, 1965, pp. 236-237; Quinton, 1965, p. 507). Firth (1965, p. 214) asserts, furthermore, that "sensations do not occur as constituents of perceptions, but at most only as complete and individual states of mind." And Quinton (1965, p. 510) says, "When we transfer our attention from objects to experience, an enormously richer awareness is obtained. We then suppose that we were in fact having experiences of as complex and detailed a kind while attending to the objects, although we were unaware of the complexity and detail. This move is not inference supported by recollection, but a convention."

It follows from this sharp dichotomizing of mental states that perceptual judgments cannot be interpreted as the results of unconscious inference from premises concerning sense data. This conclusion does not require the rejection of unconscious mental processes. Rather, it rests on a sharp distinction between causes and reasons. Whatever the causal importance of the unconscious processes, they cannot be supposed to have the status of premises in the epistemic system of a subject who is operating in the normal mode of perception. Quinton (1965, p. 525) softens this conclusion somewhat: "Perception is an intelligent activity (not an infallible reflex). . . . But not all intellectual processes are types of reasoning." Unfortunately, however, he provides no explanatory details. I suggest that one of the important ways in which empirical psychology can illuminate traditional epistemological debates is by providing some detailed information about the "intelligent activity" at which Quinton hints.

I shall now list five propositions, expressing experimental results and theoretical ideas of recent psychology, which are relevant to the problem of epistemic mediation in perception. It should be noted in advance that the five propositions are far from being straightforward matters of fact. Each of them contains one or more terms—like "information" in proposition 1 and "expectancies" in proposition 4—which should be clarified by more conceptual analysis than there will be space for in this paper. That these propositions are drawn from different schools is not merely opportunistic, for I am convinced by a synoptic view (see Neisser, 1976, p. 24) which combines elements from information-processing theories, from hypothetico-deductive models (especially Bruner and Gregory), from Campbell's (1956) analysis of perception as substitute trial and error, from theories which emphasize the richness of stimulus information (J. and E. Gibson), and from developmental psychology (especially Piaget).

1. There are stages in the process of perception which are fairly well defined temporally and describable in such terms as the selection, transformation, encoding, storing, and use of information. Especially relevant for the debate between sense data theorists and their critics is the experimental evidence for a very brief period of iconic storage, in which an image is presented as a whole, as if offered for further examination, and a longer period, in which features are analyzed and the resulting information is stored in a short-term memory (Neisser, 1967, pp. 89-90, p. 301).

2. The experimental data concerning visual and auditory pattern recognition favor a hierarchically arranged feature-analysis mechanism, rather than, for example, a template mechanism (Neisser, 1967, ch. 3).

3. Feature analysis is not merely a matter of making decisions but requires a constructive activity of synthesizing a perceptual object: "The mechanisms of visual imagination are continuous with those of visual perception" (Neisser, 1967, pp. 94-95).

4. The perceptual process is one of active search, in two different ways. First, there is an internal search for appropriate features and appropriate concepts (Bruner, 1957). Second, there is a search of the environment for additional stimulus information to confirm or disconfirm expectancies aroused by earlier perceptual activity (Gibson, 1950; Campbell, 1956; Neisser, 1976).

5. The generic trait of search for significant stimulus information is innate, since it is exhibited in newborn infants (Bower, 1966). However, the ability to search with increased control, purposefulness, and sensitivity to salient features of the physical stimuli is a skill, which is developed naturally by a growing child interacting with the environment but which can be greatly enhanced by training (Gibson, 1969).

The epistemological importance of these results and theoretical ideas goes beyond adjudicating between the sense data theorists and their critics, though one can score points for and against both sides. It seems to count against the sense data theorists that the stage of perception which is closest to the physical stimulus and least modified by subjective processing is epistemically labile: "After a tachistoscopic exposure in Sperling's . . . experiment, the subject feels that he 'saw' all the letters, but he cannot remember most of them. The uncoded ones slip away even as he tries to grasp them, leaving no trace behind. Compared with the firm clarity of the few letters he really remembers, they have only a marginal claim to being called 'conscious' at all" (Neisser, 1967, p. 301). On the other hand, the occurrence of a stage of feature detection as part of normal perception of objects is a point in favor of the *general* position of the sense data theorists that epistemic mediation occurs in perception, with the qualification that the identification of a feature may not be a conscious occurrence. In spite of this qualm, however, the evidence for a temporal fine structure to perception at least indicates greater continuity between the "phenomenological frame of mind" and the normal state of object perception than the critics of sense data theories admitted.

What empirical psychology is beginning to provide is a view of perception as an intelligent activity in a way that none of the opposing schools within empiricism explicitly formulated (though the quotation from Quinton shows that some epistemologists vaguely foresaw the possibility of an extension of the conception of rational process beyond ordinary inference). In this view, the remarkable reliability of perception under normal circumstances— reliability in the sense of agreement on the whole between any confident perceptual judgment and further judgments based on reexamination of the scene; in the sense of consilience among normal observers with similar access to a scene; and in the sense of helpful guidance in practical activity—is the result of an immense amount of critical scrutiny. The scrutiny occurs on two entirely different time scales. One is the time scale of seconds, on which whatever may be called an act of perception is consummated. (In speaking of this time scale, I make no commitment to the view that perception occurs in discrete units; this view has been justly criticized by Gibson and Neisser, and earlier by Dewey; see Ratner, 1939, p. 958 and p. 962.) It is the scale on which both the internal search for relevant features of the stimulus information occurs, and on which there is search of the environment for information which confirms or disconfirms expectations. The other is the time scale of years, during which skills in perceptual search have been developed, to some extent by neural maturation but certainly in large part as a result of trial and error.

The fact that perception involves search, with resulting confirmation or disconfirmation, and often with consequent refinement of expectancies on the short time scale and of search strategy on the long one, can be construed as a vindication of those epistemologists who maintain that perception is an epistemically mediated process. The price of the vindication, however, is a considerable transformation of the conception of mediation. Furthermore, those epistemologists who maintain the opposite view regarding mediation can find some comfort in the fact that perception, like other skills which are improved by practice, has the property that its analytic components are usually smoothly integrated and suppressed from consciousness.

One cannot stop at this point in adjudicating, however, since there is at least one other constituent in the "intelligence" of perception. The smooth integration which normally occurs when expectations are fulfilled can be interrupted by disconfirmations or by various kinds of challenges, such as the one posed by a painter who is trying to replicate an optical effect or by an epistemologist who is trying to understand the structure of his knowledge. The ease of such interruption and the ease of resumption of the integrated sequence are further evidence that the phenomenological frame of mind and the normal state of object perception are not as disjoint as the critics of sense data theories maintained. (It may be relevant to note that a common pedagogical technique for developing skills other than perception is to exhibit the elements of a temporal sequence separately—for instance, musical phrases or parts of a tennis stroke—and then to coach the student into smoothly integrating them into a larger unity.) The possibility of consciously verifying whether it is appropriate to attribute a certain feature to the stimulus information and of consciously searching the environment for confirming or disconfirming information concerning a tentative identification is one of the major ways in which perception is distinguished from reflex activities. In this way a kind of control is exercised by the subject as a whole— whatever that may mean—over the perceptual process, with some similarity to the control exercised in inference and problem solving, which are considered exemplary instances of intelligent activity.

This is a good place to point out that my invocation of empirical psychology for epistemological purposes differs in important respects from Quine's, in spite of the fact that I share with him a commitment to a naturalistic epistemology and also, it seems, to a dialectic method. According to Quine (1969, pp. 82-83), "Epistemology . . . studies a natural phenomenon, viz., a physical human subject. This human subject is accorded a certain experimentally controlled input—certain patterns of irradiation in assorted frequencies, for instance—and in the fullness of time the subject delivers as output a description of the three-dimensional external world and its history." Quine regards the observer as a black box whose perceptual processes are reliable if his reports are correct and his behavioral responses adequate, as judged by third persons who take for granted the scientific description of the physical world. This way of construing naturalistic epistemology is inseparable from a commitment to behaviorism in psychology and seems to me excessively narrow.

The cognitive psychology which I have invoked, in order to show that the reliability of ordinary perceptual judgments is well grounded in the critical scrutiny exercised by normal subjects, aims at "getting inside the black box," at least in the minimal sense of offering a "subpersonal theory" (see Dennett, 1978) analogous to microphysics, and perhaps in the further sense of linkage with introspection. In giving an account of the structure of my knowledge, I wish to be free to switch back and forth between a first- and third-person point of view. The ability to take a third-person point of view of myself is not an achievement of epistemological sophistication but, rather, the by-product of socialization and of the mastery of a public language; and it is, of course, a most remarkable thing, which we are far from completely understanding. The ability to take a first-person point of view is so fundamental that it would seem not to need a defense, except for the challenge to its scientific respectability by various forms of behaviorism; and the best defense on the same plane as the challenge is to refer to a cognitive psychology, which is a well-articulated and fruitful alternative to behaviorism.

I shall now turn to the structural problem that is raised when empirical psychology is applied to epistemology. Of the dialectical theses stated earlier, the most important one for the problem of structure is thesis 3, that epistemology and knowledge of the world should mesh and complement each other. Two claims can fairly be made on this matter. One is that at a certain level, where human cognitive processes are studied functionally, the meshing is quite well established. The other is that the areas of ignorance are enormous, both regarding details on this level and regarding matters of principle on another and presumably deeper level, where the role of consciousness is taken into account. (The importance of different levels of causal relations in cognition is stressed by Heffner, in press.)

It is a safe generalization that empirical psychologists look on a human being as a special case of a biological system and accept the following broad but nontrivial propositions: that a biological system is at least partially describable in physical terms (which, of course, leaves open the question of complete physicalistic reducibility), that it interacts causally with its environment, and that it has an array of goals (whether or not teleology is explicable in nonteleological physical terms) and of strategies for pursuing these goals. What distinguishes normal adult human beings from all other organisms is the extraordinary flexibility, farsightedness, and cumulative character of these strategies. In psychological terms, the strategies employed by human beings require the development and meshing of faculties of perception, memory, learning, and reasoning.

There is an immense wealth of results concerning human cognitive strategies. However, since these results have been obtained in subdisciplines that have not been completely integrated internally and certainly are not fully integrated with each other, they are far from being systematized. One large collection of psychological results is essentially phenomenological, showing that the normal human cognitive apparatus is capable of performing many useful tasks, whatever their mechanisms may be—tasks such as three-dimensional localization, tracking objects of interest, and perceptual identification and classification. Another collection of results consists essentially of tests of various theories about how tasks like the ones just mentioned are performed, where "how" is understood in an information-processing sense—that is, what program governs the generation of cognitive output from sensory or other input, in abstraction from the physiological means by which the program is realized. Propositions 1-5, listed earlier in this section, belong for the most part to this collection of results. Finally, there are results about the actual functioning of sense organs and the nervous system, which provide innumerable examples of functionally valuable adaptations; for example, stimulation of the outlying receptors of the retina evokes the response of turning the head and eyes so as to bring the stimulus onto the fovea. These three collections of results all exhibit in various ways that *perception is reliable because it is an activity of an organism constructed so as to extract from proximal stimuli significant information about distal objects. It is in this sense and at this level that the structural demand of meshing epistemology with knowledge of the world is satisfied.*

An example will indicate that this claim is not undermined by the fact that our understanding of human cognitive faculties is incomplete. Phenomenological investigations show that human beings are remarkably good at pattern identification, even when there is great variation among the exemplars which are recognized as instantiating a pattern, such as differences in the way a phoneme is pronounced or a letter is written. It is also well known that the efforts of artificial intelligence experts to build pattern-recognizing devices have so far fallen far short of the abilities of human observers. These failures indicate a serious lack of integration of the first and second collection of results mentioned earlier in the paragraph; for, if the program whereby a human observer recognizes a phoneme were known, its realization by an artifice would be a relatively trivial step. But as long as there is no obstacle in principle to the discovery of the mechanisms whereby the organism functions effectively, and as long as there are "progressive research programs" on this problem, it remains reasonable to assert that the reliability of perceptual judgments is grounded in the constitution of the organism.

It is appropriate to interpolate another structural remark before continuing the discussion of the meshing of epistemology with knowledge of the world. The assertion of the empirical psychologist that the subjects whom they are studying perceive reliably is obviously not presupposition free. Psychologists must themselves identify and characterize those objects which are presented to the subjects, and their own judgments are essential to their evaluation of the subjects'. We have, however, allowed for this kind of presupposition in thesis 1: that commonsense judgments about ordinary matters have a prima facie credibility, which should not be discounted without clear positive reasons. Moreover, the reliability of ordinary perceptual judgments, which is exhibited by psychological experimentation, justifies the psychologist's starting point. It should also be noted that the perceptions upon which the experimenter relies are performed typically under optimum circumstances—with plenty of time to examine the objects in question, good lighting, and so forth. These excellent circumstances are often very different from those under which the subjects of experimentation must operate—with tachistoscopic exposures, cue deprivations, unnatural contexts, and other such conditions. Consequently, it is not reasonable to use the experimenter's judgments to evaluate those of the subjects.

I wish to make two final remarks about the question of meshing epistemology with knowledge of the world, one of which expresses confidence about the views presented so far, while the other is an acknowledgment of ignorance. It was argued above that epistemic mediation in perception is to a large extent a matter of potentiality: the possibility of separating out an element which has been integrated into a complex mental process and the possibility of further search, both in the environment and in memory. Because of these possibilities, the phenomenological frame of mind and normal state of object perception are not disjoint states but are in some sense open to each other. The insistence on the role of potentiality in mental states is, in my opinion, essential to any hope for understanding the relation between consciousness and unconscious mentality. Furthermore, the recognition of potentiality in psychology is a specification of the general metaphysical view that any concrete entity in nature must be characterized not only by the properties which it actually manifests but also by those which it would or might manifest under changed conditions. This is a metaphysical view which has been strongly supported by developments in physics, since the wave function in quantum mechanics is best understood as a network of potentialities.

The second remark is that the characterization of human observers which was made above and imputed to cognitive psychologists is rather coarse. This is not to say that it is either false or unimportant, for just as a reasonable demand on a microphysical theory is that it accounts for macrophysical properties, a fine theory of the constitution of human beings should account for the characteristics which they share with other biological systems. What, above all, has been omitted in the coarse characterization of human beings, that needs to be treated in a fine characterization, is the mind-body relationship. But what a demand this is! We have begun with apparently modest questions about the reliability of perceptual judgments, and now we are confronted with the most baffling of scientific problems. Even physicalistic reductionists admit that consciousness is not yet understood as a physical collective phenomenon, although they are convinced that it is one. And, on the other hand, antireductionists (myself included) have only the vaguest of programs for a fundamental mind-matter dualism or for an interpretation of matter in protomental terms.

This confrontation with the mind-body problem is not a perversity of exposition but is inevitable in an integral epistemology of the kind I am advocating. If epistemology must make recourse to a description of knowers as entities in nature, and if consciousness is one of the crucial characteristics of these entities, then a "closing of the circle" cannot be fully accomplished without a solution to the mind-body problem. Historically, this inevitability has often been recognized, but with a diversity of accommodations. For example, one of the recurrent arguments for a phenomenalistic epistemology is that the principal alternative, a physical realism, would pose the mind-body problem in an unsolvable manner (Russell, 1921, p. 306). The pattern of such argumentation is legitimate, but the inference itself is not, because it is surely premature to say that the mind-body problem is not solvable along some realistic lines (for instance, those of Leibniz and Whitehead). A strong consideration against phenomenalism is the part of the "closing of the circle" which has already been accomplished; that is, the explanation of the reliability of perceptual judgments in terms of the constitution of organisms which are well adapted to their environment.

Evolutionary Considerations

Since the integral epistemology which I am advocating has a central naturalistic component, and since almost all epistemologists who now identify themselves as naturalists look at human cognitive faculties from an evolutionary point of view, it may seem strange that the foregoing discussion of the reliability of perception was so silent about evolution. The omission was deliberate, because it is important for clarity in the structure of epistemology to see what can be established without invoking evolutionary biology and what cannot. (See Dretske, 1971, for questions on this point.)

As emphasized in the preceding section, the discussion of the "closing of the circle" of epistemology and knowledge of the world can be carried out at different levels, with different degrees of fineness of description. I claimed that at one level—the level at which one studies a human being functionally, as an organism interacting with its environment and endowed with strategies for pursuing vital goals—the closing of the circle is quite well established. We can see, from a variety of psychological studies, *that* human cognitive strategies are effective for practical purposes and, to some extent, *how* they are effective. It is possible to discuss these questions without inquiring into the genetic question of how human beings came to be endowed with such effective cognitive strategies and with apparatus for carrying them out. The genetic question obviously lurks in the background, however; indeed, one reason why it was not obtrusively disturbing is that almost all readers, and certainly all who follow Campbell, take for granted an evolutionary answer. *Some* answer to this question—either the evolutionary one or a surrogate—is necessary if we are to know where human beings fit in nature and therefore to accomplish the "closing of the circle" at a level deeper than that of practical functioning. (I should add, however, in view of the difficulty posed by the mind-body problem, that an evolutionary answer to the genetic question does not ipso facto "close the circle" at the deepest level. Instead, we are confronted with the problem in a new guise: How is it that creatures endowed with the consciousness that we know ourselves to possess could have evolved from ancestors apparently not so endowed?)

Another reason why the question of the origin of human

cognitive faculties did not arise obtrusively in the previous section is that we were concerned with the reliability of perception, which is a practical matter, or certainly can be so understood, given the customary usage of the word *reliability*. A perceptual judgment is reliable if on the whole it is confirmed by further experience, if on the whole it agrees with the judgments of other well-placed and competent observers, and if on the whole it can serve as a basis for actions which favor the interests of the perceiver. That perception is reliable in this sense is a commonplace, shared by a number of philosophers, including some who are not naturalistic. In particular, philosophies which offer radically different answers to the genetic question may concur regarding the reliability of perception in the sense stated. Berkeley, for example, repeatedly insists that his immaterialism does not undermine the commonsense view of the world, and indeed he claims to provide a better explanation of the accessibility of the commonsense world to knowledge by perception than do the advocates of a mysterious material substratum.

Nevertheless, different answers to the genetic question are linked with different epistemological positions, but the linkage concerns other problems than those considered so far; for example, the extent to which perception reveals properties of material objects which are independent of the cognitive relation, and the extent to which the structure of experience is imposed by the mind of the subject. I agree with Campbell that, in assessing the various solutions offered to these problems, one must take into account the overwhelming evidence for an evolutionary history of human cognitive faculties. This evidence makes very implausible the Kantian thesis that the mind legislates to nature. Furthermore, the evolutionary point of view permits a naturalistic explanation of the pervasiveness of the concepts of substance and causality in experience, contrary to Kant's thesis that we can recognize the operations of the understanding but cannot give a causal explanation for these operations. (According to Kant, an attempt to give such a causal explanation would be an instance of applying the category of causality to the things in themselves.) As Lorenz has argued, with Campbell's enthusiastic concurrence (1974, pp. 441-447), a lineage of animals which evolves in a physical environment inhabited by relatively stable macroscopic objects and governed by macroscopic regularities would benefit from a propensity to apply the categories of substance and causality to the sensory content of experience. The general lines of this argument are independent of the detailed nature of the propensity—whether it is ontogenetically innate, or requires neurological maturation, or requires some empirical input to consummate the imprinting. The evolutionary point of view suggests a less rigid role of the categories of the understanding than the one to which Kant is committed because of his theory of the *constitutive* function of the understanding. Insofar as flexibility in applying the categories seems to be required not only to account for the heterogeneous intermingling of regularity and randomness in ordinary experience but also to fit the picture of the world provided by modern physics, the evidence seems to favor a naturalistic rather than a transcendental explanation of a priori elements in experience.

Our concern with the structure of epistemology, however, should make us wary of an easy victory for naturalism. It may reasonably be objected that a question has been begged when the theory of evolution and other theories with which it is linked, such as historical geology and parts of the physical sciences, are accorded a realistic interpretation. If scientific theories are interpreted nonrealistically, merely as means for "economy of thought" or as instruments for anticipating experience of the ordinary world or even one's personal sensations, then their significance for adjudicating epistemological arguments becomes negligible. Consequently, the appeal to evolutionary and other advanced scientific theories is illegitimate in epistemology without an antecedent analytic argument that these theories should be interpreted realistically. The objection can then continue that strong arguments raised by Berkeley, Hume, Mill, James, and the positivists suggest that a realistic interpretation of scientific theories is either meaningless, or else possibly meaningful but in principle undemonstrable by human beings, or else indistinguishable in cognitive content from sufficiently sophisticated phenomenalistic interpretations.

This issue is intricate, and evidently I cannot do justice to it in a part of a section of a paper which aims only at giving an overview. Fortunately, I feel that I can respectably abstain from a detailed discussion of the issue, since Skagestad's contribution to this volume is a careful discussion of the realistic interpretation of scientific theories, and since I have written about it elsewhere (Shimony, 1971). A rough outline of my position is the following.

The arguments against realistic interpretations depend on the adoption of quite narrow criteria of meaningfulness and of evidential support. There are no strict and compelling reasons for the adoption of these narrow criteria, and there is a general metaphilosophical or dialectical reason for adopting broader criteria at least at the outset of inquiry: specifically, Peirce's maxim that the road to inquiry should not be blocked (thesis 2, presented earlier). Given broader criteria, one has the possibility of bringing relevant evidence to bear on realism, instrumentalism, and other rival interpretations of scientific theories. The possibility is also kept open that the initially broad criteria of meaningfulness and evidential support will be tightened as a consequence of inquiry, perhaps for reasons of scientific fruitfulness or of coherence (the "closing of the circle" of epistemology and knowledge of the world). If such tightening should occur, it would be nonarbitrary, in contrast to an initial stringency of criteria, which has the effect of hedging a favored philosophical position against evidence. The final step in the argument is that when realism, instrumentalism, and phenomenalism are placed on the same footing as candidates for the interpretation of scientific theories, none being excluded and none favored a priori, then realism is overwhelmingly supported a posteriori.

Needless to say, the foregoing analysis is excessively condensed, but at least it illustrates a major claim of integral epistemology: that naturalistic, analytic, and dialectical considerations must be intertwined if one is to do justice to the peculiar difficulties of the central problems of epistemology.

Inductive Inference and Vindicatory Arguments

Concern for the structure of epistemology requires that at least a few statements be made about inductive inference, even though they will be inadequate for a complicated subject. Campbell explicitly states that one of the presuppositions of his descriptive epistemology is the validity of inductive inference, and he accepts Hume's skepticism with regard to the possibility of justifying this presumption. One might have expected that he would attempt, as some other naturalistic epistemologists have done, to justify this presupposition by evolutionary considerations, along the following lines: (1) Something like inductive inference has been practiced throughout the evolutionary process. (2) In particular, long before the articulation of the scientific method, the higher animals and primitive humans developed (by trial and error) the strategy of proposing hypotheses and testing them by observations. (3) The success in the biological sense of a population which has adopted this strategy shows that it is a good strategy to employ in the world as it actually is constituted, and this fact justifies inductive inference. Actually, Campbell does assert (1) and (2), but he refrains from (3). He does so, I believe, because he realizes that the transition to (3) is itself an instance of inductive inference, the validity of which is under examination. In other words, Hume's criticism of the attempt to justify inductive inference by its past successes applies equally well to arguments which appeal to evolutionary evidence.

If Hume is not to have the last word, several strategies are open. One is to make full use of the apparatus of probability theory and decision theory, which themselves are defensible on analytic or a priori grounds. Another is to concede that any kind of inductive justification of induction involves some circularity but to reason that some types of circularity are nonvicious, in the sense that they do not foreclose the issue under examination and do not hedge a favored position against adverse evidence. This strategy is clearly in the spirit of the dialectical approach to epistemology advocated as a meta-theory in an earlier section (headed "Some Dialectical Theses") and followed elsewhere in this paper. The third strategy is to employ a vindicatory argument, as sketched in thesis 4 of that section. My own opinion is that all these strategies are needed and must be used in tandem. I am not satisfied that any methodologist has worked out the details of how the meshing of these strategies is to be accomplished, but I have advocated a program along the following lines, which seems to me to be in the right direction (Shimony, 1970, 1976).

Briefly, I believe that the theory of inductive inference has an a priori component, consisting of personal probability theory and decision theory, and making free use of deductive logic. The personal probability evaluation should be qualified by a maxim of critical open-mindedness, which is called "the tempering condition"; that is, give sufficient prior probability to any seriously proposed hypothesis that it may achieve high posterior probability if favored by the data.

A vindicatory argument can be given as follows for this maxim. If we are not optimistic about human ability to make good conjectures, then seriously proposed hypotheses are coordinate with frivolously proposed ones and with merely logically possible ones that no one has ever singled out for mention. But concerning any nontrivial matter of interest, particularly concerning generalities, infinitely many rival hypotheses are logically possible, and it is hopeless to explore all of these. Our only hope of approaching the truth concerning nontrivial matters is to be optimistic about human abductive power, in the sense of permitting a seriously proposed hypothesis to be accepted after experimental scrutiny. This optimism precludes no particular method for assigning an order to hypotheses, since the proposal of an order would itself be a special case of abduction. Consequently, nothing is lost and possibly something is gained—in accordance with the general line of vindicatory arguments—if our subjective probability evaluations are made in accordance with the tempering condition. The tempering condition as stated, however, is manifestly vague, and attempts to eliminate this vagueness by a priori explications of the phrases "seriously proposed" and "sufficient prior probability" are likely to be arbitrary.

A more promising procedure is to use a posteriori considerations in explicating these phrases. By reflecting on exemplary scientific discoveries in the past, one can recognize types of hypotheses which deserve prima facie attention—meaning, of course, no more than a moderate degree of prior probability, remote from full credence. Circularity obviously enters into the proposed epistemological structure, since there is reliance on exemplary scientific achievements, and these in turn could have been identified only by inductive inference. The circularity is nonvicious, since a seriously proposed hypothesis of the past can be subjected to experimental scrutiny and as a result may acquire so low a posterior probability that it is discredited. Furthermore, there is an openness in the procedure in spite of the recourse of past achievement, for a radically new hypothesis can be accorded the status of "seriously proposed" for many reasons; it may, for example, satisfy the standards of exposition of exemplary hypotheses of the past, in spite of the novelty of its content.

This program for justifying inductive inference evidently needs much further clarification and exploration. However, space permits only two last remarks, which bear on the structure of epistemology. The first is to reiterate the indispensability of analytic considerations in an integral epistemology. Without probability theory and decision theory, and without the appeal to a vindicatory argument, there seems to be no alternative to skepticism regarding the justification of induction. The second remark is that what Campbell calls "descriptive epistemology" has a role in the justification of induction, as it does in other epistemological matters. But it appears that the term should be extended, so as to apply not only to results from the natural sciences but also to some historical information. As far as we can reconstruct the reasoning process of prescientific people, we find a mélange of critical and noncritical elements. Along with *modus tollens* and rudimentary probability theory, there is uncritical use of simple inductions and very uncritical reliance on analogies and anthropocentric explanations. If one argues from the biological success of Homo sapiens to the reasonableness of the genetic thought processes of man, then one has an undiscriminating justification of the whole package of primitive thought processes, critical and uncritical together. The a posteriori considerations which permit us to separate out different components of the package and to assess them are considerations concerning not the evolutionary development of the race but, rather, the history of science. The improvement of primitive inductive processes is thus a cultural achievement. It is, however, a very special cultural achievement, since it is the by-product of search for the truth about nature. There is no better illustration of a dialectical structure in epistemology than the refinement of inductive inferences by reflection on exemplary results of science, which themselves were achieved inductively.

References

Bower, T. G. R. "The Visual World of Infants." *Scientific American,* 1966, *215* (12), 80-92.

Bruner, J. S. "Perceptual Readiness." *Psychological Review,* 1957, *64,* 123-152.

Campbell, D. T. "Perception as Substitute Trial and Error." *Psychological Review,* 1956, *63,* 330-342.

Campbell, D. T. "Methodological Suggestions from a Comparative Psychology of Knowledge Processes." *Inquiry,* 1959, *2,* 152-182.

Campbell, D. T. "Evolutionary Epistemology." In P. A. Schilpp (Ed.), *The Philosophy of Karl Popper.* La Salle, Ill.: Open Court, 1974.

Campbell, D. T. Introductory comments to a lecture on Evolutionary Epistemology to the Boston Colloquium for the Philosophy of Science, February 22, 1977a.

Campbell, D. T. "Descriptive Epistemology: Psychological, Sociological, and Evolutionary." William James Lectures, Harvard University, 1977b.

Dennett, D. C. "Toward a Cognitive Theory of Consciousness." In C. W. Savage (Ed.), *Minnesota Studies in the Philosophy of Science.* Vol. 9: *Perception and Cognition: Issues in the Foundations of Psychology.* Minneapolis: University of Minnesota Press, 1978.

Dretske, F. "Perception from an Evolutionary Point of View." *Journal of Philosophy,* 1971, *68,* 584-591.

Firth, F. "Sense-Data and the Percept Theory." In R. J. Swartz (Ed.), *Perceiving, Sensing, and Knowing.* New York: Doubleday, 1965.

Gibson, E. J. *Principles of Perceptual Learning and Development.* New York: Appleton-Century-Crofts, 1969.

Gibson, J. J. *The Perception of the Visual World.* Boston: Houghton Mifflin, 1950.

Gregory, R. L. *The Intelligent Eye.* New York: McGraw-Hill, 1970.

Heffner, J. "Causal Relations in Visual Perception." *International Philosophical Quarterly,* in press.

Neisser, U. *Cognitive Psychology.* New York: Appleton-Century-Crofts, 1967.

Neisser, U. *Cognition and Reality.* San Francisco: Freeman, 1976.

Peirce, C. S. "Notes on Scientific Philosophy." In C. Hartshorne and P. Weiss (Eds.), *Collected Papers of Charles Sanders Peirce.* Vol. 1. Cambridge, Mass.: Harvard University Press, 1931. (Original manuscript 1897.)

Price, H. H. *Perception.* London: Methuen, 1932.

Quine, W. V. "Epistemology Naturalized." In *Ontological Relativity and Other Essays.* New York: Columbia University Press, 1969.

Quine, W. V. *The Roots of Reference.* La Salle, Ill.: Open Court, 1974.

Quinton, A. M. "The Problem of Perception." In R. J. Swartz (Ed.), *Perceiving, Sensing, and Knowing.* New York: Doubleday, 1965.

Ratner, J. (Ed.). *Intelligence in the Modern World: John Dewey's Philosophy.* New York: Random House, 1939.

Rescher, N. *Methodological Pragmatism.* Oxford: Blackwell, 1977.

Russell, B. *The Analysis of Mind.* London: Allen and Unwin, 1921.

Shimony, A. "Scientific Inference." In R. G. Colodny (Ed.), *The Nature and Function of Scientific Theories.* Pittsburgh: University of Pittsburgh Press, 1970.

Shimony, A. "Perception from an Evolutionary Point of View." *Journal of Philosophy,* 1971, *68,* 571-583.

Shimony, A. "Comments on Two Epistemological Theses of Thomas Kuhn." In R. S. Cohen, P. K. Feyerabend, and M. W. Wartofsky (Eds.), *Essays in Memory of Imre Lakatos.* Dordrecht, Netherlands: Reidel, 1976.

Shimony, A. "Is Observation Theory-Laden? A Problem in Naturalistic Epistemology." In R. G. Colodny (Ed.), *Logic, Laws, and Life: Some Philosophical Complications.* Pittsburgh: University of Pittsburgh Press, 1978.

Swartz, R. J. (Ed.). *Perceiving, Sensing, and Knowing.* New York: Doubleday, 1965.

6   *William C. Wimsatt*

Robustness, Reliability, and Overdetermination

> Philosophy ought to imitate the successful sciences in its methods, so far as to proceed only from tangible premises which can be subjected to careful scrutiny, and to trust rather to the multitude and variety of its arguments than to the conclusiveness of any one. Its reasoning should not form a chain which is no stronger than its weakest link, but a cable whose fibers may be so slender, provided they are suf-

Note: This chapter is the product of seeds planted a dozen years ago by the writings and ideas of Donald Campbell and Richard Levins. Since then these germinated ideas have received nurturance and selective pruning by many individuals. Particularly helpful have been discussions with William

ficiently numerous and intimately connected [Peirce, [1868] 1936, p. 141].

Our truth is the intersection of independent lies [Levins, 1966, p. 423].

The use of multiple means of determination to "triangulate" on the existence and character of a common phenomenon, object, or result has had a long tradition in science but has seldom been a matter of primary focus. As with many traditions, it is traceable to Aristotle, who valued having multiple explanations of a phenomenon, and it may also be involved in his distinction between special objects of sense and common sensibles. It is implicit though not emphasized in the distinction between primary and secondary qualities from Galileo onward. It is arguably one of several conceptions involved in Whewell's method of the "consilience of inductions" (Laudan, 1971) and is to be found in several places in Peirce.

Indeed, it is to be found widely among the writings of various scientists and philosophers but, remarkably, seems almost invariably to be relegated to footnotes, parenthetical remarks, or suggestive paragraphs that appear without warning and vanish without further issue. While I will point to a number of different applications of multiple determination which have surfaced in the literature, Donald Campbell has done far more than anyone else to make multiple determination a central focus of his work and to draw a variety of methodological, ontological, and epistemological conclusions from its use (see Campbell, 1958, 1966, 1969a, 1977; Campbell and Fiske, 1959; Cook and Campbell, 1979). This theme is as important a contribution as his work on evolutionary epistemology; indeed, it must be a major unappreciated component of the latter: multiple determination, because of its implications for increasing reliability, is a fundamental and universal feature of

Bechtel, Aaron Ben Ze'ev, Robert MacCauley, and Robert Richardson. Sandy Zabell and Stephen Stigler provided invaluable guidance through the literature on statistical notions of robustness which, while distantly related, have proved to be less central than I had hoped. The editors of this volume have made creative suggestions that substantially improved the paper. This work was supported by the National Science Foundation under Grant NSF SOC78-07310.

sophisticated organic design and functional organization and can be expected wherever selection processes are to be found.

Multiple determination—or *robustness,* as I will call it—is not limited in its relevance to evolutionary contexts, however. Because of its multiplicity of uses, it is implicit in a variety of criteria, problem-solving procedures, and cognitive heuristics which have been widely used by scientists in different fields, and is rich in still insufficiently studied methodological and philosophical implications. Some of these I will discuss, some I will only mention, but each contains fruitful directions for future research.

Common Features of Concepts of Robustness

The family of criteria and procedures which I seek to describe in their various uses might be called *robustness analysis.* They all involve the following procedures:

1. To analyze a *variety* of *independent* derivation, identification, or measurement processes.
2. To look for and analyze things which are *invariant* over or *identical* in the conclusions or results of these processes.
3. To determine the *scope* of the processes across which they are invariant and the *conditions* on which their invariance depends.
4. To analyze and explain any relevant *failures of invariance.*

I will call things which are invariant under this analysis "robust," extending the usage of Levins (1966, p. 423), who first introduced me to the term and idea and who, after Campbell, has probably contributed most to its analysis (see Levins, 1966, 1968).

These features are expressed in very general terms, as they must be to cover the wide variety of different practices and procedures to which they apply. Thus, the different processes in clause 1 and the invariances in clause 2 may refer in different cases to any of the following:

a. Using different sensory modalities to detect the same property or entity (in the latter case by the detection of spatiotemporal boundaries which are relatively invariant across different sensory modalities) (Campbell, 1958, 1966).

b. Using different experimental procedures to verify the same empirical relationships or generate the same phenomenon (Campbell and Fiske, 1959).

c. Using different assumptions, models, or axiomatizations to derive the same result or theorem (Feynman, 1965; Levins, 1966; Glymour, 1980).

d. Using the agreement of different tests, scales, or indices for different traits, as measured by different methods, in ordering a set of entities as a criterion for the "validity" (or reality) of the constructed property (or "construct") in terms of which the orderings of entities agree (Cronbach and Meehl, 1955; Campbell and Fiske, 1959).

e. Discovering invariance of a macrostate description, variable, law, or regularity over different sets of microstate conditions, and also determining the microstate conditions under which these invariances may fail to hold (Levins, 1966, 1968; Wimsatt, 1976, 1980b).

f. Using matches and mismatches between theoretical descriptions of the same phenomenon or system at different levels of organization, together with Leibniz's law (basically that if two things are identical, no mismatches are allowed), to generate new hypotheses and to modify and refine the theories at one or more of the levels (Wimsatt, 1976a, 1976b, 1979).

g. Using failures of invariance or matching in a through f above to calibrate or recalibrate our measuring apparatus (for a, b, or f) or tests (for d), or to establish conditions (and limitations on them) under which the invariance holds or may be expected to fail, and (for all of the above) to use this information to guide the search for explanations as to why the invariances should hold or fail (Campbell, 1966, 1969a; Wimsatt, 1976a, 1976b).

h. Using matches or mismatches in different determinations of the value of theoretical parameters to test and confirm or infirm component hypotheses of a complex theory (Glymour, 1980) and, in a formally analogous manner, to test and localize faults in integrated circuits.

One may ask whether any set of such diverse activities as would fit all these items and as exemplified in the expanded discussions below are usefully combined under the umbrella term *ro-*

bustness analysis. I believe that the answer must be yes, for two reasons. First, all the variants and uses of robustness have a common theme in the distinguishing of the real from the illusory; the reliable from the unreliable; the objective from the subjective; the object of focus from artifacts of perspective; and, in general, that which is regarded as ontologically and epistemologically trustworthy and valuable from that which is unreliable, ungeneralizable, worthless, and fleeting. The variations of use of these procedures in different applications introduce different variant tools or consequences which issue from this core theme and are explicable in terms of it. Second, all these procedures require at least partial *independence* of the various processes across which invariance is shown. And each of them is subject to a kind of systematic error leading to a kind of *illusory robustness* when we are led, on less than definitive evidence, to presume independence and our presumption turns out to be incorrect. Thus, a broad class of fallacious inferences in science can be understood and analyzed as a kind of failure of robustness.

Nonetheless, the richness and variety of these procedures require that we go beyond this general categorization to understand robustness. To understand fully the variety of its applications and its central importance to scientific methodology, detailed case studies of robustness analysis are required in each of the areas of science and philosophy where it is used.

Robustness and the Structure of Theories

In the second of his popular lectures on the character of physical law, Feynman (1965) distinguishes two approaches to the structure of physical theory: the Greek and the Babylonian approaches. The Greek (or Euclidean) approach is the familiar axiomatic one in which the fundamental principles of a science are taken as axioms, from which the rest are derived as theorems. There is an established order of importance, of ontological or epistemological priority, from the axioms out to the farthest theorems. The "Greek" theorist achieves postulational economy or simplicity by making only a small number of assumptions and deriving the rest—often reducing the assumptions, in the name of simplicity or elegance, to the minimal set necessary to derive the given theorems. The "Babylonian," in contrast, works with an approach that is much less well ordered and sees a theoretical structure that is much more richly connected:

> So the first thing we have to accept is that even in mathematics you can start in different places. If all these various theorems are interconnected by reasoning there is no real way to say "These are the most fundamental axioms," because if you were told something different instead you could also run the reasoning the other way. It is like a bridge with lots of members, and it is overconnected; if pieces have dropped out you can reconnect it another way. The mathematical tradition of today is to start with some particular ideas which are chosen by some kind of convention to be axioms, and then to build up the structure from there. What I have called the Babylonian idea is to say, "I happen to know this, and I happen to know that, and maybe I know that; and I work everything out from there. Tomorrow I may forget that this is true, but remember that something else is true, so I can reconstruct it all again. I am never quite sure of where I am supposed to begin or where I am supposed to end. I just remember enough all the time so that as the memory fades and some of the pieces fall out I can put the thing back together again every day" [Feynman, 1965, pp. 46-47].

This rich connectivity has several consequences for the theoretical structure and its components. First, as Feynman (1965, pp. 54-55) observes, most of the fundamental laws turn out to be characterizable and derivable in a variety of different ways from a variety of different assumptions: "One of the amazing characteristics of nature is the variety of interpretational schemes which is possible. It turns out that it is only possible because the laws are just so, special and delicate. . . . If you modify the laws much you find that you can only write them in fewer ways. I always find that mysterious, and I do not understand the reason why it is that the correct laws of physics seem to be expressible in such a tremendous variety of ways. They seem to be able to get through several wickets at the same time." Although Feynman nowhere explicitly says so, his own choice of examples and other considerations that will emerge later suggest another ordering principle for fundamentality among laws of nature: *The more fundamental laws will be those that are independently derivable in a larger number of ways.* I will return to this suggestion later.

Second, Feynman also observes that this multiple derivability of physical laws has its advantages, for it makes the overall structure much less prone to collapse:

> At present we believe that the laws of physics have to have the local character and also the minimum principle, but we do not really know. If you have a structure that is only partly accurate, and something is going to fail, then if you write it with just the right axioms maybe only one axiom fails and the rest remain, you need only change one little thing. But if you write it with another set of axioms they may all collapse, because they all lean on that one thing that fails. We cannot tell ahead of time, without some intuition, which is the best way to write it so that we can find out the new situation. We must always keep all the alternative ways of looking at a thing in our heads; so physicists do Babylonian mathematics, and pay but little attention to the precise reasoning from fixed axioms [Feynman, 1965, p. 54].

This multiple derivability not only makes the overall structure more reliable but also has an effect on its individual components. Those components of the structure which are most insulated from change (and thus the most probable foci for continuity through scientific revolutions) are those laws which are most robust and, on the above criterion, most fundamental. This criterion of fundamentality would thus make it natural (though by no means inevitable) that the most fundamental laws would be the least likely to change. *Given that different degrees of robustness ought to confer different degrees of stability, robustness ought to be a promising tool for analyzing scientific change.* Alternatively, the analysis of different degrees of change in different parts of a scientific theory may afford a way of detecting or measuring robustness.

I wish to elaborate and illustrate the force of Feynman's remarks arguing for the Babylonian rather than the Greek or Euclidean approach by some simple considerations suggested by the statistical theory of reliability. (For an excellent review of work in reliability theory, see Barlow and Proschan, 1975, though no one has, to my knowledge, applied it in this context.)

A major rationale for the traditional axiomatic view of science is to see it as an attempt to make the structure of scientific theory as reliable as possible by starting with, as axioms, the minimal number of assumptions which are as certain as possible and operating on them with rules which are as certain as possible (deductive rules which are truth preserving). In the attempt to secure high reliability, the focus is on total elimination of error, not on recognizing that it will occur and on controlling its effects: it is a structure in which, if no errors are introduced in the assumptions and if no errors are made in choosing or in applying the rules, no errors will occur. No effort is spared in the attempt to prevent these kinds of errors from occurring. But it does not follow that this is the best structure for dealing with errors (for example, by minimizing their effects or making them easier to find) if they do occur. In fact, it is not. To see how well it handles errors that do occur, let us try to model the effect of applying the Greek or Euclidian strategy to a real (error-prone) theory constructed and manipulated by real (fallible) operators.

For simplicity, assume that any operation, be it choosing an assumption or applying a rule, has a small but finite probability of error, p_o. (In this discussion, I will assume that the probability of error is constant across different components and operations. Qualitatively similar results obtain when it is not.) Consider now the deductive derivation of a theorem requiring m operations. If the probabilities of failure in these operations are independent, then the probability of a successful derivation is just the product of the probabilities of success, $(1 - p_o)$, at each operation. Where p_s stands for the probability of failing at this complex task (p_s because this is a serial task), then we have for the probability of success, $(1 - p_s)$:

$$(1 - p_s) = (1 - p_o)^m$$

No matter how small p_o is, as long as it is finite, longer serial deductions (with larger values of m) have monotonically decreasing probabilities of successful completion, approaching zero in the limit. *Fallible thinkers should avoid long serial chains of reasoning.* Indeed, we see here that the common metaphor for deductive reasoning as a chain is a poor one for evaluating probability of failure in reasoning. Chains always fail at their weakest links, chains of reasoning only most probably so.

When a chain fails, the release in tension protects other parts of the chain. As a result, failures in such a chain are not independent, since the occurrence of one failure prevents other failures. In this model, however, we are assuming that failures are independent of each other, and we are talking about probability of failure rather than actual failure. These differences result in a serious disanalogy with the metaphor of the argument as a chain. A chain is only as strong as the weakest link, but it *is* that strong; and one often hears this metaphor as a rule given for evaluating the reliability of arguments (see, for example, the quote from C. S. Peirce that begins this chapter). But a chain in which failure could occur at any point is always weaker than (in that it has a higher probability of failure than) its weakest link, except if the probability of failure everywhere else goes to zero. This happens when the weakest link in a chain breaks, but not when one link in an argument fails.

Is there any corrective medicine for this cumulative effect on the probability of error, in which small probabilities of error in even very reliable components cumulatively add up to almost inevitable failure? Happily there is. *With independent alternative ways of deriving a result, the result is always surer than its weakest derivation.* (Indeed, it is always surer than its *strongest* derivation.) This mode of organization—with independent alternative modes of operation and success if any one works—is parallel organization, with its probability of failure, p_p. Since failure can occur if and only if each of the m independent alternatives fails (assume, again, with identical probabilities p_o):

$$p_p = p_o{}^m$$

But p_o is presumably always less than 1; thus, for $m > 1$, p_p is always less than p_o. Adding alternatives (or redundancy, as it is often called) always increases reliability, as von Neumann (1956) argued in his classic paper on building reliable automata with unreliable components. Increasing reliability through parallel organization is a fundamental principle of organic design and reliability engineering generally. It works for theories as well as it does for polyploidy, primary metabolism, predator avoidance, microprocessor architecture, Apollo moon shots, test construction, and the structure of juries.

Suppose we start, then, with a Babylonian (or Byzantine?) structure—a multiply connected, poorly ordered scientific theory having no principles singled out as axioms, containing many different ways of getting to a given conclusion and, because of its high degree of redundancy, relatively short paths to it (see Feynman, 1965, p. 47)—and let it be redesigned by a Euclidean. In the name of elegance, the Euclidean will look for a small number of relatively powerful assumptions from which the rest may be derived. In so doing, he will eliminate redundant assumptions. The net effects will be twofold: (1) With a smaller number of assumptions taken as axioms, the mean number of steps in a derivation will increase, and can do so exponentially. This increased length of seriation will decrease reliability along any path to the conclusion. (2) Alternative or parallel ways of getting to a given conclusion will be substantially decreased as redundant assumptions are removed, and this decrease in "parallation" will also decrease the total reliability of the conclusion.

Each of these changes increases the unreliability of the structure, and both of them operating together produce a cumulative effect—if errors are possible, as I have supposed. Not only is the probability of failure of the structure greater after it has been Euclideanized, but the consequences of failure become more severe: with less redundancy, the failure of any given component assumption is likely to inform a larger part of the structure. I will elaborate on this point shortly. It has not been studied before now (but see Glymour, 1980) because of the dominance of the Cartesian-Euclidean perspective and because of a key artifact of first-order logic.

Formal models of theoretical structures characteristically start with the assumption that the structures contain no inconsistencies. As a normative ideal, this is fine; but as a description of real scientific theories, it is inadequate. Most or all scientific theories with which I am familiar contain paradoxes and inconsistencies, either between theoretical assumptions or between assumptions and data in some combination. (Usually these could be resolved if one knew which of several eminently plausible assumptions to give up, but each appears to have strong support; so the assumptions—and the inconsistencies—remain.) This feature of scientific theories has not until now (with the development of nonmonotonic logic) been modeled, because of the fear of total collapse. In first-order logic, anything whatsoever follows from a contradiction; so systems which contain contradictions are regarded as useless.

But the total collapse suggested by first-order logic (or by highly Euclidean structures with little redundancy) seems not to be a characteristic of scientific theories. The thing that is remarkable about scientific theories is that the inconsistencies are walled off and do not appear to affect the theory other than very locally— for things very close to and strongly dependent on one of the conflicting assumptions. Robustness provides a possible explanation, perhaps the best explanation, for this phenomenon.

When an inconsistency occurs, results which depend on one or more of the contradictory assumptions are infirmed. This infection is transitive; it passes to things that depend on these results, and to their logical descendants, like a string of dominoes—until we reach something that has independent support. The independent support of an assumption sustains it, and the collapse propagates no further. If all deductive or inferential paths leading from a contradiction pass through robust results, the collapse is bounded within them, and the inconsistencies are walled off from the rest of the network. For each robust result, one of its modes of support is destroyed; but it has others, and therefore the collapse goes no further. Whether this is the only mechanism by which this isolation of contradictions could be accomplished, I do not know, but it is *a* possible way, and scientific constructs do appear to have the requisite robustness. (I am not aware that anyone has tried to formalize or to simulate this, though Stuart A. Kauffman's work on "forcing structures" in binary, Boolean switching networks seems clearly relevant. See, for example, Kauffman, 1971, where these models are developed and applied to gene control networks.)

Robustness, Testability, and the Nature of Theoretical Terms

Another area in which robustness is involved (and which is bound to see further development) is Clark Glymour's account of testing and evidential relations in theories. Glymour argues systematically that parts of a theoretical structure can be and are used to test other parts of the theory, and even themselves. (His name for this is bootstrapping.) This testing requires the determination of values for quantities of the theory in more than one way: "If the data are consistent with the theory, then these different computations must agree [within a tolerable experimental error] in the value they determine for the computed quantity; but if the data are inconsistent with the theory, then different computations of the same quantity may give different results. Further and more important, what quantities in a theory may be computed from a given set of initial data depends both on the initial data and on the structure of the theory" (Glymour, 1980, p. 113).

Glymour argues later (pp. 139-140) that the different salience of evidence to different hypotheses of the theory requires the use of a variety of types of evidence to test the different component hypotheses of the theory. Commenting on the possibility that one could fail to locate the hypothesis whose incorrectness is producing an erroneous determination of a quantity or, worse, mislocating the cause of the error, he claims: "The only means available for guarding against such errors is to have a variety of evidence so that as many hypotheses as possible are tested in as many different ways as possible. What makes one way of testing relevantly different from another is that the hypotheses used in one computation are different from the hypotheses used in the other computation. Part of what makes one piece of evidence relevantly different from another piece of evidence is that some test is possible from the first that is not possible from the second, or that, in the two cases, there is some difference in the precision of computed values of theoretical quantities" (Glymour, 1980, p. 140).

A given set of data and the structure of the theory permit a test of a hypothesis (or the conjunction of a group of hypotheses) if and only if they permit determination of all of the values in the tested entity in such a way that contradictory determinations of at least one of these values could result (in the sense that it is not analytically ruled out). This requires more than one way of getting at that value. (See Glymour, 1980, p. 307.) To put it in the language of the present paper, *only robust hypotheses are testable*. Furthermore, a theory in which most components are multiply connected is a theory whose faults are relatively precisely localizable. Not only do errors not propagate far, but we can find their source quickly and evaluate the damage and what is required for an adequate replacement. If this sounds like a design policy for an automobile, followed to facilitate easy diagnostic service and repair, I can say only that there is no reason why our scientific theories should be less well designed than our other artifacts.

The same issues arise in a different way in Campbell's discussions (Campbell and Fiske, 1959; Campbell, 1969a, 1969b, 1977; Cook and Campbell, 1979) of single or definitional versus multiple operationalism. Definitional operationalism is the view that philosophers know as operationalism, that the meaning of theoretical terms is to be defined in terms of the experimental operations used in measuring that theoretical quantity. Multiple means of determining such a quantity represents a paradox for this view—an impossibility, since the means is definitive of the quantity, and multiple means means multiple quantities. Campbell's multiple operationalism is not operationalism at all in this sense but a more tolerant and eclectic empiricism, for he sees the multiple operations as contingently associated with the thing measured. Being contingently associated, they cannot have a definitional relation to it; consequently, there is no barrier to accepting that one (robust) quantity has a number of different operations to get at it, each too imperfect to have a definitional role but together triangulating to give a more accurate and complete picture than would be possible from any one of them alone.

Campbell's attack on definitional operationalism springs naturally from his fallibilism and his critical realism. Both of these forbid a simple definitional connection between theoretical constructs and measurement operations: "One of the great weaknesses in definitional operationalism as a description of best scientific practice was that it allowed no formal way of expressing the scientist's prepotent awareness of the imperfection of his measuring instruments and his prototypic activity of improving them" (Campbell, 1969a, p. 15). For a realist the connection between any measurement and the thing measured involves an often long and indirect causal chain, each link of which is affected and tuned by other theoretical parameters. The aim is to make the result insensitive to or to control these causally relevant but semantically irrelevant intermediate links: "What the scientist does in practice is to design the instrument so as to minimize and compensate for the stronger of these irrelevant forces. Thus, the galvanometer needle is as light as possible, to minimize inertia. It is set on jeweled bearings to minimize friction. It may be used in a lead-shielded and degaussed room. Remote influences are neglected because they dissipate at the rate of $1/d^2$, and the weak and strong nuclear forces dissipate even more rapidly. But these are practical minimizations, recognizable on theoretical grounds as incomplete" (1969a, pp. 14-15).

The very same indirectness and fallibility of measurement that rule out definitional links make it advantageous to use multiple links: "[W]e have only *other invalid measures* against which to validate our tests; we have no 'criterion' to check them against. . . . A theory of the interaction of two theoretical parameters must be tested by imperfect exemplifications of each. . . . In this predicament, great inferential strength is added when each theoretical parameter is exemplified in 2 or more ways, each mode being as independent as possible of the other, as far as the theoretically irrelevant components are concerned. This general program can be designated *multiple operationalism*" (Campbell, 1969a, p. 15).

Against all this, then, suppose one did have only one means of access to a given quantity. Without another means of access, even if this means of access were not made definitional, statements about the value of that variable would not be independently testable. Effectively, they would be as if defined by that means of access. And since the variable was not connected to the theory in any other way, it would be an unobservable, a fifth wheel: anything it could do could be done more directly by its operational variable. It is, then, in Margenau's apt phrase, a peninsular concept (Margenau, 1950, p. 87), a bridge that leads to nowhere.

Philosophers often misleadingly lump this "peninsularity" and the existence of extra axioms permitting multiple derivations together as redundancy. The implication is that one should be equally disapproving of both. Presumably, the focus on error-free systems leads philosophers to regard partially identical paths (the paths from a peninsular concept and from its "operational variable" to any consequence accessible from either) and alternative independent paths (robustness, bootstrapping, or triangulation) as equivalent—because they are seen as equally dispensable if one is dealing with a system in which errors are impossible. But if errors are possible, the latter kind of redundancy can increase the reliability of the conclusion; the former cannot.

A similar interest in concepts with multiple connections and a disdain for the trivially analytic, singly or poorly connected concept is to be found in Hilary Putnam's (1962) classic paper "The Analytic and the Synthetic." Because theoretical definitions are multiply connected law-cluster concepts, whose meaning is determined by this multiplicity of connections, Putnam rejects the view that such definitions are stipulative or analytic. Though for Putnam it is theoretical connections, rather than operational ones, which are important, he also emphasizes the importance of a multiplicity of them: "Law-cluster concepts are constituted not by a bundle of properties as are the typical general names [cluster concepts] like 'man' and 'crow,' but by a cluster of laws which, as it were, determine the identity of the concept. The concept 'energy' is an excellent sample. . . . It enters into a great many laws. It plays a great many roles, and these laws and inference roles constitute its meaning collectively, not individually. I want to suggest that most of the terms in highly developed sciences are law-cluster concepts, and that one should always be suspicious of the claim that a principle whose subject term is a law-cluster concept is analytic. The reason that it is difficult to have an analytic relationship between law-cluster concepts is that . . . any one law can be abandoned without destroying the identity of the law-cluster concept involved" (p. 379).

Statements that are analytic are so for Putnam because they are singly connected, not multiply connected, and thus trivial: "Thus, it cannot 'hurt' if we decide always to preserve the law 'All bachelors are unmarried' . . . because bachelors are a kind of synthetic class. They are a 'natural kind' in Mill's sense. They are rather grouped together by ignoring all aspects except a single legal one. One is simply not going to find any . . . [other] laws about such a class" (p. 384).

Thus, the robustness of a concept or law—its multiple connectedness within a theoretical structure and (through experimental procedures) to observational results—has implications for a variety of issues connected with theory testing and change, with the reliability and stability of laws and the component parts of a theory, with the discovery and localization of error when they fail, the analytic-synthetic distinction, and accounts of the meaning of theoretical concepts. But these issues have focused on robustness in existing theoretical structures. It is also important in discovery and in the generation of new theoretical structures.

Robustness, Redundancy, and Discovery

For the complex systems encountered in evolutionary biology and the social sciences, it is often unclear what is fundamental or trustworthy. One is faced with a wealth of partially conflicting, partially complementary models, regularities, constructs, and data sets with no clear set of priorities for which to trust and where to start. In this case particularly, processes of validation often shade into processes of discovery—since both involve a winnowing of the generalizable and the reliable from the special and artifactual. Here too robustness can be of use, as Richard Levins suggests in the passage which introduced me to the term:

> Even the most flexible models have artificial assumptions. There is always room for doubt as to whether a result depends on the essentials of a model or on the details of the simplifying assumptions. This problem does not arise in the more familiar models, such as the geographical map, where we all know that contiguity on the map implies contiguity in reality, relative distances on the map correspond to relative distances in reality, but color is arbitrary and a microscopic view of the map would only show the fibers of the paper on which it is printed. But in the mathematical models of population biology, it is not always obvious when we are using too high a magnification.
>
> Therefore, we attempt to treat the same problem with several alternative models, each with different simplifications, but with a common biological assumption. Then, if these models, despite their different

> assumptions, lead to similar results we have what we
> can call a robust theorem which is relatively free of
> the details of the model. Hence, our truth is the inter-
> section of independent lies [Levins, 1966, p. 423].

Levins is here making heuristic use of the philosopher's criterion of logical truth as true in all possible worlds. He views
robustness analysis as "sampling from a space of possible models"
(1968, p. 7). Since one cannot be sure that the sampled models are
representative of the space, one gets no guarantee of logical truth
but, rather, a heuristic (fallible but effective) tool for discovering
empirical truths which are relatively free of the details of the various specific models.

Levins talks about the robustness of theorems or phenomena or consequences of the models rather than about the robustness of the models themselves. This is necessary, given his view
that any single model makes a number of artifactual (and therefore nonrobust) assumptions. A theory would presumably be a
conceptual structure in which many or most of the fundamental
theorems or axioms are relatively robust, as is suggested by Levins'
statement (1968, p. 7) "A theory is a cluster of models, together
with their robust consequences."

If a result is robust over a range of parameter values in a
given model or over a variety of models making different assumptions, this gives us some independence of knowledge of the exact
structure and parameter values of the system under study: a prediction of this result will remain true under a variety of such conditions and parameter values. This is particularly important in
scientific areas where it may be difficult to determine the parameter values and conditions exactly.

*Robust theorems can thus provide a more trustworthy basis
for generalization of the model or theory* and also, through their
independence of many exact details, *a sounder basis for predictions from it.* Theory generalization is an important component of
scientific change, and thus of scientific discovery.

Just as robustness is a guide for discovering trustworthy results and generalizations of theory, and distinguishing them from
artifacts of particular models, it helps us to distinguish signal from
noise in perception generally. Campbell has furnished us with

many examples of the role of robustness and pattern matching in visual perception and its analogues, sonar and radar. In an early paper, he described how the pattern and the redundancy in a randomly pulsed radar signal bounced off Venus gave a new and more accurate measurement of the distance to that planet (Campbell, 1966).

The later visual satellite pictures of Mars and its satellite Deimos have provided an even more illuminating example, again described by Campbell (1977) in the unpublished William James Lectures (lecture 4, pp. 89-90). The now standard procedures of image enhancement involve combining a number of images, in which the noise, being random, averages out; but the signal, weak though usually present, adds in intensity until it stands out. The implicit principle is the same one represented explicitly in von Neumann's (1956) use of "majority organs" to filter out error: the combination of parallel or redundant signals with a threshold, in which it is assumed that the signal, being multiply represented, will usually exceed threshold and be counted; and the noise, being random, usually will fall below threshold and be lost. There is an art to designing the redundancy so as to pick up the signal and to setting the threshold so as to lose the noise. It helps, of course, if one knows what he is looking for. In this case of the television camera centered on Mars, Deimos was a moving target and—never being twice in the same place to add appropriately (as were the static features of Mars)—was consequently filtered out as noise. But since the scientists involved knew that Deimos was there, they were able to fix the image enhancement program to find it. By changing the threshold (so that Deimos and some noise enter as— probably smeared—signal), changing the sampling rate or the integration area (stopping Deimos at the effectively same place for two or more times), or introducing the right kind of spatiotemporal correlation function (to track Deimos's periodic moves around Mars), they could restore Deimos to the pictures again. Different tunings of the noise filters and different redundancies in the signal were exploited to bring static Mars and moving Deimos into clear focus.

We can see exactly analogous phenomena in vision if we look at a moving fan or airplane propellor. We can look through it (filtering it out as noise) to see something behind it. Lowering our threshold, we can attend to the propellor disk as a colored transparent (smeared) object. Cross-specific variation in flicker-fusion frequency indicates different sampling rates, which are keyed to the adaptive requirements of the organism (see Wimsatt, 1980a, pp. 292-297). The various phenomena associated with periodic stroboscopic illumination (apparent freezing and slow rotation of a rapidly spinning object) involve detection of a lagged correlation. Here, too, different tunings pick out different aspects of or entities in the environment. This involves a use of different heuristics, a matter I will return to later.

I quoted Glymour earlier on the importance of getting the same answer for the value of quantities computed in two different ways. What if these computations or determinations do not agree? The result is not always disastrous; indeed, when such mismatches happen in a sufficiently structured situation, they can be very productive.

This situation could show that we were wrong in assuming that we were detecting or determining the same quantity; but (as Campbell, 1966, was the first to point out), if we assume that we *are* determining the same quantity but "through a glass darkly," the mismatch can provide an almost magical opportunity for discovery. Given imperfect observations of a thing-we-know-not-what, using experimental apparatus with biases-we-may-not-understand, we can achieve both a better understanding of the object (it must be, after all, that one thing whose properties can produce these divergent results in these detectors) and of the experimental apparatus (which are, after all, these pieces that can be affected thus divergently by this one thing).

The constraint producing the information here is the identification of the object of the two or more detectors. If two putatively identical things are indeed identical, then any property of one must be a property of the other. We must resolve any apparent differences either by giving up the identification or locating the differences not in the thing itself but in the interactions of the thing with different measuring instruments. And this is where we learn about the measuring instruments. Having then acquired a better knowledge of the biases of the measuring instruments, we are in a better position not only to explain the differences but also, in the light of them, to give a newly refined estimate of the property of the thing itself. This procedure, a kind of "means-end" analysis (Wimsatt, 1976a; Simon, 1969) has enough structure to work in any given case only because of the enormous amount of background knowledge of the thing and the instruments which we bring to the situation. What we can learn (in terms of localizing the source of the differences) is in direct proportion to what we already know.

This general strategy for using identifications has an important subcase in reductive explanation. I have argued extensively (Wimsatt, 1976a, part II, 1976b, 1979) that the main reason for the productiveness of reductive explanation is that interlevel identifications immediately provide a wealth of new hypotheses: each property of the entity as known at the lower level must be a property of it as known at the upper level, and conversely; and usually very few of these properties from the other level have been predicated of the common object. The implications of these predictions usually have fertile consequences at both levels, and even where the match is not exact, there is often enough structure in the situation to point to a revised identification, with the needed refinements. This description characterizes well the history of genetics, both in the period of the localization of the genes on chromosomes (1883 to 1920) and in the final identification of DNA as the genetic material (1927 to 1953). (For the earlier period see, for example, Allen, 1979; Moore, 1972; Darden, 1974; Wimsatt, 1976a, part II. For the later period see Olby, 1974.) Indeed, the overall effect of these considerations is to suggest that *the use of identities for the deletion of error in a structured situation in the detection of error may be the most powerful heuristic known and certainly one of the most effective in generating scientific hypotheses.*

Also significant in the connection between robustness and discovery is Campbell's (1977) suggestion that things with greater entitativity (things whose boundaries are more robust) ought to be learned earlier. He cites suggestive support from language development for this thesis, which Quine's (1960) views also tend to support. I suspect that robustness could prove to be an important tool in analyzing not only what is discovered but also the order in which things are discovered.

There is some evidence from work with children (Omanson,

1980a, 1980b) that components of narratives which are central to the narrative, in that they are integrated into its causal and its purposive or intentional structure, are most likely to be remembered and least likely to be abstracted out in summaries of the story. This observation is suggestively related both to Feynman's (1965, p. 47) remark quoted above, relating robustness to forgetting relationships in a multiply connected theory, and to Simon's (1969) concept of a blackboard work space, which is maintained between successive attempts to solve a problem and in which the structure of the problem representation and goal tree may be subtly changed through differential forgetting. These suggest other ways in which robustness could affect discovery processes through differential effects on learning and forgetting.

Robustness, Objectification, and Realism

Robustness is widely used as a criterion for the reality or trustworthiness of the thing which is said to be robust. The boundaries of an ordinary object, such as a table, as detected in different sensory modalities (visually, tactually, aurally, orally), roughly coincide, making them robust; and this is ultimately the primary reason why we regard perception of the object as veridical rather than illusory. (See Campbell, 1958, 1966.) It is a rare illusion indeed which could systematically affect all of our senses in this consistent manner. (Drug-induced hallucinations and dreams may involve multimodal experience but fail to be consistent through time for a given subject, or across observers, thus failing at a higher level to show appropriate robustness.)

Our concept of an object is of something which exemplifies a multiplicity of properties within its boundaries, many of which change as we move across its boundary. A one-dimensional object is a contradiction in terms and usually turns out to be a disguised definition—a legal or theoretical fiction. In appealing to the robustness of boundaries as a criterion for objecthood, we are appealing to this multiplicity of properties (different properties detected in different ways) and thus to a time-honored philosophical notion of objecthood.

Campbell (1958) has proposed the use of the coincidence of boundaries under different means of detection as a methodological criterion for recognizing entities such as social groups. For example, in a study of factors affecting the reproductive cycles of women in college dormitories, McClintock (1971, and in conversation) found that the initially randomly timed and different-length cycles of 135 women after several months became synchronized into 17 groups, each oscillating synchronously, in phase and with a common period. The members of these groups turned out to be those who spent most time together, as determined by sociological methods. After the onset of synchrony, group membership of an individual could be determined either from information about her reproductive cycle or from a sociogram representing her frequency of social interaction with other individuals. These groups are thus multiply detectable. This illustrates the point that there is nothing sacred about using perceptual criteria in individuating entities. The products of any scientific detection procedure, including procedures drawn from different sciences, can do as well, as Campbell suggests: "In the diagnosis of middle-sized physical entities, the boundaries of the entity are multiply confirmed, with many if not all of the diagnostic procedures confirming each other. For the more 'real' entities, the number of possible ways of confirming the boundaries is probably unlimited, and the more our knowledge expands, the more diagnostic means we have available. 'Illusions' occur when confirmation is attempted and found lacking, when boundaries diagnosed by one means fail to show up by other expected checks" (1958, pp. 23-24).

Illusions can arise in connection with robustness in a variety of ways. Campbell's remark points to one important way: Where expectations are derived from one boundary, or even more, the coincidence of several boundaries leads us to predict, assume, or expect that other relevant individuating boundaries will coincide. Perhaps most common, given the reductionism common today, are situations in which the relevant system boundary is in fact far more inclusive than one is led to expect from the coincidence of a number of boundaries individuating an object at a lower level. Such functional localization fallacies are found in neurophysiology, in genetics, in evolutionary biology (with the hegemony of the selfish gene at the expense of the individual or the group; see Wimsatt, 1980b), in psychology, and (where it is a fallacy) with methodological individualism in the social sciences. In all these cases the primary object of analysis—be it a gene, a neuron, a neural tract, or an individual—may well be robust, but its high degree of entitativity leads us to hang too many boundaries and explanations on it. Where this focal entity is at a lower level, reductionism and robustness conspire to lead us to regard the higher-level systems as epiphenomenal. Another kind of illusion—the illusion that an entity is robust—can occur when the various means of detection supposed to be independent are not in fact. (This will be discussed further in the final section of this chapter.) Another kind of illusion or paradox arises particularly for functionally organized systems. This illusion occurs when a system has robust boundaries, but the different criteria used to decompose it into parts produce radically different boundaries. When the parts have little entitativity compared to the system, the holist's war cry (that the whole is more than the sum of the parts) will have a greater appeal. Elsewhere (Wimsatt, 1974), I have explored this kind of case and its consequences for the temptation of antireductionism, holism, or, in extreme cases, vitalisms or ontological dualisms.

Robustness is a criterion for the reality of entities, but it also has played and can play an important role in the analysis of properties. Interestingly, the distinction between primary and secondary qualities, which had a central role in the philosophy of Galileo, Descartes, and Locke, can be made in terms of robustness. Primary qualities—such as shape, figure, and size—are detectable in more than one sensory modality. Secondary qualities—such as color, taste, and sound—are detectable through only one sense. I think it is no accident that seventeenth-century philosophers chose to regard primary qualities as the only things that were "out there"—in objects; their cross-modal detectability seemed to rule out their being products of sensory interaction with the world. By contrast, the limitation of the secondary qualities to a single sensory modality seemed naturally to suggest that they were "in us," or subjective. Whatever the merits of the further seventeenth-century view that the secondary qualities were to be explained in terms of the interaction of a perceiver with a world of objects with primary qualities, this explanation represents an instance of an explanatory principle which is widely found in science (though seldom if ever explicitly recognized): *the explanation of that which is not robust in terms of that which is robust.* (For other examples see Wimsatt, 1976a, pp. 243-249; Feynman, 1965).

Paralleling the way in which Levins' use of robustness differs from Feynman's, *robustness, or the lack of it, has also been used in contexts where we are unsure about the status of purported properties, to argue for their veridicality or artifactuality,* and thus to discover the properties in terms of which we should construct our theories. This is the proposal of the now classic and widely used methodological paper of Campbell and Fiske (1959). Their convergent validity is a form of robustness, and their criterion of discriminant validity can be regarded as an attempt to guarantee that the invariance across test methods and traits is not due to their insensitivity to the variables under study. Thus, method bias, a common cause of failures of discriminant validity, is a kind of failure of the requirement for robustness that the different means of detection used are actually independent, in this case because the method they share is the origin of the correlations among traits.

Campbell and Fiske point out that very few theoretical constructs (proposed theoretical properties or entities) in the social sciences have significant degrees of convergent and discriminant validity, and they argue that this is a major difference between the social and natural or biological sciences—a difference which generates many of the problems of the social sciences. (For a series of essays which in effect claim that personality variables are highly context dependent and thus have very little or no robustness, see Shweder, 1979a, 1979b, 1980.)

While the natural and biological sciences have many problems where similar complaints could be made (the importance of interaction effects and context dependence is a key indicator of such problems), scientists in these areas have been fortunate in having at least a large number of cases where the systems, objects, and properties they study can be effectively isolated and localized, so that interactions and contexts can be ignored.

Robustness and Levels of Organization

Because of their multiplicity of connections and applicable descriptions, *robust properties or entities tend to be (1) more easily detectable, (2) less subject to illusion or artifact, (3) more ex-*

planatorily fruitful, and (4) predictively richer than nonrobust properties or entities. With this set of properties, it should be small wonder that we use robustness as a criterion for reality. It should also not be surprising that—since we view perception (as evolutionary epistemologists do) as an efficient tool for gathering information about the world—robustness should figure centrally in our analysis of perceptual hypotheses and heuristics (in the earlier section "Robustness, Redundancy, and Discovery" and in the next section, "Heuristics and Robustness"). Finally, since ready detectability, relative insensitivity to illusion or artifact, and explanatory and predictive fruitfulness are desirable properties for the components of scientific theories, we should not be surprised to discover that robustness is important in the discovery and description of phenomena (again, see the section on discovery) and in analyzing the structure of scientific theories (see the section "Robustness and the Structure of Theories").

One of the most ubiquitous phenomena of nature is its tendency to come in levels. If the aim of science, to follow Plato, is to cut up nature at its joints, then these levels of organization must be its major vertebrae. They have become so major, indeed, that our theories tend to follow these levels, and the language of our theories comes in strata. This has led many linguistically inclined philosophers to forgo talk of nature at all, and to formulate problems—for example, problems of reduction—in terms of "analyzing the relation between theoretical vocabularies at different levels." But our language, as Campbell (1974) would argue, is just another (albeit very important) tool in our struggle to analyze and to adapt to nature. In an earlier paper (Wimsatt, 1976a, part III), I applied Campbell's criteria for entification to argue that entities at different levels of organization tend to be multiply connected in terms of their causal relations, primarily with other entities at their own level, and that they, and the levels they comprise, are highly robust. As a result, there are good explanatory reasons for treating different levels of organization as dynamically, ontologically, and epistemologically autonomous. There is no conflict here with the aims of good reductionistic science: there is a great deal to be learned about upper-level phenomena at lower levels of organization, but upper-level entities are not "analyzed away" in the process, because they remain robustly connected with other upper-level entities, and their behavior is explained by upper-level variables.

To see how this is so, we need another concept—that of the *sufficient parameter,* introduced by Levins (1966, pp. 428, 429):

> It is an essential ingredient in the concept of levels of phenomena that there exists a set of what, by analogy with the sufficient statistic, we can call sufficient parameters defined on a given level . . . which are very much fewer than the number of parameters on the lower level and which among them contain most of the important information about events on that level.
>
> The sufficient parameters may arise from the combination of results of more limited studies. In our robust theorem on niche breadth we found that temporal variation, patchiness of the environment, productivity of the habitat, and mode of hunting could all have similar effects and that they did this by way of their contribution to the uncertainty of the environment. Thus uncertainty emerges as a sufficient parameter.
>
> The sufficient parameter is a many-to-one transformation of lower-level phenomena. Therein lies its power and utility, but also a new source of imprecision. The many-to-one nature of "uncertainty" prevents us from going backwards. If either temporal variation or patchiness or low productivity leads to uncertainty, the consequences of uncertainty alone cannot tell us whether the environment is variable, or patchy, or unproductive. Therefore, we have lost information.

A sufficient parameter is thus a parameter, a variable, or an index which, either for most purposes or merely for the purposes at hand, captures the effect of significant variations in lower-level or less abstract variables (usually only for certain ranges of the values of these variables) and can thus be substituted for them in the attempt to build simpler models of the upper-level phenomena.

Levins claims that this notion is a natural consequence of the concept of levels of phenomena, and this is so, though it may relate to degree of abstraction as well as to degree of aggregation. (The argument I will give here applies only to levels generated by aggregation of lower-level entities to form upper-level ones.) Upper-level variables, which give a more "coarse-grained" description of the system, are much smaller in number than the lower-level variables necessary to describe the same system. Thus, there must be, for any given degree of resolution between distinguishable state descriptions, far fewer distinguishable upper-level state descriptions than lower-level ones. The smaller number of distinguishable upper-level states entails that for any given degree of resolution, there must be many-one mappings between at least some lower-level and upper-level state descriptions with many lower-level descriptions corresponding to a single upper-level description. But then, those upper-level state descriptions with multiple lower-level state descriptions are robust over changes from one of these lower-level descriptions to another in its set.

Furthermore, the stability of (and possibility of continuous change in) upper-level phenomena (remaining in the same macrostate or changing by moving to neighboring states) places constraints on the possible mappings between lower-level and upper-level states: in the vast majority of cases neighboring microstates must map without discontinuity into the same or neighboring macrostates; and, indeed, most local microstate changes will have no detectable macrolevel effects. *This fact gives upper-level phenomena and laws a certain insulation from* (through their invariance over: robustness again!) *lower-level changes and generates a kind of explanatory and dynamic (causal) autonomy of the upper-level phenomena and processes,* which I have argued for elsewhere (Wimsatt, 1976a, pp. 249-251; 1976b).

If one takes the view that causation is to be characterized in terms of manipulability (see, for example, Gasking, 1955; Cook and Campbell, 1979), the fact that the vast majority of manipulations at the microlevel do not make a difference at the macrolevel means that macrolevel variables are almost always more causally efficacious in making macrolevel changes than microlevel variables. This gives explanatory and dynamic autonomy of the upper-level entities, phenomena, laws, and relations, within a view of explanation which is sensitive to problems of computational complexity and the costs and benefits we face in offering explanations. As a result, it comes much closer than the traditional hypothetico-deductive view to being able to account for whether we explain a phenomenon at one level and when we choose to go instead to a higher or lower level for its explanation. (See Wimsatt, 1976a, part III, and 1976b, particularly sections 4, 5, 6, and the appendix.)

The many-one mappings between lower- and upper-level state descriptions mentioned above are consistent with correspondences between types of entities at lower and upper levels but do not entail them. There may be only token-token mappings (piecemeal mappings between instances of concepts, without any general mappings between concepts), resulting in the upper-level properties being supervenient on rather than reducible to lower-level properties (Kim, 1978; Rosenberg, 1978). The main difference between Levins' notion of a sufficient parameter and the notion of supervenience is that the characterization of supervenience is embedded in an assumed apocalyptically complete and correct description of the lower and upper levels. Levins makes no such assumption and defines the sufficient parameter in terms of the imperfect and incomplete knowledge that we actually have of the systems we study. It is a broader and less demanding notion, involving a relation which is inexact, approximate, and admits of both unsystematic exceptions (requiring a *ceteris paribus* qualifier) and systematic ones (which render the relationship conditional).

Supervenience could be important for an omniscient Laplacean demon but not for real, fallible, and limited scientists. The notion of supervenience could be regarded as a kind of ideal limiting case of a sufficient parameter as we come to know more and more about the system, but it is one which is seldom if ever found in the models of science. The concept of a sufficient parameter, by contrast, has many instances in science. It is central to the analysis of reductive explanation (Wimsatt, 1976a; 1976b, pp. 685-689; 1979) and has other uses as well (Wimsatt, 1980a, section 4).

Heuristics and Robustness

Much or even most of the work in philosophy of science today which is not closely tied to specific historical or current scientific case studies embodies a metaphysical stance which, in effect, assumes that the scientist is an omniscient and computationally omnipotent Laplacean demon. Thus, for example, discussions of reductionism are full of talk of "in principle analyzability" or "in principle deducibility," where the force of the "in principle" claim is held to be something like "If we knew a total description of the system at the lower level, and all the lower-level laws, a sufficiently complex computer could generate the analysis of all the upper-level terms and laws and predict any upper-level phenomenon." Parallel kinds of assumptions of omniscience and computational omnipotence are found in rational decision theory, discussions of Bayesian epistemology, automata theory and algorithmic procedures in linguistics and the philosophy of mind, and the reductionist and foundationalist views of virtually all the major figures of twentieth-century logical empiricism. It seems almost to be a corollary to a deductivist approach to problems in philosophy of science (see Wimsatt, 1979) and probably derives ultimately from the Cartesian vision criticized earlier in this chapter.

I have already written at some length attacking this view and its application to the problem of reduction in science (see Wimsatt, 1974; 1976a, pp. 219-237; 1976b; 1979; 1980b, section 3; and also Boyd, 1972). The gist of this attack is threefold: (1) On the "Laplacean demon" interpretation of "in principle" claims, we have no way of evaluating their warrant, at least in science. (This is to be distinguished from cases in mathematics or automata theory, where "in principle" claims can be explicated in terms of the notion of an effective procedure.) (2) We are in any case not Laplacean demons, and a philosophy of science which could have normative force only for Laplacean demons thus gives those of us who do not meet these demanding specifications only counterfactual guidance; that is, it is of no real use to practicing scientists and, more strongly, suggests methods and viewpoints which are less advantageous than those derived from a more realistic view of the scientist as problem solver (see Wimsatt, 1979). (3) An alternative approach, which assumes more modest capacities of practicing scientists, does provide real guidance, better fits with actual scientific practice, and even (for reductive explanations) provides a plausible and attractive alternative interpretation for the "in principle" talk which so many philosophers and scientists use frequently (see Wimsatt, 1976a, part II; 1976b, pp. 697-701).

An essential and pervasive feature of this more modest alternative view is the replacement of the vision of an ideal scientist as a computationally omnipotent algorithmizer with one in which the scientist as decision maker, while still highly idealized, must consider the size of computations and the cost of data collection, and in other very general ways must be subject to considerations of efficiency, practical efficacy, and cost-benefit constraints. This picture has been elaborated over the last twenty-five years by Herbert Simon and his co-workers, and their ideal is "satisficing man," whose rationality is bounded, by contrast with the unbounded omniscience and computational omnipotence of the "economic man" of rational decision theory (see Simon, 1957, reprinted as ch. 1 of Simon, 1979; see also Simon, 1969). Campbell's brand of fallibilism and critical realism from an evolutionary perspective also place him squarely in this tradition.

A key feature of this picture of man as a boundedly rational decision maker is the use of heuristic principles where no algorithms exist or where the algorithms that do exist require an excessive amount of information, computational power, or time. I take a heuristic procedure to have three important properties (see also Wimsatt, 1980b, section 3): (1) By contrast with an algorithmic procedure (here ignoring probabilistic automata), *the correct application of a heuristic procedure does not guarantee a solution* and, if it produces a solution, does not guarantee that the solution is correct. (2) *The expected time, effort, and computational complexity of producing a solution with a heuristic procedure is appreciably less* (often by many orders of magnitude for a complex problem) *than that expected with an algorithmic procedure.* This is indeed the reason why heuristics are used. They are a cost-effective way, and often the *only* physically possible way, of producing a solution. (3) *The failures and errors produced when a heuristic is used are not random but systematic.* I conjecture that *any heuristic, once we understand how it works, can be made to fail.* That is, given this knowledge of the heuristic procedure, we can construct classes of problems for which it will always fail to produce an answer or for which it will always produce the wrong answer. This property of systematic production of wrong answers will be called the *bias* of the heuristic.

This last feature is exceedingly important. Not only can we

work forward from an understanding of a heuristic to predict its biases, but we can also work backward from the observation of systematic biases as data to hypothesize the heuristics which produced them; and if we can get independent evidence (for example, from cognitive psychology) concerning the nature of the heuristics, we can propose a well-founded explanatory and predictive theory of the structure of our reasoning in these areas. This approach was implicitly (and sometimes explicitly) followed by Tversky and Kahneman (1974), in their analysis of fallacies of probabilistic reasoning and of the heuristics which generate them (see also Shweder, 1977, 1979a, 1979b, 1980, for further applications of their work; and Mynatt, Doherty, and Tweney, 1977, for a further provocative study of bias in scientific reasoning). The systematic character of these biases also allows for the possibility of modifications in the heuristic or in its use to correct for them (see Wimsatt, 1980b, pp. 52-54).

The notion of a heuristic has far greater implications than can be explored in this chapter. In addition to its centrality in human problem solving, it is a pivotal concept in evolutionary biology and in evolutionary epistemology. It is a central concept in evolutionary biology because any biological adaptation meets the conditions given for a heuristic procedure. First, it is a commonplace among evolutionary biologists that adaptations, even when functioning properly, do not guarantee survival and production of offspring. Second, they are, however, cost-effective ways of contributing to this end. Finally, any adaptation has systematically specifiable conditions, derivable through an understanding of the adaptation, under which its employment will actually decrease the fitness of the organism employing it, by causing the organism to do what is, under those conditions, the wrong thing for its survival and reproduction. (This, of course, seldom happens in the organism's normal environment, or the adaptation would become maladaptive and be selected against.) This fact is indeed systematically exploited in the functional analysis of organic adaptations. It is a truism of functional inference that learning the conditions under which a system malfunctions, and how it malfunctions under those conditions, is a powerful tool for determining how it functions normally and the conditions under which it was designed to function. (For illuminating discussions of the problems, techniques, and fallacies of functional inference under a variety of circumstances, see Gregory, 1962; Lorenz, 1965; Valenstein, 1973; Glassman, 1978.)

The notion of a heuristic is central to evolutionary epistemology because Campbell's (1974, 1977) notion of a vicarious selector, which is basic to his conception of a hierarchy of adaptive and selective processes spanning subcognitive, cognitive, and social levels, is that of a heuristic procedure. For Campbell a vicarious selector is a substitute and less costly selection procedure acting to optimize some index which is only contingently connected with the index optimized by the selection process it is substituting for. This contingent connection allows for the possibility—indeed, the inevitability—of systematic error when the conditions for the contingent concilience of the substitute and primary indices are not met. An important ramification of Campbell's idea of a vicarious selector is the possibility that one heuristic may substitute for another (rather than for an algorithmic procedure) under restricted sets of conditions, and that this process may be repeated, producing a nested hierarchy of heuristics. He makes ample use of this hierarchy in analyzing our knowing processes (Campbell, 1974). I believe that this is an appropriate model for describing the nested or sequential structure of many approximation techniques, limiting operations, and the families of progressively more realistic models found widely in progressive research programs, as exemplified in the development of nineteenth-century kinetic theory, early twentieth-century genetics, and several areas of modern population genetics and evolutionary ecology.

To my mind, Simon's work and that of Tversky and Kahneman have opened up a whole new set of questions and areas of investigation of pragmatic inference (and its informal fallacies) in science, which could revolutionize our discipline in the next decade. (For a partial view of how studies of reduction and reductionism in science could be changed, see Wimsatt, 1979.) This change in perspective would bring philosophy of science much closer to actual scientific practice without surrendering a normative role to an all-embracing descriptivism. And it would reestablish ties with psychology through the study of the character, limits, and biases of processes of empirical reasoning. Inductive procedures in science are heuristics (Shimony, 1970), as are Mill's methods and other methods for discovering causal relations, building models, and generating and modifying hypotheses.

Heuristics are also important in the present context, because the procedures for determining robustness and for making further application of these determinations for other ends are all heuristic procedures. Robustness analysis covers a class of powerful and important techniques, but they are not immune to failures. There are no magic bullets in science, and these are no exception.

Most striking of the ways of failure of robustness analysis is one which produces illusions of robustness: the failure of the different supposedly independent tests, means of detection, models, or derivations to be truly independent. This is the basis for a powerful criticism of the validity of IQ scales as significant measures of intelligence (see McClelland, 1973). Failures of independence are not easy to detect and often require substantial further analysis. Without that, such failures can go undetected by the best investigators for substantial lengths of time. Finally, the fact that different heuristics can be mutually reinforcing, each helping to hide the biases of the others (see Wimsatt, 1980b, sections 5 and 8), can make it much harder to detect errors which would otherwise lead to discovery of failures of independence. The failure of independence in its various modes, and the factors affecting its discovery, emerges as one of the most critical and important problems in the study of robustness analysis, as is indicated by the history of the group selection controversy.

Robustness, Independence, and Pseudorobustness:
A Case Study

In recent evolutionary biology (since Williams' seminal work in 1966), group selection has been the subject of widespread attack and general suspicion. Most of the major theorists (including W. D. Hamilton, John Maynard Smith, and E. O. Wilson) have argued against its efficacy. A number of mathematical models of this phenomenon have been constructed, and virtually all of them (see Wade, 1978) seem to support this skepticism. The various mathematical models of group selection surveyed by Wade all admit of the possibility of group selection. But almost all of them predict that group selection should only very rarely be a significant evolutionary factor; that is, they predict that group selection should have significant effects only under very special circumstances—for extreme values of parameters of the models—which should seldom be found in nature. Wade undertook an experimental test of the relative efficacy of individual and group selection—acting in concert or in opposition in laboratory populations of the flour beetle, *Tribolium*. This work produced surprising results. Group selection appeared to be a significant force in these experiments, one capable of overwhelming individual selection in the opposite direction for a wide range of parameter values. This finding, apparently contradicting the results of all of the then extant mathematical models of group selection, led Wade (1978) to a closer analysis of these models, with results described here.

All the models surveyed made simplifying assumptions, most of them different. Five assumptions, however, were widely held in common; of the twelve models surveyed, each made at least three of these assumptions, and five of the models made all five assumptions. Crucially, for present purposes, the five assumptions are biologically unrealistic and incorrect, and each independently has a strong negative effect on the possibility or efficacy of group selection. It is important to note that these models were advanced by a variety of different biologists, some sympathetic to and some skeptical of group selection as a significant evolutionary force. Why, then, did all of them make assumptions strongly inimical to it? Such a coincidence, radically improbable at best, cries out for explanation: we have found a systematic bias suggesting the use of a heuristic.

These assumptions are analyzed more fully elsewhere (Wade, 1978; Wimsatt, 1980a). My discussion here merely summarizes the results of my earlier analysis, where (in section 5) I presented a list of nine reductionistic research and modeling strategies. Each is a heuristic in that it has systematic biases associated with it, and these biases will lead to the wrong answer if the heuristic is used to analyze certain kinds of systems. It is the use of these heuristics, together with certain "perceptual" biases (deriving from thinking of groups as "collections of individuals" rather than as robust entities analogous to organisms), that is responsible for the widespread acceptance of these assumptions and the almost total failure to notice what an unrealistic view they give of group selection. Most of the reductionistic heuristics lead to a dangerous over-simplification of the environment being studied and a dangerous underassessment of the effects of these simplifications. In the context of the perceptual bias of regarding groups as collections of individuals (or sometimes even of genes), the models tend systematically to err in the internal and relational structure they posit for the groups and in the character of processes of group reproduction and selection.

The first assumption, that the processes can be analyzed in terms of selection of alternative alleles at a single locus, is shown to be empirically false by Wade's own experiments, which show conclusively that both individual and group selection is proceeding on multilocus traits. (For an analysis of the consequences of treating a multilocus trait erroneously as a single-locus trait, see Wimsatt, 1980b, section 4.) The fifth assumption, that individual and group selection are opposed in their effects, also becomes untenable for a multilocus trait (see Wimsatt, 1980b, section 7).

The second assumption is equivalent to the time-honored assumption of panmixia, or random mating within a population, but in the context of a group selection model it is equivalent to assuming a particularly strong form of blending inheritance for group inheritance processes. This assumption is factually incorrect and, as R. A. Fisher showed in 1930, effectively renders evolution at that level impossible. The third assumption is equivalent to assuming that groups differ in their longevity but not in their reproductive rates. But, as all evolutionary biologists since Darwin have been aware, variance in reproductive rate has a far greater effect on the intensity of selection than variance in longevity. So the more significant component was left out in favor of modeling the less significant one. (The second and third assumptions are discussed in Wimsatt, 1980b, section 7.) The fourth assumption is further discussed and shown to be incorrect in Wade (1978).

The net effect is a set of cumulatively biased and incorrect assumptions, which, not surprisingly, lead to the incorrect conclusion that group selection is not a significant evolutionary force. If I am correct in arguing that these assumptions probably went unnoticed because of the biases of our reductionistic research heuristics, a striking analogy emerges. The phenomenon appeared to be a paradigmatic example of Levinsian robustness. A wide variety of different models, making different assumptions, appeared to show that group selection could not be efficacious. But the robustness was illusory, because the models were not independent in their assumptions. The commonality of these assumptions appears to be a species of method bias, resulting in a failure of discriminant validity (Campbell and Fiske, 1959). But the method under consideration is not the normal sort of test instrument that social scientists deal with. Instances of the method are reductionistic research heuristics, and the method is reductionism. For the purposes of problem solving, our minds can be seen as a collection of methods, and the particularly single-minded are unusually prone to method bias in their thought processes. This conclusion is ultimately just another confirmation at another level of something Campbell has been trying to teach us for years about the importance of multiple independent perspectives.

References

Allen, G. E. *Thomas Hunt Morgan: The Man and His Science.* Princeton, N.J.: Princeton University Press, 1979.

Barlow, R. E., and Proschan, F. *The Mathematical Theory of Reliability and Life Testing.* New York: Wiley, 1975.

Boyd, R. "Determinism, Laws, and Predictability in Principle." *Philosophy of Science,* 1972, *39,* 431-450.

Campbell, D. T. "Common Fate, Similarity, and Other Indices of the Status of Aggregates of Persons as Social Entities." *Behavioral Science,* 1958, *3,* 14-25.

Campbell, D. T. "Pattern Matching as an Essential in Distal Knowing." In K. R. Hammond (Ed.), *The Psychology of Egon Brunswik.* New York: Holt, Rinehart and Winston, 1966.

Campbell, D. T. "Definitional Versus Multiple Operationalism." *et al.,* 1969a, *2* (1), 14-17.

Campbell, D. T. "Prospective: Artifact and Control." In R. Rosenthal and R. Rosnow (Eds.), *Artifact in Behavioral Research.* New York: Academic Press, 1969b.

Campbell, D. T. "Evolutionary Epistemology." In P. A. Schilpp (Ed.), *The Philosophy of Karl Popper.* La Salle, Ill: Open Court, 1974.

Campbell, D. T. "Descriptive Epistemology: Psychological, Sociological, and Evolutionary." William James Lectures, Harvard University, 1977.

Campbell, D. T., and Fiske, D. W. "Convergent and Discriminant Validation by the Multitrait-Multimethod Matrix." *Psychological Bulletin*, 1959, *56*, 81-105.

Cook, T. D., and Campbell, D. T. *Quasi-Experimentation: Design and Analysis for Field Settings*. Chicago: Rand McNally, 1979.

Cronbach, L. J., and Meehl, P. E. "Construct Validity in Psychological Tests." *Psychological Bulletin*, 1955, *52*, 281-302.

Darden, L. "Reasoning in Scientific Change: The Field of Genetics at Its Beginnings." Unpublished doctoral dissertation, Committee on the Conceptual Foundations of Science, University of Chicago, 1974.

Feynman, R. P. *The Character of Physical Law*. Cambridge, Mass.: M.I.T. Press, 1965.

Fisher, R. A. *The Genetical Theory of Natural Selection*. New York: Clarendon Press, 1930.

Gasking, D. A. T. "Causation and Recipes." *Mind*, 1955, n.s. *64*, 479-487.

Glassman, R. B. "The Logic of the Lesion Experiment and Its Role in the Neural Sciences." In S. Finger (Ed.), *Recovery from Brain Damage: Research and Theory*. New York: Plenum, 1978.

Glymour, C. *Theory and Evidence*. Princeton, N.J.: Princeton University Press, 1980.

Gregory, R. L. "Models and the Localization of Function in the Central Nervous System" [1962]. Reprinted in C. R. Evans and A. D. J. Robertson (Eds.), *Key Papers: Cybernetics*. London: Butterworth, 1967.

Kauffman, S. A. "Cellular Gene Control Systems." In A. A. Moscona and others (Eds.), *Current Topics in Developmental Biology*. Vol. 6. New York: Academic Press, 1971.

Kim, J. "Supervenience and Nomological Incommensurables." *American Philosophical Quarterly*, 1978, *15*, 149-156.

Laudan, L. "William Whewell on the Consilience of Inductions." *The Monist*, 1971, *55*, 368-391.

Levins, R. "The Strategy of Model Building in Population Biology." *American Scientist*, 1966, *54*, 421-431.

Levins, R. *Evolution in Changing Environments*. Princeton, N.J.: Princeton University Press, 1968.

Lorenz, K. Z. *Evolution and Modification of Behavior.* Chicago: University of Chicago Press, 1965.

McClelland, D. D. "Testing for Competence Rather Than for 'Intelligence.' " *American Psychologist,* 1973, *29,* 107.

McClintock, M. K. "Menstrual Synchrony and Suppression." *Nature,* January 22, 1971, *229,* 244-245.

Margenau, H. *The Nature of Physical Reality.* New York: McGraw-Hill, 1950.

Moore, J. A. *Heredity and Development.* (2nd ed.) New York: Oxford University Press, 1972.

Mynatt, C. R., Doherty, M. E., and Tweney, R. D. "Confirmation Bias in a Simulated Research Environment: An Experimental Study of Scientific Inference." *Quarterly Journal of Experimental Psychology,* 1977, *29,* 85-95.

Olby, R. *The Path to the Double Helix.* Seattle: University of Washington Press, 1974.

Omanson, R. C. "The Narrative Analysis: Identifying Central, Supportive and Distracting Content." Unpublished manuscript, 1980a.

Omanson, R. C. "The Effects of Centrality on Story Category Saliency: Evidence for Dual Processing." Paper presented at 88th annual meeting of the American Psychological Association, Montreal, September 1980b.

Peirce, C. S. "Some Consequences of Four Incapacities." In C. Hartshorne and P. Weiss (Eds.), *Collected Papers of Charles Sanders Peirce.* Vol. 5. Cambridge, Mass.: Harvard University Press, 1936. (Originally published 1868.)

Putnam, H. "The Analytic and the Synthetic." In H. Feigl and G. Maxwell (Eds.), *Minnesota Studies in the Philosophy of Science.* Vol. 3. Minneapolis: University of Minnesota Press, 1962.

Quine, W. V. O. *Word and Object.* Cambridge, Mass.: M.I.T. Press, 1960.

Rosenberg, A. "The Supervenience of Biological Concepts." *Philosophy of Science,* 1978, *45,* 368-386.

Shimony, A. "Statistical Inference." In R. G. Colodny (Ed.), *The Nature and Function of Scientific Theories.* Pittsburgh: University of Pittsburgh Press, 1970.

Shweder, R. A. "Likeness and Likelihood in Everyday Thought: Magical Thinking in Judgements About Personality." *Current*

Anthropology, 1977, *18,* 637-648; reply to discussion, pp. 652-658.

Shweder, R. A. "Rethinking Culture and Personality Theory. Part I." *Ethos,* 1979a, *7,* 255-278.

Shweder, R. A. "Rethinking Culture and Personality Theory. Part II." *Ethos,* 1979b, *7,* 279-311.

Shweder, R. A. "Rethinking Culture and Personality Theory. Part III." *Ethos,* 1980, *8,* 60-94.

Simon, H. A. "A Behavioral Model of Rational Choice." In P. Nash, *Models of Man.* New York: Wiley, 1957.

Simon, H. A. *The Sciences of the Artificial.* Cambridge, Mass.: M.I.T. Press, 1969.

Simon, H. A. "The Structure of Ill-Structured Problems." *Artificial Intelligence,* 1973, *4,* 181-201.

Simon, H. A. *Models of Thought.* New Haven, Conn.: Yale University Press, 1979.

Tversky, A., and Kahneman, D. "Decision Under Uncertainty: Heuristics and Biases." *Science,* 1974, *185,* 1124-1131.

Valenstein, E. *Brain Control.* New York: Wiley, 1973.

von Neumann, J. "Probabilistic Logic and the Synthesis of Reliable Organisms from Unreliable Components." In C. E. Shannon and J. McCarthy (Eds.), *Automata Studies.* Princeton, N.J.: Princeton University Press, 1956.

Wade, M. J. "A Critical Review of the Models of Group Selection." *Quarterly Review of Biology,* 1978, *53* (3), 101-114.

Williams, G. C. *Adaptations and Natural Selection: A Critique of Some Current Evolutionary Thought.* Princeton, N.J.: Princeton University Press, 1966.

Wimsatt, W. C. "Complexity and Organization." In K. F. Schaffner and R. S. Cohen (Eds.), *Proceedings of the Meetings of the Philosophy of Science Association, 1972.* Dordrecht, Netherlands: Reidel, 1974.

Wimsatt, W. C. "Reductionism, Levels of Organization, and the Mind-Body Problem." In G. G. Globus, G. Maxwell, and I. Savodnik (Eds.), *Consciousness and the Brain: Scientific and Philosophical Strategies.* New York: Plenum, 1976a.

Wimsatt, W. C. "Reductive Explanation: A Functional Account." In C. A. Hooker, G. Pearse, A. C. Michalos, and J. W. van Evra (Eds.), *Proceedings of the Meetings of the Philosophy of Science Association, 1974.* Dordrecht, Netherlands: Reidel, 1976b.

Wimsatt, W. C. "Reduction and Reductionism." In P. D. Asguith and H. Kyburg, Jr. (Eds.), *Current Problems in Philosophy of Science*. East Lansing, Mich.: Philosophy of Science Association, 1979.

Wimsatt, W. C. "Randomness and Perceived-Randomness in Evolutionary Biology." *Synthese*, 1980a, *43* (3), 287-329.

Wimsatt, W. C. "Reductionistic Research Strategies and Their Biases in the Units of Selection Controversy." In T. Nickles (Ed.), *Scientific Discovery*. Vol. 2: *Historical and Scientific Case Studies*. Dordrecht, Netherlands: Reidel, 1980b.

PART II

The final chapter in the preceding section illustrates most clearly the mutual relevance of philosophy and methodology of science. It is in this arena that Donald Campbell has published his most well-known and widely cited papers. Through applications ranging from the construction of psychological tests to the design of large-scale social experiments, he has repeatedly demonstrated how the methods of social science can be informed by an explicit recognition of underlying epistemological assumptions.

Internal Validity

In one of the most cited papers in the social sciences, Campbell and Stanley (1963; also Campbell, 1957b; Cook and Campbell,

Methodological
Applications

1979) sought to explicate the rationale for true experimentation in social settings in terms of the logic and validity of causal inference. The sine qua non of good experimental design, they argued, lies in its "internal validity"—the extent to which variations in the outcome variable can be legitimately attributed to experimentally controlled variation in the treatment variable, as opposed to other, extraneous sources. The internal validity of a research design is threatened to the extent that "plausible rival hypotheses," other than the treatment manipulation of interest, could account for observed effects. The role of random assignment in the true experiment is to rule out (render implausible) a multitude of common threats to the validity of a cause-effect inference.

165

Quasi-experimental Designs

The internal validity of a true experiment (one involving random assignment) provides a standard of comparison for alternative research approaches. In the absence of randomization, the adequacy of "quasi-experimental" designs is to be evaluated in terms of the extent to which features of the research design (other than random assignment) can serve to rule out those same threats to internal validity. Campbell and Stanley suggested a number of alternative quasi-experimental research designs that could—under ideal conditions of implementation—compare favorably with true experiments with respect to the control of threats to validity. In later papers and addresses, Campbell (1969a, 1971) extended the logic of quasi-experimental design and internal validity to the political arena by arguing for the application of systematic scientific methodology to the assessment of the effects of social programs. These arguments had considerable influence on the development of program evaluation research as a field of applied social science and on the commitment to evaluation efforts that now accompanies major social programming in the United States and elsewhere.

Construct Validity and Operational Definitions

Another major application of epistemological concepts to social science methodology lies in Campbell's explication of construct validity as a central notion in his methodology. The relationship between scientific practice and the themes of hypothetical realism, triangulation, and entitativity is most clearly represented in his contributions to the psychological literature on testing and assessment (see Campbell, 1950, 1957a, 1960; Campbell and Fiske, 1959). In his approach to measurement theory, Campbell has consistently eschewed the notion of definitional operationalism (1969b) in favor of a position that recognizes the inevitable fallibility of any single measure or observation as a representation of the psychological construct it is intended to tap:

Measurements involve processes which must be specified in terms of many theoretical parameters. For

any specific measurement process, we know on theoretical grounds that it is a joint function of many scientific laws. Thus, we know on scientific grounds
that the measurements resulting cannot purely reflect a single parameter of our scientific theory. . . .
Let us consider in detail . . . a single meter, the galvanometer. The amount of needle displacement is certainly a strong function of the electrical current in
the attached wire. But it is also a function of friction
in the needle bearings, of inertia due to the weight of
the needle, of fluctuations in the earth's and other
magnetic and electrical fields, of the strong and weak
nuclear interactions in adjacent matters, of photon
bombardment, and so on. We know on theoretical
grounds that the needle pointer movements cannot be
independent of all of these, that is, that the pointer
movements cannot be *definitional* of a single parameter. . . . Analogously, for a tally mark on a census-
taker's protocol indicating family income, or rent
paid, or number of children, we know on theoretical
grounds that it is only in part a function of the state
of the referents of the question. It is also a function
of the social interaction of the interview, of the interviewer's appearance, of the respondent's fear of similar strangers . . . and so on. A manifest anxiety questionnaire response may be in part a function of
anxiety, but it is also a function of vocabulary comprehension, of individual and social class differences
in the use of euphoric and dysphoric adjectives,
[and] of idiosyncratic definitions of the key terms
[1969b, pp. 14-15].

Consistent with his view that all knowing is indirect, Campbell suggested that our only access to psychological phenomena
(such as social attitudes, personality traits, and decision rules) lies
in what he calls "outcroppings": "Any given theory has innumerable implications and makes innumerable predictions which are
unaccessible to available measures at any given time. The testing of
the theory can only be done at the available outcroppings, those
points where theoretical predictions and available instrumentation
meet. Any one such outcropping is equivocal, and all types avail-

able should be checked. The more remote or independent such checks, the more confirmatory their agreement" (Webb, Campbell, Schwartz, and Sechrest, 1966, p. 28).

The relationship between methodological triangulation and construct validity is best represented in Campbell and Fiske's (1959) discussion of the "multitrait-multimethod matrix." In that paper they argued that the adequacy of any measure of a hypothetical construct must be demonstrated in terms of two types of validity—convergent validity (the agreement between *different* methods of measuring the same construct) and discriminant validity (differentiation of results when the same method is used to assess hypothetically different qualities of the same objects). The results of analyses of the interrelationships among multiple independent measures of multiple constructs provide, simultaneously, information on methods variance (the extent to which results from similar methods converge when they are supposed to be assessing different constructs) and information on trait variance (the extent to which results from different measures of the same construct converge in their ordering of a set of objects). The relative size of trait variance to method variance, in turn, tells us something about the validity or entitativity of the psychological constructs under study.

Although Campbell realized that some instruments or techniques of measurement may well be more fallible than others, his emphasis on sources of systematic error and bias in all modes of knowing precludes reliance on any single method of assessment. Similarly, his advocacy of quantitative, experimental methods of hypothesis testing and program evaluation is tempered by a consistent appreciation for alternative research designs and more qualitative approaches as sources of cross-validation:

> Not only does science depend upon common-sense knowing as the trusted grounding for its elaborate esoteric instrumentation and quantification, in addition many products and achievements of science are cross-validated in ways accessible to common sense . . . the cross-validation of the quantitative by the qualitative. . . . For many experimental program evaluations, especially quasi-experimental ones, a ma-

jor category of threats to validity are the many other events, other than the experimental program, which might have produced the measured changes. Generating these specific alternative explanations and estimating their plausibility are matters of commonsense knowing and require a thorough familiarity with the specific local setting. . . . The polarity of quantitative versus qualitative approaches to research on social action remains unresolved, if resolution were to mean a predominant justification of one over the other. . . . Each pole is at its best in its criticisms of the other, not in the invulnerability of its own claims to descriptive knowledge. . . . I cannot recommend qualitative social science . . . as substitutes for the quantitative. But I have strongly recommended them both as needed cross-validating additions [Campbell, 1974].

Overview of Chapters

The chapters included in the second section of this volume build directly on Campbell's writings on methodological triangulation and the logic of experimental design. The first three chapters deal with the potential convergence of the quantitative, comparative approach to research with the qualitative, case-study approach. Robert LeVine discusses how ethnographic research can be informed by Campbell's epistemological contributions, with particular emphasis on the notion of triangulation from multiple vantage points. The chapters by Paul Rosenblatt and Louise Kidder both illustrate how qualitative and quantitative research methodologies can be made comparable when assessed in a Campbellian framework. Rosenblatt's chapter builds on an earlier Campbell (1975) paper, in which he suggested a sense in which a "degrees-of-freedom" analysis could be applied to conclusion drawing from case-study research. Kidder extends the logic of internal validity to qualitative research methods, arguing that the causal hypotheses implicit in typical qualitative research reports render them comparable to experimental or quasi-experimental research in the need to rule out standard threats to the validity of conclusions.

Thomas Cook expands more directly on Campbell's application of evolutionary models to the realm of social experimentation, seeking to develop stances for evaluation that are appropriate to the unique features of complex, ongoing social programs. Starting from the assumptions that large-scale programs rarely if ever die but that local projects representing such programs are heterogeneous in conception and implementation, Cook concludes that the appropriate goal of evaluation research efforts in such contexts is the identification of local project types that are successful, consistent with a variation-and-selective-retention view of social change.

The final chapter in this section, by David Kenny, grapples directly with the central problem of measurement identified by Campbell; namely, that all observations are in part a function of what is observed and in part a function of the characteristics of the observer. The paper revisits the setting of the earlier work by Campbell and his associates (1964) on response biases in interpersonal perception, and proposes some methodological innovations with particular application to the assessment of interpersonal attraction. Kenny demonstrates how a more sophisticated multivariate, analytic approach to the types of data collected by Campbell and associates can be used to assess (1) the relative contribution of observer response biases and observee differences to social perceptions and (2) the emergent properties of particular dyadic relationships—that is, the extent to which specific perceptions derive from the interaction between two individuals rather than from the characteristics of the two considered separately.

References

Campbell, D. T. "The Indirect Assessment of Social Attitudes." *Psychological Bulletin,* 1950, *47,* 15-38.

Campbell, D. T. "A Typology of Tests, Projective and Otherwise." *Journal of Consulting Psychology,* 1957a, *21,* 207-210.

Campbell, D. T. "Factors Relevant to the Validity of Experiments in Social Settings." *Psychological Bulletin,* 1957b, *54,* 297-312.

Campbell, D. T. "Recommendations for APA Test Standards Regarding Construct, Trait, or Discriminant Validity." *American Psychologist,* 1960, *15,* 546-553.

Campbell, D. T. "Reforms as Experiments." *American Psychologist,* 1969a, *24,* 409-429.

Campbell, D. T. "Definitional Versus Multiple Operationalism." *et al.*, 1969b, *2*, 14-17.

Campbell, D. T. "Methods for the Experimenting Society." Paper presented at 79th annual meeting of the American Psychological Association, Washington, D.C., Sept. 5, 1971.

Campbell, D. T. "Qualitative Knowing in Action Research." Kurt Lewin Award Address presented at 82nd annual meeting of the American Psychological Association, New Orleans, Sept. 1, 1974.

Campbell, D. T. " 'Degrees of Freedom' and the Case Study." *Comparative Political Studies*, 1975, *8*, 178-193.

Campbell, D. T., and Fiske, D. W. "Convergent and Discriminant Validation by the Multitrait-Multimethod Matrix." *Psychological Bulletin*, 1959, *56*, 81-105.

Campbell, D. T., and Stanley, J. C. "Experimental and Quasi-Experimental Designs for Research on Teaching." In N. L. Gage (Ed.), *Handbook of Research on Teaching*. Chicago: Rand McNally, 1963.

Campbell, D. T., Miller, N., Lubetsky, J., and O'Connell, E. J. "Varieties of Projection in Trait Attribution." *Psychological Monographs*, 1964, *78* (entire issue 592).

Cook, T. D., and Campbell, D. T. *Quasi-experimentation: Design and Analysis for Field Settings*. Chicago: Rand McNally, 1979.

Webb, E. J., Campbell, D. T., Schwartz, R. D., and Sechrest, L. B. *Unobtrusive Measures: Nonreactive Research in the Social Sciences*. Chicago: Rand McNally, 1966.

7   *Robert A. LeVine*

Knowledge and Fallibility in Anthropological Field Research

For many academic psychologists, there is a wide gulf between data obtained through the formal methods of empirical science and data obtained through other means, not fully specifiable in advance, which they deride as impressionistic and anecdotal. Ethnographic data are usually placed in the latter category. Donald Campbell has long propounded a radically different view, which emphasizes the continuity of all knowledge-gathering processes—from those of the amoeba to those of modern physical science. He sees all devices for obtaining information as simultaneously privileged and restricted by their distinctive structures and locations. The very features that give a sense organ or a research instrument privileged access to certain specialized information of adaptive value are the same ones that limit its capacity to detect other relevant information, thus generating biased inferences about the world. Strategies for reducing this fallibility involve multiplying sources of information, thus diversifying the biases in order to transcend them. Examples of different sources of knowledge include trial-and-error learning, often studied in animals; binocular vision in humans; triangulation in astronomy and navigation; and the multitrait-multimethod matrix in scientific psychological assessment. In adopting triangulation or the diversification of biases as a strategy for overcoming the fallibility of their research instruments, scientists are in effect imitating, or using the same mechanisms of knowing as, nature. For Campbell, ethnography, like other forms of data collection, exemplifies this evolution in scientific methods of knowing, and his perspective is an important one for the understanding and improvement of anthropological method.

Campbell's Contributions to Ethnographic Methodology

In 1955 Campbell published "The Informant in Quantitative Research," in which he attempted to demonstrate that the use of key informants, derived from anthropology, can "produce findings of validity and generality" (p. 342). The specific context was a study of morale in ten submarine crews; he discovered that rankings of the crews on morale made by land-based informants at the squadron headquarters correlated highly (.9) with rankings resulting from an anonymous morale questionnaire filled out by all crew members. Informants who had more contact and communication with crew members made more accurate rankings, which were also more accurate than rankings based on reputational assessments by the crew members themselves—though all informant and reputational rankings were significantly correlated. Campbell concluded not only that the reports of informants were valid but also that their validity seemed to vary with their degree of access to communication with those on whom they were reporting. The informants had made no deliberate or systematic research on the relative morale of the ten submarine crews, but their judgments based on informally acquired knowledge were almost as good as if they had. Tapping *their* knowledge was an indirect, unobtrusive, and less expensive way of getting the same information. The informants were to be seen as experts in the literal sense; that is, persons whose experience with a phenomenon reached an intensity and duration sufficient to endow their reports with a validity matched only by expensive scientific investigation. Anthropology was to be credited with inventing a method for using the indigenous knowledge of informants rather than merely duplicating it (although duplication could demonstrate its transsituational generality, or convergent validity), a method to be recommended to the other social sciences. The central lesson was that the stabilized interaction imposed by social roles gave the participants privileged access to certain kinds of information about each other.

The access to special knowledge in a social role, however, involves a particular angle of vision, an inevitably biased perspective, which makes generalizations highly fallible. The concomitant veridicality and fallibility of social perception is a major theme of Campbell's work on group stereotypes. His defense (1967) of the "grain of truth" in ethnic stereotypes is based on the idea that limited social contexts, such as employment and commerce, form the conditions for interaction and hence mutual observation between ethnic populations in certain settings; group stereotypes of each other result from overgeneralization of attributes inherent in their most frequent role relationships. When white Anglo-Americans report that Jews are crafty and work too hard and that Negroes are ignorant and lazy, they are telling us that they know Jews as shopkeepers and Negroes as laborers and are basing general characterizations on observations defined by and inaccurately generalized from those limited occupational contexts. Proof of this can be found in the variety of groups to which these characterizations are attributed. In Southern California during the 1930s and 1940s, when Campbell was growing up there, it was the Japanese-Americans who were stereotyped as working too hard and the "Okies" (poor white migrants from Oklahoma) who were seen as ignorant and lazy. Earlier generations in his family had informed him that the Irish were seen as ignorant and lazy in nineteenth-century America, a bit of folk wisdom amply confirmed by the historical evidence assembled by Solomon (1972). Similarly, in the twentieth-century midwestern United States, Swedes and Poles were stereotyped as stupid when they were visibly concentrated in manual occupations. Thus, skin color and distinctive traits of language, culture, and national origin were largely irrelevant to these stereotype contents, a conclusion one could reach only by comparison of diverse ethnic situations.

Campbell saw in cross-cultural comparison the inferential power of a correlational method—one that, by means of a sifting or winnowing procedure, systematically separates factors generally related to each other from those confounding factors that incidentally occur with them. Like statistical correlation, and sometimes *with* statistical correlation, cross-cultural comparison depends on what Campbell (in Campbell and Fiske, 1959) called "the heterogeneity of irrelevancies"; that is, the notion that confounding factors differ from one instance to another in a population of instances, and therefore the more frequently co-occurring factors can be identified as indicative of causal, functional, or structural relationship (Campbell, 1961). When one examines ethnic stereotypes comparatively, the diversity of groups and situations makes it possible for the most salient sources of stereotype content to stand out as common denominators, "signals" in a welter of "noisy" group characteristics. What have Jews in medieval Europe in common with Hindus in East Africa, Lebanese in West Africa, Chinese in Southeast Asia? Not skin color, religion, language, or national origin—only their socioeconomic positions as alien trading communities of urban sophistication in largely rural societies; hence, the apparent similarity in stereotypes of them as cunning and greedy by the indigenous peoples. Only a comparative perspective encompassing otherwise heterogeneous ethnic groups could identify which features were salient and which irrelevant to stereotype formation.

Another means devised by Campbell to identify the grain of truth in fallacious and ethnocentric group stereotypes is the examination of mutual intergroup attributions: If the English say that Americans are too forward and Americans say that the English are unfriendly, they are agreeing on the nature and direction of the difference while setting different values on it. An anthropological example concerns the Hidatsa Indians in Montana, who regard Anglos as stingy and are, in turn, regarded by the Anglos as improvident (Campbell and LeVine, 1961; Bruner, 1956). Intergroup agreement on such behavioral differences, in the face of discrepancies in values, provides convergent validation for that part of the stereotype that reflects observed behavior produced in accordance with different standards. In other words, the implicit agreement in reciprocal stereotypes indicates an accurate perception of difference in certain social contexts that is inaccurately generalized as a population attribute and endowed with opposite evaluations.

Decentering and Back-Translation. In the use of culturally diverse instances and reciprocal stereotypes, the investigator transcends the ethnocentric biases of groups and achieves a decentered position outside the perspective of any particular group. In comparative research the multiplicity of cases reveals to the investigator (but not to the groups themselves) the sources of veridicality in stereotype content. Similarly, reciprocal stereotypes place the investigator in the Godlike position of eavesdropping on both sides of a group boundary. The social scientist frees himself from the biases inherent in group membership and role relationships by diversifying the sources of bias in his observations; for instance, by studying diverse groups with differing values in contact situations (Americans and British, Hidatsa and Anglos) or diverse groups with similar values in similar but distantly located situations (Hindus in East Africa, Chinese in Southeast Asia). This deliberate diversification of sources of bias in ethnic research puts the observer in a decentered position, from which the winnowing of truth from falsity in intergroup perception is possible. This view of Campbell's concerning intergroup stereotypes has implications for ethnography which will be spelled out in a later section.

At this point the process of decentering deserves further attention. If, as Campbell argues, group membership and roles structure the observations on which we base social cognition, conferring ethnocentric bias on accurate judgments, the ethnographer is constantly striving to reduce the fallibility of field observations by diversification (cross-checking informants' accounts, comparing ideals and behavior, and so forth) to avoid the bias of any single perspective. But decentering is most problematic in the case of translation, for in rendering vernacular words, phrases, and texts (and hence ideas, concepts, and attitudes) in another language, the ethnographer is susceptible to the biases inherent in that procedure, which can distort meanings while attempting to convey them. How does the ethnographer transcend the biases of the language into which he or she is translating material from another language? This problem came to Campbell's attention in 1964, when he and I were conducting fieldwork among the Gusii of western Kenya to try out a preliminary version of our field manual for the cross-cultural study of ethnocentrism (LeVine and Campbell, 1972). Many Gusii terms have no exact equivalents in English; for example, *ogoita* means "to injure or to kill" (in a fight), *okobea* means "to be sad or disappointed," but with the understanding that one is angry. English terms for aggression and hostility cannot do justice to the Gusii concepts without contextual qualification; furthermore, English terms carry their own connotations and other contextual presumptions not present in the Gusii original. Each language represents a distinctive organization of meanings, one that is well adapted to the most salient situations and concerns of its native speakers and therefore poorly adapted to render accurately terms with meanings based on other situations and concerns. To Campbell this difficulty meant that the investigator had to move beyond the constraints of source language and target language alike, to find a decentered position from which the distinctive meanings and blind spots of both languages could be systematically explored. One had to develop a strategy for exposing failures of communication in order to remedy them.

Campbell joined with the linguist Oswald Werner to attack this problem in detail (Werner and Campbell, 1970). As they put it: "Translation is . . . not simply code-switching, where one code is unambiguously retrievable if the other is given. The world of different speakers is not just the same world with different labels attached. Exact translation is therefore impossible in principle. The crucial problems are methods, techniques, etc., by which the quality or approximation of translation may be improved" (1970, p. 403). More specifically: *"There is no one correct translation of a sentence into another language.* There are for every sentence in the source language many possible appropriate sentences in the target language. . . . If translation is viewed as a special kind of paraphrase across two languages, then, as in all paraphrases, there are an indeterminately large number of sentences in the target language which are appropiate paraphrases or translations of a sentence in the source language. . . . Obviously, the many sentences of the target language have multiple equivalents and near equivalents

in back-translation into the source language. A *decentered* translation is then viewed as a *set* of equivalent or near-equivalent sentences of the source language corresponding to a similar set of
sentences in the target language" (p. 402; italics in original).

The research strategy they adopted is one of "multiple iteration," mapping many acceptable sentence paraphrases of the
source language onto many acceptable sentences of the target language. When more context is provided, "Every new sentence in the
appropriate set allows for pinpointing more precisely the intended
'meaning' and facilitates translation" (1970, p. 407). Back-translation is offered as a special case of mapping equivalent sets of sentences in one language onto a set in another. When one bilingual
person translates a text from source to target language, and another bilingual independently translates it back into the source
language, the investigator has available for comparison his original
text and the twice-translated one in the investigator's native language. In exploring discrepancies with the translators, the investigator discovers what contexts are considered relevant to the topic
in each language, how competent the translators are in considering
and selecting among appropriate alternative phrases, and which
other biases each translator alone might not have been willing or
able to reveal. When there are no discrepancies between the original and a twice-translated text, the two near-equivalent versions
constitute a form of "triangulation"; that is, convergent evidence
that an unbiased or decentered translation has occurred and that
the translators are competent. The first-round back-translation,
however, usually contains informative discrepancies: "Back-translation will instructively inform the investigator of what part of
his content can be successfully asked and what part of his social
science interest is uncommunicable, at least with the translation
talent available. It will force a realistic abandonment of many
subtle distinctions that cannot be communicated. It will further an
active revision of the English-language 'original' " (1970, pp. 412-
413). The investigator goes through as many rounds of back-translation as are necessary to eliminate discrepancies between a text
devised and revised by the investigator and a derived source-language
text that has been translated into and out of the target language
by two translators working independently.

The Werner and Campbell chapter (1970) is rich in concrete

suggestions for using back-translation as an ethnographic tool. Their approach illustrates Campbell's treatment of epistemological problems in social research. First, he diagnoses the difficulty in acquiring knowledge as being inherent in the medium of knowing itself. In the case of translation, the object to be known is the meaning of a term in one language, but the medium of knowing (or vehicle for conveying that knowledge) is another language with a differing lexicosemantic organization. As the original communication is decoded it is encoded in the other language, which biases the original meanings toward its own referential framework. This generates mismatches and errors in translation. Second, Campbell locates a potential solution to the problem in the diversity of errors, in this case the multiplicity of paraphrases always available for rendering the meaning of a sentence in a single language. Third, he proposes a procedure of winnowing, or "multiple iteration," which involves trial and error (and is based on analogies with natural selection and statistical correlation); formal application of this procedure identifies the errors as such, permitting their elimination. In this case it is multiple back-translation, with its successive approximations, that constitutes the method for identifying and eliminating mismatches. (Those acquainted with Campbell's work will recognize here the continuity with his work on pattern matching, variation and selection, and the multitrait-multimethod matrix.) Fourth, Campbell recommends the procedure as a form of decentering, permitting the investigator to transcend the particular errors inherent in each method of acquiring or conveying information (in this case, each language). Fifth, he nevertheless maintains that even this relatively decentered knowledge is subject to biases, which more advanced iterative procedures may later identify and eliminate. In translation, as in scientific description generally, the accuracy of knowledge is as fallible as the devices used to acquire it, and the possibility always exists that better devices, taking account of more biases and therefore attaining a higher level of decentration, will be constructed to render an external reality more faithfully.

Summary of Campbell's Contributions. Donald Campbell is a psychologist who has done relatively little anthropological fieldwork but has taken a serious interest in ethnography and its methods for the last twenty-five years. His contributions have emphasized both the validities and the pitfalls in ethnographic inquiry, explicating some of its procedures in great detail, proposing certain improvements, and above all setting ethnography in a coherent epistemological framework. Campbell's unique descriptive epistemology, which encompasses both "natural" and "scientific" processes of knowing, includes ethnography as one of many scientific activities but one which has special problems and special solutions worthy of attention by anthropologists and nonanthropologists alike.

Like other scientists (and other animals), the ethnographer struggles for knowledge with fallible instruments for knowing. His distinctive sources of information come equipped with distinctive biases, and his task is to devise data collection strategies for superseding these biases, usually by finding additional sources with different biases. The guiding metaphor for this process is visual perception, in which a particular angle of vision provides unique information with unique limitations: binocular vision compensates for the limitations of one eye with a second one, identical except for its angle of vision, adding information and diversifying the biases. The brain decenters the information from the perspective of either eye, giving a synthesized perception in depth. In ethnography the anthropologist finds a key informant who is close enough to a social or cultural phenomenon to know it well, but far enough removed to be able to report on it, often better than those currently most absorbed in it. This location of the key informant endows his report with an accuracy and depth of great value for social science investigation, but the informant's location is only one of several possible locations and therefore embodies potential biases in itself. The ethnographer compensates for the bias of one informant by asking the same questions of others, whose biases (based on their social positions and experience) are likely to be different; in synthesizing the accounts of different informants and allowing for their distinctive biases, the ethnographer strives for a decentered description that supersedes any particular account in terms of the amount of information and supersedes all accounts in terms of its relatively unbiased point of view.

The description from one culture's point of view is unbiased only relative to an informant or group of informants; it is likely to be biased by the ethnocentric perspective all the informants of a culture share. One standard solution is for the ethnographer to make independent observations, biased no doubt, but differently, providing another perspective on the same patterns of behavior. Campbell's solution is to interview neighboring groups of other cultures who were able to observe the first group's behavior but did not share their ethnocentric perspective. (This solution is limited in principle to those customs available for outgroup observation and in practice by the tendency of contemporary anthropologists to work within a single cultural group.) Since each outgroup brings its own ethnocentric perspective to the task of ethnographic description, those biases must also be dealt with by diversification (through a panel of several outgroups or in the reciprocal stereotypes of pairs of groups).

In the task of ethnographic translation, it is the language itself that occupies a position simultaneously privileged (relative to decoding communications within its speech community) and biased (relative to representing translations accurately). The capacity of each language to generate a variety of acceptable paraphrases, each with its own inaccuracy, makes it possible for the ethnographer (through back-translation, multistage translation, and other trial-and-error methods) to develop a decentered translation that at least avoids the worst distortions one language might impose on communications from another.

By explicating the processes of inference involved in ethnographic data collection in terms of his unified epistemological approach, exposing methodological capabilities and limitations and proposing strategies for dealing with the limitations, Campbell has contributed to our understanding of ethnography as a form of scientific inquiry. In the next sections I shall explore certain applications of his general methodological principles to anthropological fieldwork.

Principles of Ethnography

In this section I use the terms of Campbell's descriptive epistemology to explicate principles of ethnographic method recognized by anthropologists generally and to recommend principles of research design that would improve certain kinds of ethnographic investigations. Campbell's analysis of knowledge processes suggests that all instruments for knowing are fallible; that their fallibility resides primarily in their situation-specific points of view, which bias transsituational judgments; and that methodological advance consists of generating strategies for overcoming their particular restrictions in order to approximate "objectivity." This analysis applies to all knowledge processes, from sensation in the one-celled organism to data collection in modern physical science and including the practice of social policy in an "experimenting society." It covers short-term processes like visual perception and long-term ones like natural selection, in which variation and selection collaborate in adapting a knowledge structure (perceptual judgment, sense organ, scientific generalization) to diverse environmental information, creating a novel structure superior in inferential scope to any previous variant.

In applying this model of knowing to ethnography, we are dealing with the conditions under which statements about the customs and behavior of a people can be accepted as valid. I submit that most anthropologists are prepared to accept, albeit tentatively, such statements as valid when they are made by a professionally qualified ethnographer on the basis of an adequate field experience.

A "professionally qualified ethnographer" is a person who (1) has learned to suspend moral judgment and ingroup loyalty while obtaining knowledge about cultures other than his own, (2) has had experience in generalizing from an array of specific ethnographic facts reported by others to a pattern characteristic of a group, (3) has had experience in comparing characteristic patterns of one group to those of another with respect to a common dimension of culture or social structure. The first qualification is supposed to reduce the tendency to make loyalistic misperceptions of foreign cultures. In these, the observer makes incorrect generalizations about the other group because he wants to prove his own group's superiority (through exaggerating differences or overemphasizing the outgroup's deviation from his own group's standards) or to justify the opposition or oppression by his own group to the outgroup. The professionally qualified ethnographer is presumed to be free of this bias, motivated by ingroup loyalty, and is in this respect deemed a priori a more accurate observer of the outgroup behavior than an average person from the ethnographer's culture. The second and third qualifications are supposed to reduce the tendencies to conclude mistakenly that some particular act or practice indicates a general pattern and to overlook relevant acts or practices—"errors of hasty judgment," characteristic of untrained observers. Thus, we trust the ethnographer's attribution of a behavioral disposition to a group more than that of a nonethnographer, in part because we assume that the ethnographer's training has reduced his susceptibility to the biases referred to as "loyalistic misperception" and "errors of hasty judgment."

"Adequate field experience" is extended exposure to diverse aspects of the functioning and thought of a group through observation, conversation, and "participation." This kind of experience is alleged to reduce further (though not eliminate) the possibility of the ethnographer's concluding that certain acts or practices form a general pattern when in fact they do not or of omitting in the formulation of a general pattern certain acts or practices which are actually related to the ones included—the "errors of hasty judgment" mentioned above. Just as the "professionally qualified ethnographer" is contrasted with the untrained observer, so "adequate field experience" is contrasted with a casual visit in terms of the probability of forming accurate generalizations. Furthermore, anthropologists are aware that field experience does not allow for perfect accuracy and that such experiences vary widely in accuracy. The established professional consensus is that the longer the ethnographer's residence in the field situation and the greater his command of the local language, the more adequate is the field experience. The effects of these variables on ethnographic judgments of group behavior have been shown in cross-cultural study (Naroll, 1962). The general point is that the greater the exposure of the ethnographer to the particularities of a group's acts and beliefs, the less likely is that ethnographer to make incorrect general statements concerning the patterning of those acts and beliefs. But the likelihood is not reduced to nil.

Now, the ethnographer in the field situation depends in part on another type of observer: the informant. What is an "informant"? The informant is usually a group member who, due either to generally superior inductive and verbal skills or to a life history which has brought exposure to the types of behavioral contrasts in which we are interested (for example, "cultural marginality"), is capable of drawing generalizations from the particularities of his or her social environment and communicating them to an ethnographer. The good informant is less likely than the ethnographer to misjudge ingroup behavior because of inadequate duration of residence or inadequate knowledge of the local language—biases to which ethnographers as outsiders are susceptible. However, the informant is so close to the events on which he or she is reporting that other biases enter: his or her position in society and individual life history may cause a misperception of the larger group and its functioning, both as error resulting from a parochial point of view (with concomitant lack of awareness that other points of view exist) and as the motivated result of subgroup loyalties. The ethnographer reduces the net bias in informants' accounts by diversification of informants; that is, by selecting informants who vary in social position, aspects of life history, and subgroup loyalty. Those points in which such diverse informants agree are accepted as probably valid unless the ethnographer has good reason to believe that a common bias is operating. One such common bias is the loyalistic misperception of ingroup behavior by informants, their tendency to glorify or justify the entire ingroup. The ethnographer has two defenses against this: a request for details and particulars, the response to which may reveal inconsistencies in generalization; and independent observations of the events concerned.

This somewhat tortuous reconstruction of ethnographic epistemology suggests that it is widely, if implicitly, recognized by anthropologists that biased human observers (such as ethnographers and informants) can produce acceptable generalizations concerning group behavior—as long as their biases are known and there is a strategy available for eliminating, reducing, or controlling these biases. The biases of ethnographers are controlled through strategies of training (rehearsal of the processes of ethnographic generalization and comparison) and of criteria for exposure to the culture under study (length of residence in the field, knowledge of the local language).

Working with informants, the ethnographer employs a strategy which can be called the *diversification of biases,* in which misperceptions not widely shared cancel each other out, leaving a net information yield of generalizations supported by agreement among informants. If such generalizations are not regarded by the informants as particularly favorable to the group, the ethnographer may be inclined to accept them and to offer them to professional colleagues as validated statements concerning group behavior. If the generalizations appear to involve ingroup glorification, the ethnographer will suspect his informants of shared loyalistic misperception and may have recourse to strategies for eliminating this bias; namely, the elicitation of fine-grain detail, which can be examined for consistency with the reported general pattern and ultimately with independent first-hand observations by the ethnographer or others who are known not to share the loyalistic biases of the ingroup informants.

In addition, informants are likely to share another bias in perception—a bias whereby they fail to report existing behavior patterns that appear to them universal and obvious, because they are unacquainted with cultural contexts in which these behavior patterns are absent. This bias can be eliminated only by the use of an outside observer, the ethnographer or some other exotic person, preferably from a cultural context dissimilar to the one under study.

Ethnography, then, does not presume to be an infallible knowledge-gathering process but is most accurately seen as a series of operational strategies for reducing the biases known to affect human judgments of group behavior. Elsewhere (LeVine, 1966), I have gone so far as to suggest that the judgments of outsiders untrained in anthropology could be used to generate valid information about the customs and behavior of a group if one knew their biases and were able to obtain judgments from outsiders of diverse biases. This was an effort to salvage ethnographic information from the same processes of intercultural observation that produce the "grains of truth" Campbell (1967) saw in ethnic stereotypes. It illustrates that the strategy of diversifying biases in order to transcend them can apply not only within the conventions of ethnographic research but in other cultural inquiries as well.

Anthropologists in recent years have been moving beyond ethnographic conventions in a number of directions; one such direction is hypothesis-testing research as conducted in psychology and sociology. To help guide these efforts, I proposed some principles of research design in anthropological fieldwork (LeVine,

1970), applying Campbell's notions to the field conditions under which anthropologists work. In deciding on a hypothesis testable in the field, the anthropological investigator is often in the position of making an irreversible personal commitment to a field site, or at least the society in which the field site will be located, before knowing in detail the kinds of cultural variation to be found there.

The general situation can be conceptualized in terms of pattern matching (Campbell, 1966), a multiple-iteration procedure in which the investigator finds a match between one of many potentially testable anthropological hypotheses and those cultural factors discovered to be locally variable at the research site. This usually has to be done during the exploratory phase of fieldwork or just after its completion, when the investigator is unable to move to another society where a particular hypothesis might be more effectively tested. At that point, with a limited number of variables available, the investigator is advised to change hypotheses, rather than field sites, to devise an effective research design for hypothesis testing. The principle that emerges is that the anthropologist should develop and be prepared to test an array of hypotheses relevant to a major research focus, so that alternative hypotheses can be selected in the field if the cultural setting proves unsuitable for testing the first-choice hypothesis.

A second principle that emerges from this analysis concerns the selection of an area or society for field study. Although advance knowledge of the field site in the detail needed to design hypothesis-testing research is often unavailable, increasing ethnographic coverage of the world makes it possible for anthropologists to judge one area more likely than another for certain types of hypotheses. In making the selection, the anthropologist is advised to avoid a society or area that is homogeneous on the dimensions of concern in the study and opt for maximum variation on them. While the investigator might be tempted to conduct research where a particular cultural phenomenon—for example, the centralized state or antagonism between the sexes—is highly developed, a lack of variation in these institutions would preclude a hypothesis-testing design and result in observations in which the dimensions of interest are confounded. Where these institutions vary locally, however, a correlational research design becomes possible to sort "irrelevant" from invariant co-occurrences and help identify the causal and functional relationships of the cultural phenomenon. This is the maximum-variation principle.

A third, closely related, principle applies to that phase of fieldwork when the collection of ethnographic background data has been completed and a final hypothesis-testing research design is being developed. Invoking the "multiplicity criterion" from Campbell's general strategy of fallibility reduction, I proposed five guidelines for the field worker: multiplicity of instances, indexes, tests, hypotheses, and phase-specific conditions.

Multiplicity of instances means operationalizing the variables of central interest so that there are as many instances of them as possible in the field setting. This often requires reducing a society-wide concept—such as centralized political system, market economy, or marital stability—to a dimension on which there is observable variation across territorial or other group divisions, local communities, domestic groups, or individuals. If the investigation is to test hypotheses rather than generate them, a multiplicity of instances is essential.

Multiplicity of indexes refers to the use of several measures for each variable under study, to diversify the biases inherent in all measures and achieve convergent validation. Thus, to compare several communities on a global variable, such as social disorganization or disintegration, the field worker would be well advised to find several indicators (such as crime rates, marital separation, litigation, amount of factionalism or interpersonal tension) to make their covariation a matter for empirical research rather than theoretical assumption.

Multiplicity of tests means replication; for example, testing the same hypothesis in different settings, in different samples within a common setting, or at different levels of subgroup differentiation. The more diverse tests the hypothesis survives, the more confidence it deserves. The field worker is advised to seek opportunities for such replication within the field location.

In suggesting a multiplicity of hypotheses, I was proposing that data be collected to test hypotheses that could be viewed as plausible rivals or alternatives to those hypotheses the field worker might be attached to. This is particularly important in anthropological fieldwork because the investigator might not be able to return to the field location to collect more information at a later date. This suggestion deals with the investigator's theoretical bias and introduces the general model of testing plausible rival hypotheses into the process of field work.

Finally, multiplicity of phase-specific conditions refers to the possibility of finding in one field setting ongoing processes that have affected some individuals or groups more than others and permit cross-sectional or repeated-measures designs. These include processes of individual development, institutional change, and immigration; their investigation could involve the adoption of quasi-experimental research designs (Campbell and Stanley, 1966).

These guidelines and recommendations serve only to illustrate the possibilities for using the principles and perspectives of Campbell's epistemology to improve the hypothesis-testing effectiveness of research carried out as part of anthropological fieldwork.

Multicultural Perspectives on Group Differences in Behavior

Campbell's epistemological notions encourage the ethnographer to experiment with new methods of reducing the cultural biases inherent in anthropological research. In our work on the cross-cultural study of ethnocentrism, Campbell and I were impressed with the fact that standard ethnography is limited to the perspectives of the anthropological visitor and the people whose culture is under study, though these do not exhaust the perspectives available for exploring the validity of statements about their customs and behavior. Like individuals in a work or school setting, each culturally defined population acquires a "peer reputation" among its neighbors, and such reputations represent potential sources of ethnographic data on ingroup culture and behavior from another series of perspectives. They provide opportunities for comparing data from different perspectives in order to examine their validity. In 1965 we conducted a methodological experiment in East Africa, going beyond the usual ethnographic perspectives to obtain reputational data from groups in the region. In this section I shall show how some results from that experiment compare with data collected in the framework of ethnographic fieldwork.

We conducted an attitude survey of thirty contiguous ethnic

groups in Uganda, Kenya, and Tanzania, interviewing fifty respondents from each concerning their images of their own and other ethnic groups. Ten groups were sampled in each country, and each respondent was interviewed concerning the ten intranational groups and four well-known or bordering groups from the other countries. One part of the interview involved naming which of the fourteen groups was highest on each of forty-eight traits. Since only one group could be named for each trait, the scatter of choices was considerable, even within ethnic samples. For purposes of this analysis, I regard group A as having attributed a particular trait to group B if more than one third of the group-A respondents made the attribution and if group B received more such attributions than any other outgroup.

Two examples from the Kenya results illustrate how reputational data might be related to ethnographic description. The first example concerns differences between the Kikuyu and Luo in thriftiness. At the time of the survey, the Kikuyu and Luo were the most publicly visible ethnic groups in Kenya and were presented in the mass media as competing for power. For our trait description "thrifty, like to save money," eight of the ten Kenya groups named the Kikuyu as highest. For the description "lacking in thrift, like to spend money," five Kenya groups named the Luo. No other group received more than one attribution from a Kenya sample on either of these traits. Since the ethnic groups making those attributions vary in proximity to Kikuyu and Luo and in other factors likely to affect the images they might form of them, this degree of agreement seems quite substantial and suggests that the image of the thrifty Kikuyu and spendthrift Luo is common to a diverse panel of neighboring groups. Furthermore, the Kikuyu and Luo themselves were among the groups making both attributions. Despite their respective ingroup biases, the Kikuyu and Luo respondents tended to agree with the outgroup consensus concerning where their groups stand on this dimension. Thus, the attributions cannot be due to any of the group-centered biases that distinguish one group from another. What is responsible for this level of consensus?

Ethnographic work among the Gusii, one of the other Kenya groups sampled, sheds light on this question. In Kisii District, where the Gusii live, the Kikuyu were visible during the

1960s, primarily as shopkeepers and independent bus drivers, and
the Gusii dealt with them primarily in commercial transactions.
Furthermore, the Kikuyu were at that time virtually the only African entrepreneurs in the district (most were Indian), so they stood
out as models of commercial activity and investment. The Luo, on
the other hand, had long occupied positions as clerks and school-
teachers in Kisii, and were seen as prototypes of white-collar bureaucrats. Gusii informants claimed that Luos were particularly
interested in spending their money on clothes to enhance their appearance as Westernized civil servants and professionals elevated
above the level of the rural population. The Gusii data suggest that
the survey attributions of Kikuyu thrift and Luo spending originated with the actual roles that persons from these groups were
seen to perform outside their home areas. Thus, the images might,
if the Gusii experience were replicated elsewhere, reflect rural
stereotypes of entrepreneurs and white-collar clerks in their midst.
The agreement of the Kikuyu and Luo samples with these stereotypes suggests that they view the difference between themselves
and others in similar terms and perhaps for similar reasons; namely, that the stereotypes are grounded in the socioeconomic conditions of their interactions with other peoples. From an ethnographic viewpoint, the reputational data provide leads for further
research on the history of Kikuyu and Luo economic adaptations
in the postcolonial period.

The second example concerns preoccupation with witchcraft among the Kamba. At the time of the survey, the Kamba
were not as visible nationally as the Kikuyu and Luo, and were
known outside their own area primarily as soldiers and policemen,
roles into which they had been recruited during the colonial
period. The mass media carried occasional stories concerning
witchcraft among the Kamba; but belief in witchcraft, sorcery,
and other magical procedures is not limited to any particular
group in East Africa. Beliefs and practices concerning witchcraft
are much less public than economic behaviors and hence less likely
to enter into national reputations. In our survey, the Kikuyu and
Embu samples, both adjacent to the Kamba, named the Kamba as
the group most "interested in using witchcraft and poison," and
the Kamba sample agreed with them. Since this is generally regarded as a malicious comment about a person or group in Kenya,

the Kamba self-attribution is particularly important as an indication of something more than an expression of hostility by neighboring groups. Perhaps they have accepted an outgroup stereotype of them; perhaps they are referring to different customs and beliefs than their neighbors are; or perhaps this is, however incompletely described, a reflection of consensus about a social reality.

In this case we can check the published data by Edgerton (1971) on the Kamba studied in 1961-62 for the Culture and Ecology Projects against the reputational attributions. In that project pastoral and agricultural samples of Kamba were compared with pastoral and agricultural samples of three other (nonadjacent) East African peoples. When the extensive interviews were coded for the mentioning of witchcraft, the Kamba pastoral sample showed the highest mean number of mentions per person of witchcraft, but the Kamba farmers were lowest of the eight samples (Edgerton, 1971, p. 184). Edgerton found this anomalous and is not able to explain it. In his ethnographic report on their distinctiveness, however, he says:

> In 1913, Charles Dundas wrote: "I have heard Akamba say that they dislike all foreigners, but hate each other." And "the Mkamba is much too mistrustful and suspicious to make it possible for him to ever work to any extent in combination with others." The Kamba themselves often made similar comments in 1962: "I must hide my feelings and my plans from everyone; otherwise, I can do nothing, because other Kamba will always try to thwart me if they can." Or, "Never tell anyone what you plan to do. If you do, they will try to stop you; someone may even try to kill you." This sort of mistrust and concealment of motives is *not* unique to the Kamba. It exists in all four of these tribes. What is distinctive is the extent to which the Kamba conceal their feelings, and the frequency with which they say that it is proper to do so [Edgerton, 1971, pp. 115-116].

As in the first example, here, too, the reputational data and interviews with ingroup respondents themselves raise questions for further ethnographic research. The research that is called for, now informed by reputational and ingroup perspectives, must be focused not only on the social facts of witchcraft and interpersonal behavior among the Kamba but also on the ways in which the relevant beliefs and practices become socially visible and hence available for the building of ingroup and outgroup reputations. Stimulated by Campbell's conception of knowledge, ethnographers should investigate the underlying social processes that connect the perspectives of the ethnographer, the ingroup, and neighboring outgroups and that account for their divergence and convergence.

References

Bruner, E. M. "Primary Group Experience and the Process of Acculturation." *American Anthropologist,* 1956, *58,* 605-623.

Campbell, D. T. "The Informant in Quantitative Research." *American Journal of Sociology,* 1955, *60,* 339-342.

Campbell, D. T. "The Mutual Methodological Relevance of Anthropology and Psychology." In F. L. K. Hsu (Ed.), *Psychological Anthropology: Approaches to Culture and Personality.* Homewood, Ill.: Dorsey Press, 1961.

Campbell, D. T. "Pattern Matching as an Essential in Distal Knowing." In K. R. Hammond (Ed.), *The Psychology of Egon Brunswik.* New York: Holt, Rinehart and Winston, 1966.

Campbell, D. T. "Stereotypes and the Perception of Group Differences." *American Psychologist,* 1967, *22,* 817-829.

Campbell, D. T., and Fiske, D. W. "Convergent and Discriminant Validation by the Multitrait-Multimethod Matrix." *Psychological Bulletin,* 1959, *56,* 81-105.

Campbell, D. T., and LeVine, R. A. "A Proposal for Cooperative Cross-Cultural Research on Ethnocentrism." *Journal of Conflict Resolution,* 1961, *5,* 82-108.

Campbell, D. T., and Stanley, J. C. *Experimental and Quasi-experimental Designs for Research.* Chicago: Rand McNally, 1966.

Edgerton, R. E. *The Individual in Cultural Adaptation.* Berkeley: University of California Press, 1971.

LeVine, R. A. "Outsiders' Judgments: An Ethnographic Approach to Group Differences in Personality." *Southwestern Journal of Anthropology,* 1966, *22,* 101-111.

LeVine, R. A. "Research Design in Anthropological Field Work."

In R. Naroll and R. Cohen (Eds.), *A Handbook of Method in Cultural Anthropology*. New York: Natural History Press/Doubleday, 1970.

LeVine, R. A., and Campbell, D. T. *Ethnocentrism: Theories of Conflict, Ethnic Attitudes, and Group Behavior*. New York: Wiley, 1972.

Naroll, R. *Data Quality Control*. New York: Free Press, 1962.

Solomon, B. M. *Ancestors and Immigrants: A Changing New England Tradition*. Chicago: University of Chicago Press, 1972.

Werner, O., and Campbell, D. T. "Translating, Working Through Interpreters, and the Problem of Decentering." In R. Naroll and R. Cohen (Eds.), *A Handbook of Method in Cultural Anthropology*. New York: Natural History Press/Doubleday, 1970.

8 *Paul C. Rosenblatt*

Ethnographic Case Studies

This chapter is an extended commentary on Donald Campbell's treatment of the ethnographic case study. It reflects my experience in consuming ethnographic case studies and in doing exploratory fieldwork in Mexico and Bali. It also reflects the profound influence that Campbell has had on my own thinking; where the chapter deviates from Campbell or suggests extensions of his thinking, it does so from the perspective of Campbellian epistemology.

I believe that Campbell's published evaluation of the case

Note: The impetus for the development of ideas in this chapter was a series of discussions with Donald T. Campbell during the spring of 1978. I am indebted also to Michael Baizerman and Popie M. Mohring for many stimulat-

study began within the context of his discussion of experimental treatments and experimental observations in "Factors Relevant to the Validity of Experiments in Social Settings" (Campbell, 1957). In his initial view of the case study, he did not single out ethnographic case studies; instead, he evaluated all case studies as though they were based on the same research operations, specific observations, and experimental treatments similar to those contained in an experiment. The goal of the case study was assumed to be knowledge of the effect of "the treatment" (some event or experience); a presumed subsidiary goal was the marshaling of evidence bearing on a general theory, for which "the treatment" and the observations are appropriate indicators. In his initial analysis, Campbell criticized the case study primarily because of its lack of a basis of comparison, a lack of some sort of "control" group against which to evaluate the threats to validity in Campbell's immensely useful list—maturation, history, and so on.

As his analysis of research design and design options became more elaborate, Campbell retained his very critical view of the case study (Campbell, 1961; Campbell and Stanley, 1963; Campbell and LeVine, 1970; Cook and Campbell, 1976). However, in a relatively recent article, " 'Degrees of Freedom' and the Case Study," he observed that in the ethnographic case study there are opportunities for validity testing (Campbell, 1975). He drew an analogy between the "degrees of freedom" in two types of studies: one involving statistical analysis of data from a number of independent cases; the other involving analyses of data from a single case. In a study aggregating information from a number of independent cases in order to perform an analysis of variance, each additional case provides another degree of freedom, another observation or set of observations with which to probe one's hypotheses. In a study of a single case, each additional piece of information provides another instance with which to probe one's hypotheses. One's analyses in a case study may be statistical, but they may often be qualitative.

ing discussions of epistemology and to Roxanne M. Anderson, Gael Cheek, Barry E. Collins, Cyndi Englund, Alletta Hudgens, Patricia A. Johnson, Riv-Ellen Prell-Foldes, and especially Popie M. Mohring for useful comments on earlier drafts of this chapter.

In his new analysis, Campbell continued to view the case study as relatively vulnerable to researcher self-deception, as having a low signal-to-noise ratio, and as fallible. However, he pointed out that he had been working in the past with a caricature of the case study that did not fit the ethnographic case study. In particular, he pointed out that he had been writing as though the observer had noted and interpreted "a single striking characteristic of a culture," rather than a pattern of conceptually related characteristics. Although he had previously condemned the detailed data gathering of all such studies as involving "the error of misplaced precision" (Campbell and Stanley, 1963, p. 177), he now claimed some epistemic value for ethnographic work that made good use of diverse theory-probing opportunities. The value, he argued, lay in the possibility of testing numerous implications of a theoretical viewpoint, of exploring many different challenges of theory. Campbell saw evidence of such exploration in ethnographic reports in which the investigator, after careful research, said that some phenomenon could not be explained or discarded early explanations. Campbell argued that if there had been no empirical challenging of explanations, the investigator would have no problem in producing an explanation for a single case; any conceivable explanation might be offered. Similarly, if such challenging had not occurred, there would be no discarding of early explanations. Though still seeing the fallibility of ethnographic studies, he acknowledged in his "Degrees of Freedom" paper that such studies have potential for epistemological discipline.

Assumptions in Campbell's Analysis of Ethnographic Case Studies

Assumption That Explanation Is the Goal of Case Studies. A substantial proportion of ethnographic writing, perhaps more in the past than in the present, has had descriptive rather than explanatory goals. Should the same epistemological standards be applied to studies with descriptive goals that would be applied to explanatory studies? One could argue that different standards must be applied, that with descriptive goals the crucial standard to apply is one of accuracy. How reliable and how well supported are the assertions made in reports of the fieldwork? Are the assertions based on careful, redundant observation and on reports of trustworthy informants?

One could also argue, however, that all ethnographic work is based on theoretical premises (Beattie, 1959; Herskovits, 1954; Hollander, 1967; Jarvie, 1970). Even the concepts around which a descriptive study is organized (kinship system, religion, division of labor) embody theory about what is important and what is unimportant and about what observations should be linked to what. From this perspective, all ethnographic work requires the disciplined winnowing that Campbell called for in his "Degrees of Freedom" paper. For example, if there seems to be a system of kinship important in people's lives, people will gossip about kinship, and their children will learn about kinship when the children are relatively young. People will dream about kinship and seem to understand each other when they are talking about it. If there is a kinship system, informants will all give similar descriptions of it and, if given information about some aspects of kin relations of an unknown ingroup member, will be able to predict other aspects of those relations.

It can also be argued, as Geertz (1973) has, that the phenomena dealt with in ethnographic work are so complicated, multiplex, interconnected, and inexplicit that one must always engage in analysis and interpretation if one is to report anything at all. From that perspective, too, a case may be made that ethnographic work with descriptive goals deserves the discipline Campbell asks for in explanatory studies.

It seems fallacious to claim accuracy for a study with descriptive goals. One's measures cannot be accurate in an absolute sense. They are inevitably selective as to focus and basis of comparison. Moreover, one has to choose between literal interpretation of whatever information one gets and interpretation in terms maximally sensible to one's professional audience. Thus, the question of whether a descriptive study is accurate is ambiguous, and claim of accuracy is not a sign that the study is bias free, not a sign that the study lacks interpretive elements, and not a sign that much sense can be made of uninterpreted information. So Campbell's assumption of an interpretive element in all studies seems sensible. There may be no ethnographic studies, perhaps no case studies at all, that lack theoretical footing or interpretation and that could be useful without analysis and interpretation.

The fieldwork of many contemporary ethnographers has explicit theoretical links. In fact, there are a number of "schools" of ethnography (for example, the Manchester school and the Geertz school) that provide comprehensive, integrated, well-thought-through views of culture illustrated richly in a substantial number of published monographs. However, much ethnographic activity is still not explicitly linked to any theoretical focus but is defined as general ethnography or the gathering of background data. For general ethnographic or background work, an ethnographer often finds ways, after returning from the field, to link the background or general ethnographic data to theory not considered (and perhaps not even extant) during the time in the field (see LeVine, 1970). Perhaps most often the linkage is either to a theoretical framework that is in fashion (for example, structural analysis in the style of Lévi-Strauss) or to a theoretical statement that the researcher thinks is incomplete or wrong. Such analyses almost certainly do not include disciplined search in the field setting for data challenging the analysis.

In a sense, ex post facto reports are only illustrative rather than theory probing. They show that one can find illustration of a theory in one's own field material or that one can produce data contrary to the theory. But illustrations may be of some scientific value; metaphysical or incomprehensible theories may be impossible even to illustrate. (In that sense, critical exchanges in scholarly journals concerning whether a published investigation or challenge of Theory X was based on a proper understanding of Theory X may be a sign that Theory X is not clearly stated or not scientific in the sense of being falsifiable.) Theories that can be falsified with examples may be more likely to be falsified in other ways. However, an ex post facto analysis allows for an uncomfortably large amount of room for biased recall and biased marshaling of data. People can be assumed to make more errors of reporting for recalled than for freshly observed events, and they may not effectively take into account the greater unreliability of recalled information when they use it (Trope, 1978). There is also reason to wonder whether the data were gathered with measures and methods appropriate to the theory being challenged.

Although Campbell seems to assume in all his discussions of research design that research is done ultimately in order to probe a general theory, the research goal of ethnographic accounts is often particularistic explanation, and often that explanation is historical. By definition, historical explanation is ex post facto. Like other types of ex post facto work, historical work gives one some sort of premeasures, perhaps even a series of them. An ethnographer may be able to use measures drawn from archeological data, from earlier written accounts, and from events or stories recalled by contemporary informants. The calibration of measures on the past and on the present and the assessment of methods variance require great skill. From Campbell's degrees of freedom perspective, one would demand the same probing of historical explanation as for a more social scientific explanation. To the extent that those probes are impossible or very weak because one has inadequate sources of data and cannot go back in time, historical explanation would rank with other ex post facto explanation as weak epistemologically. In general, ex post facto case analyses may be of some use, but they are obviously weaker than analyses that involve use of Campbell's degrees of freedom perspective in the field.

Assumption That Case Studies Involve Implicit Comparison. Campbell argued that case studies inevitably include implicit comparison. From his perspective, before one can say that the Samoans have a tranquil period of adolescence or that matters of personal status are very important in Korea, one must make some sort of implicit comparison of Samoa and Korea with other parts of the world. The comparative element may well be present, but it is not clear whether the comparisons are generally unstated. Many ethnographers use surprise or the unexpected as one source of focus for their work. The gathering of anecdotes is not strong evidence, but if one looks at descriptions by anthropologists of their field experiences, it seems quite common that the initial reaction included comparison embodied in surprise or attention to the unexpected. A Japanese anthropologist who came to India reported, for example, discrepancies from the expectations she had from

Japan in how people talked about households, in how privacy was handled, in sibling relations, in verbal restraint, and in relations of the younger brother to the wife of the elder brother (Nakane, 1975). An Igbo anthropologist studying the Navajo "became quite aware" of the difference in the treatment of strangers between his own culture and the Navajo (Uchendu, 1970). An anthropologist from India, studying Zuni Indians, was struck by the ease and freedom enjoyed by Zuni women and by the "misconduct" of Zuni children (Pandey, 1975). An American anthropologist was struck by differences between U.S. family life and family life in India (Gould, 1975). One need not even be a cultural stranger to find surprises in one's fieldwork. Mohring (1980), a Greek immigrant studying Greek immigrants, was surprised at how little the social science assumptions about human nature fit the reality and everyday interpersonal interaction of her informants, and her informants were surprised that anyone so Greek could ask questions so discordant with common sense and obvious truths.

Naroll and Naroll (1963) have argued that attention to the exotic is a bias of ethnographic work that leads to inadequate attention to what is expected and obvious. But one could also argue that alertness to disconfirmed expectations is at the core of the epistemology that Campbell has advocated, an epistemology involving disciplined challenge of theoretical assertions and expectations. Some works on how to do ethnography advise one to use surprise, the unexpected, or a sense of differentness as cues to what to study (see, for instance, LeVine, 1970; Mead, 1970; Richards, 1939). Mead went so far as to recommend that the researcher cherish first impressions, when the sense of difference is greatest. However, sensitization to the unusual may lead one to ignore or to fail to report much valuable information, and sensitization and awareness stemming from a sense of the unusual may still leave much to be desired when it comes to accurate assessment and reporting or even to conceptualizing at all sensibly.

Although sensitization to the unexpected and different makes comparison somewhat explicit, the comparison is still weak if one lacks the same kinds of data on one's own culture that one has on the culture one studied formally. When I returned from Bali, I attempted to gather data in the United States to bolster one of the assumptions I had made about what was true in the United

States, to document an "implicit" baseline for evaluating the Balinese on a particular dimension. I found that my assumption, which dealt with crying of infants and young children, was false; my basis for comparison was faulty.

In Bali my focus was on etiquette and the expression of emotions. One specific emotional behavior I looked at was crying by infants and young children. During my wanderings in several Balinese *bandjars* (neighborhood communities), I never saw a Balinese infant or child cry tears (with one possible exception, but the child had such a bad cold that I thought I might be seeing nasal mucus on his cheeks). I rarely saw or heard Balinese infants or young children making crying sounds, and these were never forceful or persistent. I thought that Balinese children were much harder to upset than American children; on looking for comparable data in the United States, however, I was amazed to discover that things there are not so different. I attempted to measure in the same way I had in Bali—wandering in my neighborhood, in the Minnesota equivalent of the marketplace, and at other places where parents and children might be seen. As in Bali, I almost never saw infants or young children cry, and some of the crying I saw was not accompanied by tears. There may be a difference between the United States and Bali, but it is not one that I can demonstrate with the measurement I made (counts of occurrences but not counts of number and duration of nonoccurrences). So in this one instance, an attempt to make the U.S. baseline explicit proved that the assumed, implicit baseline was in error. The phenomenon in the United States was found to be not radically different, if different at all, from what was observed and presumed to be different in another culture.

The mechanisms that underlie misperception of one's baseline culture are worth considering. One may overgeneralize from a small number of observations (Tversky and Kahneman, 1971). One may attend to vivid, striking events much more than to bland events and rely too much on the vivid events when searching memory for background information. It is the vivid events, the events initially attended to, that are likely to be recalled (see Borgida and Nisbett, 1977). Campbell might speak of enhancement of contrast (Campbell, 1956): the disposition to overestimate differences when two entities—or, as in the study of crying, two sets of data—

are being compared. From that perspective—one that views enhancement of differences as a normal part of the perception and categorizing of one's world—one might speculate that such enhancement occurs commonly in the absence of precise measurement on aspects of one's own baseline culture.

Whatever the mechanisms, the experience I had in studying crying suggests that the issue may not be one of implicit comparison but of comparability of measures. Whether one overgeneralizes from small numbers, attends too much to the vivid, or enhances contrast, one needs at least somewhat comparable measures on the target culture and on the baseline culture. Achieving comparability may often be difficult; it may require far more knowledge than one has when one begins measuring. Even establishing that something like comparability has been achieved may be quite difficult. Nonetheless, errors seem too likely if one merely assumes one knows what is true of a comparison culture and generates no information with which to bolster the assumption. Even if one turns to the vast ethnographic literature to find a baseline for comparison, the measurement procedures underlying statements in published ethnographies are typically unclear and probably of necessity somewhat different from what one would use in one's own field setting.

Are there noncomparative assertions? There is a distinction in anthropology between etic analyses—analyses using an imposed frame of reference—and emic analyses—analyses working within the conceptual frameworks of the people being studied. People who do emic analyses often argue that for such analyses comparison is by definition inappropriate. The test of validity of assertions, in their view, is informant judgment, not comparisons against alternatives generated by the ethnographer. For example, in a culture that counts some unborn people and some no-longer-breathing people as persons, it would be inappropriate, from the viewpoint of those who view emic methodology as noncomparative, to apply a comparative framework in order to understand who is a person. However, there are implied comparisons in the choice of what to report and in the use of terms in one's own language in reporting one's findings (see Jarvie, 1975; Kobben, 1970; Pelto, 1970). If, for example, I write that some of the dead of Culture Z are defined as alive, I am reflecting the fact that the issue can be delineated in my own language and the likelihood that the issue is noteworthy because studies of other cultures have dealt with the same issue or because the issue is somehow significant in my own culture. Both the usefulness of a nonindigenous language and the relevance of the topic to another culture give an emic analysis a comparative element. Beyond those comparative elements, explicit comparison with comparable measures may add to the understanding of the culture of interest. By making comparable measurements, one may find that the culture of interest is not so different (or very different) from conceivable baseline cultures. A number of studies, for example, have indicated that, in some instances, dead people in the United States and England are not counted dead by their closest relatives, at least for a while after the death (Rosenblatt, Walsh, and Jackson, 1976, ch. 3). One may also, in an emic investigation, be eliciting a comparative element in the actors' conceptual framework. That comparative element may also need probing if one hopes to understand what one is learning. Emic investigations (without any explicit comparison) can be of value in yielding understanding of how the people in a given culture understand things. But the claim of being noncomparative seems mistaken, and formalizing comparative aspects of the analysis may make the analysis much more useful.

If one can generate a maxim out of the discussion of "implicit" comparison, it is that it is best to assume one is quite ignorant of one's own culture. It seems that Campbell was correct that there is a comparison problem in the ethnographic case study. It is not necessarily a problem of *implicit* comparison; often the comparison is quite explicit. The problem is rather that the comparison is made without comparable measures. One may often know much more about focal study topics in the culture one has been investigating than one knows about one's own culture. This is not to say that making comparable measures on different cultures is a simple matter; it often is very difficult to establish comparability. But at the very least an attempt at comparable measurement is preferable to comparison involving formal measurement on a studied culture and nonmeasurement on another culture one presumes one knows.

Assumption That Observer Can Be Distinguished from Observed. Campbell's analysis of the case study may assume an

observer-observed distinction—the observer administers the measures and evaluates the treatments in a case study. Although
Campbell certainly has had a sense of observer-observed interaction in his writings about reactive measurement, he may not have
indicated in his analysis of ethnographic case studies how much
the ethnographic report is a product of interaction between ethnographer and people studied. One asks questions in the field that
informants never considered; the attempt of the ethnographer to
understand pushes both the ethnographer and the informant to
new awarenesses, perceptions, and understandings. Initial ethnographic questions, moreover, are often grounded on assumptions
about people and cultures that do not fit the people and the culture one is studying (Mohring, 1980, offers rich documentation of
this point from her work with Greek immigrants to the United
States).

       If one can learn from the boredom, shoulder shrugging,
terseness, and seemingly irrelevant answers of one's informants, if
one can learn from their direct and indirect evaluation of one's
questions, one's ethnographic work will be much stronger. Moreover, some of the phenomena to report come out of the observer's
observation of those others observing the observer—Balinese wondering about my deviance as a tourist or Mexicans struggling to
understand what it was I wanted to write about their religious beliefs and practices and why I wanted to write it. Perhaps the key
sign of observer-observed blending is that the ethnographer often
has to negotiate observations and interpretations with the people
observed. Myerhoff (1978), reporting a study of elderly Jews in
Venice, California, provides long and fascinating transcripts from
such negotiations with a wise and perceptive informant who repeatedly chided her for exaggerating. He also pushed Myerhoff to
see how much her observation and interpretation was a result of
her own personality and personal history. I can remember a conversation in English with two young Balinese men about "cremation." For them and for me, there was a painful stretching to try
to grasp what the other meant by "cremation." (The literature
seemed to me to imply that Balinese dead are cremated, but there
appeared to be only a few cremations a week on all of Bali.) We
finally reached the understanding that "cremation" referred to a
ritual event following a death, not to a burning of the remains.
Aside from the perspective that interaction gave me on ethno-

graphic work with people who seem to speak my language, it seemed to me that all three of us came out of the interaction with a negotiated reality that we had lacked prior to the negotiation.

Even with a blurred distinction between observer and observed, it seems appropriate to engage in the challenging of theory which Campbell recommended. One must only expect that much of the data for a degrees of freedom analysis of a case study will come out of interaction of "observer" with "observed." This is not to argue that the separation of observer and observed in a controlled laboratory study or in a questionnaire study is somehow superior to the more complex investigation in which observer and observed interact with each other. On the contrary, such interaction may provide much more powerful probing of theoretical assumptions, much more rich development of alternative interpretations of responses to apparently sound measuring instruments, and much shorter half-lives for theories that are faulty. There are great risks in not interacting with the persons studied. Putting laboratory manipulations, formal measurements, and research assistants between oneself and the people studied may leave one ignorant of a vast amount of useful information and not even knowledgeable enough to recognize one's ignorance.

Assumption That There Are "Degrees of Freedom" in the Case Study. Campbell's recommendation to seek information relevant to implications of one's theories is not new to fieldwork in anthropology. Anthropologists have been urged, for example, by Bateson (1941) and by Pelto (1970, pp. 314-315), and fieldwork sociologists by Geer (1967) to seek such information. But Campbell has placed the seeking of such information in a persuasive epistemological framework, a framework which makes it easy to conceptualize and engage in the process of tracking down theoretical implications. The key concept in his analysis is the analogy he draws between degrees of freedom in studies based on a number of cases and the "degrees of freedom" quality of case studies in which a number of observations relevant to a specific theory can be made. Just as the stability of a given F in an analysis of variance is greater if it rests on a larger number of cases, so the analogy runs, the trustworthiness of a theoretical assertion grounded in a field study increases if it rests on a larger number of empirical tests developed to test the theoretical assertion.

In support of the notion that "degrees of freedom" are used

in the ethnographic case study, Campbell notes that field workers often report that, in their early days in the field, they made many errors of understanding and fact, which they subsequently corrected. Although such correcting processes are easy to document, they may infrequently be corrections of relevance to theory. They may instead typically deal with "facts" (X is not actually Y's grandmother; the ceremony for twin infants is actually not the same for upper-caste infants as for other twin infants) or interpretations of "facts." To challenge broader theoretical structures requires a sophisticated deduction of implications of one's theory and a hunt for data relevant to those implications—a hunt that many ethnographers have not been educated to undertake and may not consider undertaking. I wanted to study emotional control in Bali and planned systematic data gathering before I arrived: attending funerals, watching films in company with Balinese to compare my reactions with that of the Balinese, playing competitive games like chess with the Balinese, talking with them about emotional control, watching the treatment of infants and children who were or who might be crying or angry, learning Balinese religious beliefs concerning emotional control, watching bargaining in the marketplace, watching competitive games, seeing how people reacted when I intentionally insulted them (a measurement activity about which I had strong ethical reservations; I tried it only once with a young Javanese hustler of tourists and decided I was neither brave enough nor crass enough to gather systematic data in that area). But I hardly ever see ethnographic reports that marshal such an array of data.

What I am arguing is that even if one's initial impressions of some things are corrected, one's theories and implicit metatheories (see Madan, 1975) may continue to operate unchallenged. One does not often spell out theoretical implications. One continues to ignore many things that are present but deemed unimportant by one's metatheories, and one may continue to pursue topics which one thinks are important but which informants have little or no interest in discussing. The reasons why many people in field settings have difficulty coming to new theoretical implications, difficulty in developing a powerful degrees of freedom analysis of the sort Campbell recommends, are undoubtedly complex. A substantial number of anthropologists who exploit the degrees of freedom potential of the field setting, who track down data relevant to important theoretical deductions, say that major new ideas about data sources to pursue and about previously unrealized theoretical implications came out of interaction with other ethnographers then in the field, or came while they were off at a nearby location for a rest (Maxwell, 1970), or came after they had returned home from the field (Silverman, 1972), or came out of correspondence with colleagues. Apparently it takes actual or imaginary interaction to develop new perspectives and to recognize the vulnerability of what one has achieved so far.

The assumption of "degrees of freedom" in the case study includes a conviction that there is sufficient statistical power (to maintain the analogy between the ethnographic case study and studies of multiple cases) in an ethnographic case study to make disciplined winnowing of theories possible. Although Campbell has suggested that pattern-matching tests of theories or alternative theories can be done within the context of ethnographic case studies and that multiple measures of crucial factors are desirable, especially comparison between several observers looking at the same culture, he offers no rule of thumb for deciding when one knows enough. Enough, in the context of Campbell's previous work, would not, of course, be enough for one to be certain that one was right, but it might be enough so that one can feel that one has something worth submitting for publication.

A fascinating observation in the Campbell and LeVine (1970) chapter on field manual ethnography is that, when they circulated a draft of their ethnocentrism field manual among colleagues, only psychologists and experienced survey researchers, not anthropologists, criticized their suggestion that the instrument be administered to as few as two to five individuals (p. 379). Campbell has been aware, therefore, of the tolerance many anthropologists have for assertions based on a small number of observations, but he did not address that issue in any detail in his brief "Degrees of Freedom" paper. How much information in a case study is enough? It is clear from other things Campbell has written that he is concerned that there be a "heterogeneity of irrelevancies" in measurement, that one's multiple measures remove the effect of the biases of each measure in one's assessment of reality (see, for example, Campbell and LeVine, 1970, p. 381). But the application of that concept to the ethnographic case study is not obvious. One may know enough to know that on one topic irrelevancies are minimal whereas on another they are numerous and important. Thus, on the former topic a single informant may suffice, whereas on the latter one may need many. But how can one know which topics require only a single informant? Honigmann (1970, p. 68) asserted that, although most things in the Eskimo community he studied were in flux, "the kin term system, qualities of interpersonal relations, and some physical features in the town apparently change only imperceptibly from year to year."

Although there may be a latent consensus among ethnographers about the aspects of culture that are likely to be invariant across time, across informants, and across occurrences, one might need data from many sources in a culture in order to be sure that an area of culture was so invariant that very few sources would suffice. However, it is not obvious that ethnographers usually use multiple sources for most of what they report. In fact, there are deterrents to gathering much documentation. One deterrent is that it is impossible to provide much range of information if one must report substantial documentation on each point. Another deterrent is the inconvenience or near impossibility of gathering data on rare occurrences or on events that are intended by the people being studied to be private. An ethnographer might try to use multiple informants but find that is very difficult to do, as the following reports illustrate: Powdermaker (1966) received pressure from informants not to gather redundant information; Honigmann (1970) found it impossible to keep any Eskimo interpreter interested in repeatedly asking the same questions; an informant took offense that Kobben (1967) was checking one of the informant's statements; and in Maxwell's (1970) experience in a Samoan community, the first respondents he used in psychological testing were the most easily recruited ones, while each subsequent respondent was more resistant to being recruited. So an ethnographer trying not to antagonize informants may not do a great deal of cross-checking of information obtained from informants. Then, too, ethnographers typically invest much more time and energy in the recruitment of each informant than does a survey researcher or an experimental social psychologist. The role of the research subject is well enough known in Western cultures and the topics of inquiry

in survey research or experimental social psychology so well delineated that there need be little investment in the development of
rapport. Thus, there are pressures in ethnographic work for basing
assertions on a small number of observations.

For me the most difficult aspect of fieldwork was deciding
when I knew enough—I never thought that I knew enough about
anything. Perhaps that was an artifact of my short stays in the
field, but I could not extrapolate to a time when I would be in the
field long enough. And that may be a common feeling among ethnographers. Silverman refers to "the three-months rule: If I only
had three months more" (1972, p. 205). To illustrate the problem,
when I was studying spiritist healers in Baja California, Mexico, in
company of Michael Kearney, an anthropologist who had done
substantial previous work in Mexico, I had the opportunity to
gather some data on the apparent affects of spiritist ceremony. We
had a vehicle and an interest in attending religious services. Our
first Sunday in Mexico, an elderly man who walked with a cane
told us about a service that was about to begin nearby. We offered
him a ride, which he accepted. I was startled, when we arrived at
the *templo* where the service was to be held, to see that he had a
great deal of difficulty walking. A ride of perhaps half a mile
seemed to have led to a serious loss of function. About an hour
and half later, after the service, he seemed to walk as he did originally.

During the course of the fieldwork, we subsequently had
the opportunity to bring our favorite informant, Mica, to the same
services. Several times she seemed fine earlier in the morning and
when she got into our vehicle, but when we arrived, she seemed in
dreadful shape, having trouble getting about, seeming preoccupied
and agitated, behaving as people did who felt possessed by bad
spirits or who felt threatened by such spirits, and seeming very depressed. Each time she seemed better after the service.

With two cases, and with several measures on one of them,
it seemed to me that the spiritist service had led to a "cure" of a
problem present at the onset of the service but not present half an
hour or more before the onset. Those observations made me wonder about events in the offices of psychiatrists, social workers,
counselors, and psychologists. Might there sometimes be the illusion that somebody had been helped who seemed distressed at the

beginning of the session when in reality the person was distressed *only* at the beginning of the session, not before arriving at the healer's office? The problem might not have appeared had the session not been scheduled. I could speculate further that problems are absent or not out of control without healers, or that some people, when they have arrived at a place where they can receive some help, let go of control that they could have continued to exercise. Some people may feel cured or helped even though the observed improvements have arisen only as artifacts of the dynamics of the treatment situation. I can imagine that such a finding and the conclusions that may be derived from it may have important mental health implications, but I have never tried to present my observations to people who do healing because I don't think I know enough.

To know enough, I would have to follow through on Campbell's recommendations, develop my theoretical alternatives, and hunt for evidence. For example, assume that the theoretical alternatives were (1) a playing of the sick role or (2) a disinhibition of problems actually present. The role theory would imply that the people I observed were playing a role that was appropriate on the occasion, that they were not intrinsically disturbed and would not behave in a disturbed way in other situations; by contrast, the disinhibition theory would imply that the problems expressed at the ceremonies were present but inhibited at other times. The role-play theory would imply that the sick behavior might be more common if one lacked another role to play in the situation, which was true for Mica, who at another *templo* would sometimes function as an auxiliary healer. I never saw her break down at the *templo* where she had a healer role in the way she broke down at the *templo* where she had no such role. Sometimes, however, at the *templo* where she occasionally broke down, she did not break down.

The theory of the sick role would also imply that sick behavior would be responsive to reinforcements such as attention and past cures. The reinforcement notion may be valid, but the behavior of Mica persisted even though the first time I saw her break down she was told by the spirit possessing the healer to sit down and be quiet. The disinhibition theory might imply a leakage of symptoms at other times than at the ceremonies, and I must say that there were other occasions when Mica seemed to be fighting spirits and to be depressed. These mental ex post facto reviews of field experiences are no real substitute for systematic gathering of data, though Campbell asserted in the "Degrees of Freedom" paper that even mental experiments can be of some value in winnowing theories. Obviously, one would need to develop more thoroughly the theoretical alternatives for explaining the breakdown phenomenon, but at the crude level of my example the score is 1 for and 1 against the sick-role explanation and 1 for the disinhibition theory.

I suppose the proper way to keep score would be on an implication-by-implication basis, with enough data on at least some implications to provide some sense of stability of measures. LeVine (1970) has urged the study of phenomena with multiple instances; and some phenomena that provide multiple instances should, in my opinion, be part of one's data base. But I think it would be a mistake to ignore, when one challenges various theoretical alternatives, the rare phenomenon. For example, I saw Mica only once at Todos Santos, at a time when she and other people who attended the *templos* believed that many spirits were about and, in some cases, in search of someone to possess. But her behavior at that one time provided evidence relevant to a role explanation of the breakdown, since it was role-appropriate to have problems with spirits, and she did.

Campbell's methodological analyses have always emphasized triangulation, the convergence of independent sources of information. In the "Degrees of Freedom" paper, Campbell considered ideal a situation in which several ethnographers worked independently on the same culture, with independence promoted by their working in different communities and hence with different informants and different community peculiarities. However, it seems highly unlikely that an anthropologist could work independently of previous reports (see Silverman, 1972). Modern education in anthropology demands that one be aware of previous research on a culture before one enters the field.

The degrees of freedom in statistical analyses of an appropriately ample number of cases rest, by definition, on the independence of those cases. Although the separate individuals studied in a laboratory experiment may be nonindependent in that they share a common biology and a common cultural background, the individuals are supposedly independent in that they respond independently of one another. The analogy of Campbell's analysis of degrees of freedom in the case study would lead us to look for comparable independence among separate tests of theoretical implications. For example, consider separate explorations of the hypothesis that the Balinese are emotionally unreactive. An investigation of the nonreactivity of infants to loud noises should be independent of an investigation of the nonreactivity of adults to violations of etiquette. And the separate infants studied in one investigation and the separate adults studied in the other should be independent respondents, just as the college sophomores in an experimental social psychology laboratory study are.

There are, of course, the same risks to nonindependence among separate observations in ethnographic work as there are in the laboratory study. Just as one's laboratory "subjects" may have gossiped about the study with prospective "subjects," so one's ethnographic "subjects" may gossip. As I walked through the dark after leaving the house compound of a Balinese professor who had made some effort to educate me about the etiquette of being a guest, I passed near a trio of women who were speaking rapidly. I was able to catch my name and place of origin. What startled me was that word of me was running ahead of me. Even if I had gone directly to another dwelling in the neighborhood, I could not assume that the people there were ignorant of what I was up to.

Perhaps even more bothersome than the potential nonindependence of subjects within a specific exploration of theoretical implications is the potential nonindependence of the separate explorations of the implications of a given theoretical assertion. The observer is the same in each case; it is I who observe infants and who look at adult responses to my boorishness. Moreover, the separate explorations may not be conceptually independent. Of all the possible explorations of theoretical implications, one may explore only a few, and those few may be closely linked. One obvious nonindependence in my explorations of Balinese emotional reactivity was that I looked almost entirely at "negative" emotions like fear, disgust, and anger. Another obvious nonindependence was that I looked at public behavior, behavior with others present. It therefore seems appropriate in a degrees of freedom analysis of an ethnographic case study to attend to the independence of respondents and to the independence of separate tests of theoretical implications. To some extent, the degree of independence may be impossible to evaluate; one may not be able to determine, for example, the extent of ethnographer bias. But there is certainly some hope of assessing the relatedness of supposedly separate explorations of implications of a given theoretical assertion.

Assumption That Capitalizing on Chance Is Possible in a Case Study. The final assumption that I want to discuss is that the notion of capitalizing on chance is applicable to the case study. On one level, the notion makes perfect sense. One can overgeneralize a chance finding. On the other hand, how can it be capitalizing on chance to say that whatever one found is what one found? Perhaps the issue is one of reporting. Capitalizing on chance in case studies is not a matter of saying you found something when you found nothing; it may be a matter of obscuring the scarcity of instances or the variability of one's data. Alternatively, it may be a matter of selective blindness to the context of the "fact" and to its multiple interpretations. It would thus be capitalizing on chance for me to say that I saw no Balinese crying at the one funeral I attended in Bali if I failed to note that the Balinese in attendance were all employees of the person who had died, a Chinese merchant; that they were all young males; that the funeral decorations and activities followed a traditionally Chinese pattern; and that I saw the noncrying Balinese only in one setting (visitation at the merchant's store, with the coffin on display) during the course of events from the occurrence of the death through the initial body disposal. It is true that the noncrying might represent a low level of emotionality, but it might represent a distant relationship, public self-presentation in one particular setting, inhibition as a result of the Chinese cultural context, or a response to my presence.

Further Analysis of Strengths of Ethnographic Case Studies

Among Campbell's many contributions to the philosophy of research is a relativistic epistemology, one that justifies the valuing of knowledge from many sources. His placing of research designs within a common epistemic framework enables both a more rational use of them and the construction of supplementary procedures to guard against threats to validity even in the case of rather weak designs. His contributions in such areas as quasi-experimental design, unobtrusive measures, and the knowledge embodied in traditions represent that relativistic valuing of knowledge processes. It seems to me to be consistent with Campbell's epistemological relativism to consider possible strengths of ethnographic case studies beyond those considered in Campbell's "Degrees of Freedom" paper.

Assume that people seem often to think in terms of the specific case, to be persuaded by a specific case, and to search for a specific case in order to explore the validity of a proposition. (I believe that people relate to cases in these ways, but I can at this point provide little documentation beyond my own experiences and informal observations.) From the perspective of a comparative psychology of knowledge processes (Campbell, 1959), one can assume that there is some value to dealing with specific instances if people are strongly inclined to do so. The value may not, however, lie in the characterization of the case, which might often be primarily of entertainment value (Rosenblatt and Anderson, 1981), but in something else. The following is a list of something elses.

My students often seem to value the vividness and detail of a case. I think there are four major reasons for that valuing. One reason is that it makes the applications of definitions and principles clearer. Until a specific instance is presented, there is too much uncertainty about the meanings of terms and principles. Even when definitions and principles are precisely stated, the normal ambiguity of the English language entails some uncertainty. A case may not eliminate uncertainty, but it may reduce it. Second, concreteness may increase memorability. With an example in mind, at least some students are more confident that they can remember definitions and principles. Although the underlying dynamic may again be a matter of clarifying meaning, for some students memory is better organized to recall specific events than to recall social science abstractions. Alternatively, memory may be better for case material because case material is stimulating; it leads to more thought and hence to easier accessibility of the case material when recall is appropriate (Borgida and Nisbett, 1977). Third, students may like a case example because it is an aid to the searching of memory for comparable material (Borgida and Nisbett, 1977; Nisbett, Borgida, Crandall, and Reed, 1976) in the student's own experiences. Finally, students may value a case because, congruent with Campbell's degrees of freedom analysis, a case can be probed for internal consistency, probed in a way that a generalization cannot. If generalizations and principles are being taught, the skeptic has no basis for evaluation until there is some evidence to winnow. Even a single unconvincing or faulty illustration can provide one with a sense of the invalidity of a proposition. There is typically much more information with which to do a degrees of freedom analysis, more material with which to do conceptual challenging and pattern matching, in a case report than in a summary of a multivariate statistical analysis on a large sample.

From the viewpoint of a person doing ethnographic research, the ethnographic case study has significant strengths. Statements of theory, generalizing or theory-probing comparative studies, and written accounts of cases are inevitably selective. One needs selectivity and filtering to perceive, even to perceive new "knowledge," but selectivity and filtering also interfere with the development of new applications of theory and of new theory. Long-term contact with an ethnographic case can lead to the perception of occurrences and of connections that one could never see in the short run and that a team of individuals, each looking at different things, would miss. Ingredients for discovery as a result of long-term contact with a situation include increased exposure to information incongruent with one's initial theories; acquisition of linguistic competence; the accumulation of disparate pieces of information, which have the potential to provide solutions to a puzzle or awareness of a puzzle that one can then solve; and greater likelihood of getting backstage information (Berreman, 1962) or catching people in unguarded moments (Langness, 1970).

Long-term work provides a time-series baseline against which to detect change and the impact of events. Long-term contact with a situation may also increase the value of hunch, feeling, and other second-rate difficult-to-document data that help to give one a sense of which theories may be appropriate and which theories may not be, which data are worth reporting, which classifications and interpretations are invalid and which are valid. Perhaps these are among the reasons why studies of data quality in ethnographic work often show length of stay in the field to be a good indicator of data quality, measured by fit of the data reported in an ethnography to functional theories that have so far survived empirical winnowing (Naroll, 1962; Witkowski, 1978). In order to acquire the highest expertise in a culture, there may be no alternative but to do a long-term ethnographic case study.

It is consistent with this perspective on long-term fieldwork to expect that the theories one eventually favors will often come relatively late in the experience. A consequence of the delayed development of theory is that one will often have opportunity for systematic use of degrees of freedom to probe one's eventual theory during only a relatively brief portion of one's fieldwork. Although one's initial theories lead one to information, they also may make it difficult to see contrary information, to see the inappropriateness of one's methods, and to think in alternative ways (Mohring, 1980). To guard against too great rigidity of initial theoretical position, some people (for example, Myerhoff, 1978, ch. 1) argue against initial preconceptions, although it seems to me impossible to work without preconception. Other people (for example, Glaser and Strauss, 1967) argue for a flexible interaction between phenomena and theory, an expectation that one's initial observations will be as much a source of theoretical focus as a result of it.

From a degrees of freedom perspective, the key ideas would be to generate theoretical interpretations quickly, to track them down relentlessly, and to expect to revise or augment theory and begin a new round of degrees of freedom analyses. Initial theoretical interpretations are often not sustained by relentless tracking down of implications, but atheoretical "immersion" in a culture may often be a waste of time. From the degrees of freedom perspective, one needs to move quickly into error correction processes. It may be that people who advocate a delay in theorizing are from the beginning doing the same testing of theoretical implications that I advocate, that the difference of opinion concerns when one applies the word *theory* to one's ideas. But if there is a difference of opinion here on when one should begin active data gathering on multiple implications, it seems to me that moving quickly into theory challenging will lead to much more insight. I do not advocate the continued pursuit of information concerning implications of a theory one believes to be inappropriate. There may be substantial amounts of time spent trying to conceptualize a situation or trying to choose theoretical foci that would be maximally interesting and productive. But from the perspective of this chapter, blundering ahead with a theory that one suspects is naive is far more productive than waiting patiently but passively for a strong theory to come to one.

Compiling Cases

In addition to the strengths of ethnographic case studies listed in the preceding section, ethnographic case studies provide the data for generalizing studies. Campbell has argued strongly over the years for criticism as a necessity for scientific progress. Until the preparation of his "Degrees of Freedom" paper, however, he did not consider the case study worthwhile to criticize; he considered all case studies outside the pale in a scientific sense. There are at present no widely agreed-upon standards for evaluation of ethnographic case studies, for documentation in reports of ethnographic case studies, or for processing ethnographic anecdotes.

Given the chaos in ethnographic epistemology, in ethnographic reporting, in standards for doing fieldwork, and in conceptions of what knowledge is, how can one possibly do cross-cultural work, compiling data from many different ethnographic case studies? In order to compare with a large sample, one is pushed to hone theory to basic conceptual elements and to be wise about the narrow range of topics commonly reported in ethnographies. Given the heterogeneity of ethnographic studies, any comparative study which sets out to use fine-grain measurement or to measure variables alien to most ethnographies is doomed to an enormous loss of cases or to a high level of measurement error.

Thus, not all things one can imagine in comparative study are things for comparative study. What can be studied fruitfully? My opinion is that the safest topics to study are the ones that are cross-culturally universal and that are built into the species—things related to sexual intercourse, birth, subsistence, group membership, living in close proximity to others, and death (see Winch, 1964). Not only are these universal, but issues in these areas also receive great attention from ethnographers and apparently from many people they study.

I do not, however, think that topics currently out of fashion in ethnographies should be neglected because one cannot generate a large number of cases. Many aspects of culture go unmentioned in most ethnographic reports. Examples include the medical knowledge that people obtain from feces and other physical signs, the fate of children born deaf or otherwise handicapped, how visitors know when to go home, etiquette and grammar for talking to oneself; how people interact with kin who are related to them in several different ways—for example, with a person who is at once a mother's brother and father-in-law (see Keesing, 1968; van Velsen, 1967; I owe my sensitivity on this issue also to correspondence with Karl Eggert), what midwives consider to be knowledge about childbirth and how the validity of their knowledge is established, and social definitions of spouse beating. Here and there, one may find mention of one of these things, but it is impossible to find even a handful of detailed descriptions. As LeVine (1970, p. 190) has indicated, most ethnographers prefer to follow a conservative (my word, not his) course in developing background data—looking at standard language, economy, religion, and social organization topics. In fact, given the need for ethnographic interpretation to rest on a comparative base, ethnographers are probably much better off being conservative than developing novel theory and methods. Mead, in a letter written to Franz Boas from Bali, reported distress that, after many months of applying new observational methods in the study of the Balinese, she was having difficulty thinking about her findings because there were no materials from other cultures to serve as a basis of comparison (Mead, 1977, p. 213). Although Cole and Gay (1972), in their study of memory among the Kpelle of Liberia, have shown that work with novel methods can generate within-culture comparative material of great value in interpreting findings, the possibilities for within-culture comparisons are limited. So the need of ethnographers for comparative material may restrict the range of topics that comparativists doing large-sample, content analytic studies can investigate. Nonetheless, even though comparative work with a handful of strong descriptions available on an unconventional topic is less persuasive than a study with a large sample, it may provide some disciplining of conceptualizations and may stimulate further ethnographic work.

There are, of course, other limits to what can be studied. Certain phenomena are peculiar to specific cultural contexts or can be studied only within a particular cultural context (Collins, 1974). In addition, there are the ontological questions which the critics of comparative work have been raising at least since the time of Franz Boas. These critics argue that it is pointless to study aspects of culture out of their context. Campbell (1961) has argued that there is an empirical question involved, that whether taking something out of context is a mistake depends on whether one's comparative analysis produces systematic findings or random findings. However, obtaining random findings is an expensive way to discover that one has made a mistake. It may be that Campbell's recommendation of degrees of freedom analyses will lead to ethnographic reports that can be used in comparative work with minimal out-of-context errors. Out-of-context errors might arise most often when a comparativist is coding an ethnographic report containing a great deal of shallow, disorganized description. As description becomes organized and theory probing, it becomes clearer what a given piece of information means.

The vividness, memorability, and other attractive features of the case study can still be retained, to some extent, in large-sample studies if one reports case-study anecdotes, as my co-authors and I did in our study of grief and mourning (Rosenblatt, Walsh, and Jackson, 1976). The provision of case material clarifies meanings for readers, stimulates further theorizing, and can give readers some power to do their own degrees of freedom analysis.

It is not obvious how the Campbellian degrees of freedom program for the case study blends with an interest in large-sample comparative studies. If ethnographer A has carefully tracked down evidence challenging alternative interpretations of the high rate of marital breakup in culture A' and ethnographer B has done careful work in culture B' on adoption and fosterage, how can one use their work to study either topic comparatively, let alone unrelated topics? Even if A and B studied the same topic, to the extent that different theories guided their work they could have reported noncomparable information. Information from the two instances could be noncomparable if the different theories called for different methodologies or called for very different descriptions of phenomena (for example, phenomenological versus structural). To do comparative work involving secondary analyses of the ethnographic work of others, one must assume that the variance due to theoretical orientations is not extremely large or that the theory variance can be dealt with.

Using Available Degrees of Freedom

One obvious need, in order for people to do degrees of freedom analyses, is tolerance for theory failure and an openness to evidence that one has made errors, perhaps enormous errors. Along with this need is a willingness to fall short of the ideals offered in this methodological essay. If, after one has spent nine of one's twelve months in the field, one finds that one has been asking wrongheaded questions, one may not have sufficient time to do a strong degrees of freedom analysis on one's new theoretical hunches. Ideals of rich data gathering should never block the turning away from richly documented nonsensical theory to sparsely documented theory that may be stronger.

A second obvious need, in order for people to do the degrees of freedom analyses which Campbell has suggested, is training in how to think. Even the brightest of humans may not be very good at seeing alternative interpretations and thinking through their implications (see Einhorn and Hogarth, 1978). People may not even be very good at identifying phenomena to explain, let alone at thinking up a single explanation. Many field workers have talked about the value, for generating new ideas, of corresponding with colleagues; of chatting with colleagues about the fieldwork; of leaving the field periodically; of trying to write, in the midst of fieldwork, articles or dissertation chapters based on their work so far. Even members of a well-educated social science elite do not automatically think through things very well; they gain from the reactions of others and from thinking through how others will react to systematic presentations they are preparing. The gains from actual, prospective, or imaginary interaction with colleagues include a sense of what one has learned that one's colleagues do not know and a negotiation of reality that clarifies what one thinks and links these thoughts more richly to the frameworks colleagues can understand.

I believe that people can learn to be self-critical, to back away from theories they are invested in, and to generate alternative interpretations. Analysis of the development of theoretical fluency would require at least another chapter in this book. Ingredients for theoretical fluency include a sense of the relativity of theory and method; specific theories and methods are useful in some situations, harmful in others. I am sure that one of the reasons the Campbellian analyses of design and measurement have had such an enormous impact is that they give people easily remembered, easily applied tools for generating alternative explanations. They are as much tools for generating theory as they are technical, methodological tools. One of the great strengths of my education with Campbell was that he was continually generating explanations and alternative explanations and objectifying phenomena both in the scholarly literature and the everyday world. Perhaps every ethnographic training program needs a Campbell to model that sort of thing. But in the absence of a Campbell, one could learn habits of generating explanations and of tracking down data relevant to one's explanations by working with the conceptual tools provided by his degrees of freedom analysis.

References

Bateson, G. "Experiments in Thinking About Observed Ethnological Material." *Philosophy of Science,* 1941, *8,* 53-68.

Beattie, J. H. M. "Understanding and Explanation in Social Anthropology." *British Journal of Sociology,* 1959, *10,* 45-60.

Becker, H. S. "Social Observation and Social Case Studies." In D. L. Sills (Ed.), *International Encyclopedia of the Social Sciences.* Vol. 11. New York: Macmillan and Free Press, 1968.

Berreman, G. D. "Behind Many Masks: Ethnography and Impression Management in a Himalayan Village." *Society for Applied Anthropology Monographs,* 1962, no. 4.

Borgida, E., and Nisbett, R. E. "The Differential Impact of Abstract vs. Concrete Information on Decisions." *Journal of Applied Social Psychology,* 1977, 7, 258-271.

Campbell, D. T. "Enhancement of Contrast as Composite Habit." *Journal of Abnormal and Social Psychology,* 1956, *53,* 350-355.

Campbell, D. T. "Factors Relevant to the Validity of Experiments in Social Settings." *Psychological Bulletin,* 1957, *54,* 297-312.

Campbell, D. T. "Methodological Suggestions from a Comparative

Psychology of Knowledge Processes." *Inquiry,* 1959, *2,* 152-182.

Campbell, D. T. "The Mutual Methodological Relevance of Anthropology and Psychology." In F. L. K. Hsu (Ed.), *Psychological Anthropology: Approaches to Culture and Personality.* Homewood, Ill.: Dorsey Press, 1961.

Campbell, D. T. "Pattern Matching as an Essential in Distal Knowing." In K. R. Hammond (Ed.), *The Psychology of Egon Brunswik.* New York: Holt, Rinehart and Winston, 1966.

Campbell, D. T. "Natural Selection as an Epistemological Model." In R. Naroll and R. Cohen (Eds.), *A Handbook of Method in Cultural Anthropology.* New York: Natural History Press/Doubleday, 1970.

Campbell, D. T. " 'Degrees of Freedom' and the Case Study." *Comparative Political Studies,* 1975, *8,* 178-193.

Campbell, D. T., and LeVine, R. A. "Field Manual Anthropology." In R. Naroll and R. Cohen (Eds.), *A Handbook of Method in Cultural Anthropology.* New York: Natural History Press/Doubleday, 1970.

Campbell, D. T., and Stanley, J. C. "Experimental and Quasi-Experimental Designs for Research on Teaching." In N. L. Gage (Ed.), *Handbook of Research on Teaching.* Chicago: Rand McNally, 1963.

Cole, M., and Gay, J. "Culture and Memory." *American Anthropologist,* 1972, *74,* 1066-1084.

Collins, P. W. "The Present Status of Anthropology as an Explanatory Science." *Boston Studies in the Philosophy of Science,* 1974, *11,* 337-348.

Cook, T. D., and Campbell, D. T. "The Design and Conduct of Quasi-Experiments and True Experiments in Field Settings." In M. D. Dunnette (Ed.), *Handbook of Industrial and Organizational Psychology.* Chicago: Rand McNally, 1976.

Einhorn, H. J., and Hogarth, R. M. "Confidence in Judgment: Persistence of the Illusion of Validity." *Psychological Review,* 1978, *85,* 395-416.

Geer, B. "First Days in the Field." In P. E. Hammond (Ed.), *Sociologists at Work.* New York: Doubleday, 1967.

Geertz, C. "Thick Description: Toward an Interpretative Theory of Culture." In *The Interpretation of Cultures: Selected Essays by Clifford Geertz.* New York: Basic Books, 1973.

Glaser, B. G., and Strauss, A. L. *The Discovery of Grounded Theory: Strategies for Qualitative Research.* Chicago: Aldine, 1967.

Gould, H. A. "Two Decades of Fieldwork in India—Some Reflections." In A. Beteille and T. N. Madan (Eds.), *Encounter and Experience: Personal Accounts of Fieldwork.* Delhi: Vikas, 1975.

Herskovits, M. J. "Some Problems of Method in Ethnography." In R. F. Spencer (Ed.), *Method and Perspective in Anthropology.* Minneapolis: University of Minnesota Press, 1954.

Hollander, A. N. J. den. "Social Description: The Problem of Reliability and Validity." In D. G. Jongmans and P. C. W. Gutkind (Eds.), *Anthropologists in the Field.* Assen, Netherlands: Van Gorcum, 1967.

Honigmann, J. J. "Fieldwork in Two Northern Canadian Communities." In M. Freilich (Ed.), *Marginal Natives: Anthropologists at Work.* New York: Harper & Row, 1970.

Jarvie, I. C. "Understanding and Explanation in Sociology and Social Anthropology." In R. Borger and F. Cioffi (Eds.), *Explanation in the Behavioral Sciences.* Cambridge, England: Cambridge University Press, 1970.

Jarvie, I. C. "Epistle to the Anthropologists." *American Anthropologist,* 1975, 77, 253-266.

Keesing, R. M. "Nonunilineal Descent and Conceptual Definition of Status." *American Anthropologist,* 1968, 70, 82-84.

Kobben, A. J. F. "Why Exceptions?" The Logic of Cross-Cultural Analysis." *Current Anthropology,* 1967, 8, 3-19.

Kobben, A. J. F. "Comparativists and Noncomparativists in Anthropology." In R. Naroll and R. Cohen (Eds.), *A Handbook of Method in Cultural Anthropology.* New York: Natural History Press/Doubleday, 1970.

Langness, L. L. "Unguarded Moments." In R. Naroll and R. Cohen (Eds.), *A Handbook of Method in Cultural Anthropology.* New York: Natural History Press, 1970.

LeVine, R. A. "Research Design in Anthropological Fieldwork." In R. Naroll and R. Cohen (Eds.), *A Handbook of Method in Cultural Anthropology.* New York: Natural History Press/Doubleday, 1970.

Madan, T. N. "On Living Intimately with Strangers." In A. Beteille and T. N. Madan (Eds.), *Encounter and Experience: Personal Accounts of Fieldwork.* Delhi: Vikas, 1975.

Maxwell, R. J. "A Comparison of Field Research in Canada and Polynesia." In M. Freilich (Ed.), *Marginal Natives: Anthropologists at Work*. New York: Harper & Row, 1970.

Mead, M. "The Art and Technology of Fieldwork." In R. Naroll and R. Cohen (Eds.), *A Handbook of Method in Cultural Anthropology*. New York: Natural History Press/Doubleday, 1970.

Mead, M. *Letters from the Field, 1925-75*. New York: Harper & Row, 1977.

Mohring, P. M. "Life, My Daughter, Is Not the Way You Have It in Your Books." Unpublished doctoral dissertation, University of Michigan, 1980.

Myerhoff, B. *Number Our Days*. New York: Dutton, 1978.

Nakane, C. "Fieldwork in India—A Japanese Experience." In A. Beteille and T. N. Madan (Eds.), *Encounter and Experience: Personal Accounts of Fieldwork*. Delhi: Vikas, 1975.

Naroll, R. *Data Quality Control*. New York: Free Press, 1962.

Naroll, R., and Naroll, F. "On Bias of Exotic Data." *Man*, 1963, *53*, 24-26.

Nisbett, R. E., Borgida, E., Crandall, R., and Reed, H. "Popular Induction: Information Is Not Always Informative." In J. Carroll and V. Payne (Eds.), *Cognition and Social Behavior*. Hillsdale, N.J.: Erlbaum, 1976.

Pandey, T. N. " 'India Man' Among American Indians." In A. Beteille and T. N. Madan (Eds.), *Encounter and Experience: Personal Accounts of Fieldwork*. Delhi: Vikas, 1975.

Pelto, P. J. *Anthropological Research: The Structure of Inquiry*. New York: Harper & Row, 1970.

Powdermaker, H. *Stranger and Friend: The Way of an Anthropologist*. New York: Norton, 1966.

Richards, A. I. "The Development of Field Work Methods in Social Anthropology." In F. C. Bartlett, M. Ginsberg, E. J. Lindgren, and R. H. Thouless (Eds.), *The Study of Society*. London: Routledge and Kegan Paul, 1939.

Rosenblatt, P. C., and Anderson, R. M. "Sexual Attraction in Cross-Cultural Perspective." In M. Cook (Ed.), *The Bases of Human Sexual Attraction*. London: Academic Press, 1981.

Rosenblatt, P. C., Walsh, R. P., and Jackson, D. A. *Grief and Mourning in Cross-Cultural Perspective*. New Haven, Conn.: Human Relations Area Files, 1976.

Silverman, M. G. "Ambiguities and Disambiguities in Field Work." In S. T. Kimball and J. B. Watson (Eds.), *Crossing Cultural Boundaries: The Anthropological Experience.* San Francisco: Chandler, 1972.

Trope, Y. "Inferences of Personal Characteristics on the Basis of Information Retrieved from One's Memory." *Journal of Personality and Social Psychology,* 1978, *36,* 93-106.

Tversky, A., and Kahneman, D. "Belief in the Law of Small Numbers." *Psychological Bulletin,* 1971, *76,* 105-110.

Uchendu, V. A. "A Navajo Community." In R. Naroll and R. Cohen (Eds.), *A Handbook of Method in Cultural Anthropology.* New York: Natural History Press/Doubleday, 1970.

van Velsen, J. "The Extended-Case Method and Situational Analysis." In A. L. Epstein (Ed.), *The Craft of Social Anthropology.* London: Tavistock, 1967.

Winch, P. "Understanding a Primitive Society." *American Philosophical Quarterly,* 1964, *1,* 307-324.

Witkowski, S. "Ethnographic Fieldwork: Optimal Versus Non-optimal Conditions." *Behavior Science Research,* 1978, *13,* 245-253.

Louise H. Kidder

Qualitative Research and Quasi-Experimental Frameworks

Qualitative and quantitative research occupy different worlds, and Donald Campbell has educated us in both. His writings about degrees of freedom in the case study (Campbell, 1975), about the mutual methodological relevance of psychology and anthropology (Campbell, 1961), and about qualitative knowing in action research (Campbell, 1974) analyze qualitative research as a legitimate form of scientific knowing. His writing about experiments and quasi-experiments (Campbell and Stanley, 1966; Cook and Campbell, 1979) demonstrates his high standards for what constitutes science. While Campbell has informed us about both worlds, the two remain separate in his writings. This chapter is an attempt to bridge the gap between the two.

This is an immodest venture, because the division between qualitative and quantitative research is entrenched in several disciplines. In psychology it appears as a tension between experimental and clinical methods. In sociology it appears in the separation of fieldwork and statistical work. In the logic of scientific inquiry, it appears as a difference between hypothetico-deduction and analytic induction (see, for example, Becker, 1963; Glaser and Strauss, 1967; Lindesmith, 1968; McCall and Simmons, 1969; Znaniecki, 1934). The differences between qualitative and quantitative researchers appear also in caricatures; the former are considered "soft," and the latter "hard," or the former called "navel gazers" and the latter "number crunchers" (see Sherif, 1979). Without denying some of the real differences, this chapter explores the similarities. The exploration is one-sided; it applies the quantitative researcher's criteria for reliability and validity to qualitative work, using the typology that Campbell and his colleagues have developed (Campbell and Stanley, 1966; Cook and Campbell, 1979).

Not all qualitative research is as different from quantitative work as appears on the surface. Although field workers or participant observers do not use the language of quasi-experimental designs, we can discern designs in their studies. By studying naturally occurring phenomena, participant observers never become experimenters in the literal sense. They can seldom control what happens to whom, but they can *observe* what happens to whom when, and their work has an implicit quasi-experimental design. ("Participant observation" is here used in the broad sense, in which actual participation need not be involved.)

In the pages that follow, I will make explicit the quasi-experimental designs of several examples of qualitative research and assess their internal, external, construct, and statistical conclusion validity. I will apply the rules of one to the other, to answer the question posed by Zelditch (1962, p. 566): "Quantitative data are often thought of as 'hard,' and qualitative as 'real and deep'; thus, if you prefer 'hard' data you are for quantification and if you prefer 'real, deep' data you are for qualitative participant observation. What to do if you prefer data that are real, deep, *and* hard is not immediately apparent."

Causal Analysis in Qualitative Work

Internal validity as defined by Campbell (Campbell and Stanley, 1966; Cook and Campbell, 1979) is an issue only for research that reports causal relationships. Some qualitative work is purely descriptive and therefore not concerned with internal validity. The descriptive character of some fieldwork may contribute to its image as "soft," for it is not engaged in the difficult enterprise of causal analysis. Some qualitative researchers *do* make causal assertions, however, handicapped though they are without experimental designs.

Whenever qualitative research presents a career analysis—the career of a marijuana user, an embezzler, a parole officer—it includes causal assertions. Careers are socialization patterns or paths, and the steps necessary to become socialized are links in a causal chain. Participant observation rarely contains only one or two causal statements of the "*A* causes *B*" variety. Instead, it contains a set of propositions about the causal chain. The following are examples of causal statements from participant observation research:

> Instead of the deviant motives leading to the deviant behavior, it is the other way around; the deviant behavior in time produces the deviant motivation. Vague impulses and desires—in this case, probably most frequently a curiosity about the kind of experience the drug will produce—are transformed into definite patterns of action through the social interpretation of a physical experience [Becker, 1963, p. 42].

> Routine career contingencies reward parole officers for underreporting deviant behavior. *Ceteris paribus,* parole officers will ignore most of the crimes, incidents, and violations they observe in their caseloads. The exceptions to this rule are situations where the parole officer realizes some benefit from reporting an incident. . . . Ordinarily, the parole officer will report an incident and thereby create a record only when he can use that record to enhance his work environment [McCleary, 1977, p. 576].

Many research students and professionals are reluctant to say "this caused that" when they report the results of their work. Instead, they claim that theirs is not causal research but a description of a process, or an exploration of a phenomenon, or a report of a relationship. For instance, I did not say that my own participant observation study of sojourners in India was a causal analysis; instead, I claimed to "describe the socialization of aliens into their role" (Kidder, 1977, p. 48). In fact, I wished to say that being in India for a long time and occupying a high status caused sojourners to become more alien rather than more Indian. Without a true experiment, however, I did not make such a bold claim. This same problem besets quantitative researchers with nonexperimental designs. In the same study, I had quantitative data on attitudes, and I did not make causal assertions about those data either. Instead, I reported a relationship: "Favorability toward the host country . . . [was] examined in relation to the respondent's occupational status and elapsed time in India" (Kidder, 1977, p. 48). Reports of relationships and descriptions of processes are retreats to a safe position, a position that cannot be attacked because the writer does not claim to know what caused what.

There are many reasons for the reluctance to be so direct as to assert "this caused that," including measurement error and bias. But one problem plagues qualitative research in particular, because qualitative researchers work inductively rather than deductively. They do not begin with a hypothesis; instead, they generate hypotheses from their data. Analytic induction proceeds in the direction opposite from hypothetico-deduction. Hypothetico-deduction researchers work from the top down; they begin with theoretical premises, predict a pattern of results, and examine the data to test the theoretical deduction. Analytic inductive researchers work from the bottom up—beginning with data and developing theoretical categories, concepts, and propositions from the data (Glaser and Strauss, 1967). By forming hypotheses to fit the data, qualitative researchers violate some of the assumptions of statistical testing in quantitative research, and this too may inhibit them from making claims about causes.

If we listen carefully to the conclusions of participant observation research, however, we hear causal assertions even if they are not as simply stated as "this caused that." In his study of how parole officers report violations, McCleary (1978, pp. 132-135) listed five career contingencies that *cause* parole officers to ignore crimes: (1) Full reporting cuts into parole officers' "free" time. (2) Full reporting places parole officers in jeopardy because they may have to defend their decisions in a hearing. (3) Parole officers believe their performance is evaluated on the basis of the good behavior of their parolees. (4) Full reporting creates "busy work" for parole officers. (5) Parole officers limit their ability to counsel or give a man a "second chance" when they report violations.

In his study of marijuana users, Becker (1963, p. 49) also made causal assertions and identified a series of events that must take place if a person is to use marijuana for pleasure: (1) learning the technique of smoking to produce effects, (2) learning to perceive the effects, (3) learning to enjoy the effects.

And in his study of embezzlers, Cressey (1953) identified four conditions that must be present if embezzlement is to occur. The person must (1) be in a position of financial trust, (2) have a nonshareable financial problem, (3) recognize embezzlement as a possible solution for the problem, (4) develop a way to rationalize embezzlement without calling it that, so that it seems like a "loan" or a justifiable use of someone else's money. Cressey found no case of embezzlement that did not have all these elements, and he concluded that these four conditions as a set are necessary and sufficient causes of embezzlement.

The five career contingencies that McCleary listed as reasons for underreporting crimes, the three steps that Becker listed as necessary for becoming a marijuana user, and the four conditions Cressey listed as necessary for embezzlement to occur are all causes. In each case, the researchers have said "This leads to that." To make such definite statements about causes and effects, they must each have ruled out rival explanations. None of these studies was a true experiment, because the researchers did not control what happened to whom under conditions of random assignment. Instead, they each recorded what happened, to whom, when, and created implicit quasi-experimental designs.

Quasi-Experimental Designs in Participant Observation

Both Cressey and Becker confined their studies to people who exhibited the phenomenon in question—embezzlement in one

case and marijuana use in the other. They did not study populations of nonembezzlers and nonusers of marijuana, but they interviewed users and embezzlers about previous times in their lives
when they had not indulged in marijuana use or embezzlement.
They formulated hypotheses about the conditions necessary for
marijuana use or embezzlement to occur by noting the absence of
those conditions during times of nonuse or nonembezzlement.
Cressey and Becker were able to study the full range of the phenomenon—embezzlement and nonembezzlement of funds, use and
nonuse of marijuana—by obtaining retrospective reports from their
samples of users and embezzlers. They could have reported their
results in a cross-tabulation or covariance table like that in
Table 1.

Table 1. Phenomenon (Embezzlement or Marijuana Use)

		Present	*Absent*
Prior Conditions	Present	100% of the cases	0 cases
	Absent	0 cases	100% of the cases

The right-hand column is a phantom; it contains retrospective reports of people who currently exhibit the phenomenon, not
interviews with people who have never embezzled or used marijuana for pleasure. It is a within-subjects variable, based on recall.
Had it been possible for either Cressey or Becker to be there before some of the convicts had embezzled or before some marijuana
users learned to perceive the effects, they could have gathered on-
site observations of the absence of those conditions. They relied
on retrospective reports as a substitute. They did not interview
samples of nonembezzlers or nonusers about why they had not indulged, because such interviews would yield reasons that have
little bearing on the propositions they were testing. For instance,
people may say they have never used marijuana for pleasure because they have not had the opportunity or because they consider
it immoral or illegal (Becker, 1963). Such interviews would give information about the causes for not becoming a marijuana user,
which would be a separate study.

We can depict both Becker's study of marijuana users and

Cressey's study of embezzlers as time series designs. We borrow the following notation from Campbell and Stanley (1966):

X = treatment or cause

O = observation or effect

The causes are the necessary conditions for the marijuana use or embezzlement to occur. The effects are the reports that marijuana use or embezzlement did or did not occur. There are two differences between a traditional time-series design and the design of these two studies: (1) the X's are not single causes but a combination of necessary conditions, and (2) the X's and O's are measured through retrospective self-reports rather than the investigators' on-site observations or archives.

In his analysis of what causes people to become marijuana users, Becker identified three steps or causes that work in sequence. The three steps combined we will represent as follows: X_1-X_2-X_3. The first step, X_1, is learning to smoke "properly" to produce physical symptoms. This is not sufficient to cause marijuana use alone, however, because the physical symptoms are not always obvious. The novice must learn to detect them. Having cold hands or feet, developing an intense hunger, and having fits of laughter are signs known to initiates, but novices often need to have their own symptoms pointed out before they recognize them as signs of being high. This is the second step, X_2. Having learned to produce and perceive the effects, a novice becomes a continual user only if he or she also learns to enjoy those effects. The effects are not obviously pleasant, as the following example shows: "A new user had her first experience of the effects of marijuana and became frightened and hysterical. She 'felt like she was half in and half out of the room,' and experienced a number of alarming physical symptoms" (Becker, 1963, p. 56). Experienced users help novices reinterpret such feelings as pleasant, exciting, and sought after. This is the final step, X_3.

Becker's evidence consisted of retrospective reports from people who had tried smoking and either continued or quit as a result of the presence or absence of conditions X_1, X_2, and X_3. When people first begin smoking, they try it a second and third time, even if they do not experience all the conditions (X_1-X_2-X_3) during their first try, either because their friends encourage them or because they believe there is more to the experience than what they felt. For instance, the following report describes a progressive accumulation of experience until all three conditions were met:

> "I didn't get high the first time. . . . The second time I wasn't sure, and he [smoking companion] told me, like I asked him for some of the symptoms or something, how would I know, you know. . . . So he told me to sit on a stool. I sat on—I think I sat on a bar stool—and he said, 'Let your feet hang,' and then when I got down my feet were real cold, you know.
>
> "And I started feeling it, you know. That was the first time. And then about a week after that, sometime pretty close to it, I really got on. That was the first time I got on a big laughing kick, you know. Then I really knew I was on" [Becker, 1963, pp. 49-50].

We can diagram this report as a time series, in which the effect (becoming a regular user) appeared only after all three necessary conditions were present:

O	X_1	O	X_2-X_2	O	X_1-X_2-X_3	O
not regular user	smoked, felt no effects	not regular user	smoked, felt effects	not regular user	smoked, felt and enjoyed effects	became regular user

As evidence that all three conditions must be present for a person to become a regular user, Becker reports cases where one or more conditions were absent and a person did not become a marijuana user. For instance:

> "It was offered to me and I tried it. I'll tell you one thing. I never did enjoy it at all. I mean, it was just nothing that I could enjoy. [Well, did you get high when you turned on?] Oh, yeah, I got definite feelings from it. But I didn't enjoy them. I mean I got plenty of reactions, but they were mostly reactions of fear" [Becker, 1963, p. 54].

In another instance, a person who had been a regular user quit because he had some unpleasant and frightening experiences which made him discontinue use for three years:

> "It was too much, like I only made about four tokes, and I couldn't even get it out of my mouth, I was so high, and I got real flipped . . . I walked outside, and it was five below zero, and I thought I was dying . . . I fainted behind a bush. I don't know how long I laid there . . . all weekend I started flipping, seeing things there and going through hell, you know, all kinds of abnormal things . . . I just quit for a long time then" [Becker, 1963, p. 57].

If we draw the time series for this person, it includes the onset, offset, and onset of marijuana use, as follows:

$$O\ O\ O\ O\quad X_1\text{-}X_2\text{-}X_3\ \ O\ \ X_1\text{-}X_2\text{-}X_3\ \ O\quad X_1\text{-}X_2\ \ O\ \ X_1\text{-}X_2\ \ O$$
$$\text{nonuse}\qquad\qquad\text{onset and use}\qquad\qquad\text{offset and nonuse}$$

$$X_1\text{-}X_2\text{-}X_3\ \ O$$
$$\text{onset}$$

Not all participant observation relies on retrospective reports alone to deduce the pattern of the career. McCleary, in his study of the socialization of parole officers, write field notes while he observed parole officers. He used both his own observations of their behavior and their retrospective reports of what happened earlier. His work provides a useful illustration of how qualitative researchers can rule out threats to internal validity.

Threats to Internal Validity in Qualitative Research

McCleary's research on parole officers describes both the causes for underreporting and the causes for full reporting among parole officers. Parole officers do not always ignore crimes or violations by their parolees. If parole officers can benefit from reporting incidents, they do so.

McCleary (1977, p. 576) makes a straightforward causal assertion: "Routine career contingencies reward parole officers for underreporting deviant behavior. . . . The exceptions to this rule are situations where the parole officer (PO) realizes some benefit from reporting an incident."

In addition to the five causes of underreporting listed earlier, McCleary identified three causes or occasions for full reporting:

1. *To threaten a parolee.* Parolees know their dossiers can be used against them, so a PO may threaten to report an incident to control a particular parolee.
2. *To get rid of a parolee.* By reporting incidents that show a parolee is in need of special treatment, a PO can have that parolee transferred from his caseload to a special treatment program.
3. *To protect himself.* A PO can write full reports to demonstrate he is doing his job, following the rules, and keeping track of "dangerous men."

To diagram the implicit time-series designs for McCleary's study, we adopt the following notation:

O = observations of parole officers' reports of incidents
X_U = routine career contingencies that reward underreporting
X_F = special conditions that reward full reporting

Some of the observations of full reporting of incidents were made directly by McCleary. Some of the observations were told to him by parole officers. Since full reporting was the exception to the rule, McCleary observed fewer of those instances and had to rely on informants. For instance, when two newspapers began an investigation of the department of corrections, parole officers panicked and began putting their files in order to protect themselves and their supervisors. One parole officer told McCleary: "That's the only time I ever saw all nine POs in the office at the same time. We wrote a lot of paper that week. When the story hit, we were all protected" (McCleary, 1978, p. 145).

The research, therefore, contains a mixture of McCleary's direct observations of X's and O's and indirect observations given to him by his respondents. For instance, a time series for a new recruit may look like the following:

$$O \quad O \quad O \quad X_U O \quad X_U O \quad X_U O$$

The initial O's are stories about that recruit rather than McCleary's firsthand observations. He was told that this new PO first wrote overlong reports and had to learn to underreport: "His first reports give vivid descriptions of the 'personal, social, financial, family, employment, and psychological problems' of the clients he has contacted. He is quite proud of these reports, so he is shocked when his supervisor hands them back to him. All are rejected. All require 'further investigation' or 'more work' " (McCleary, 1978, p. 159). Contrary to the novice's belief that his reports were too brief, he gradually learned that they were too long. He learned that if he wanted to make statements about a client's financial problems, he had to back up his claims by contacting the client's employer, family, friends, and landlord. "Each sentence of a three-page report will require ten pages of corroboration. . . . When he seeks his trainer's help with the reports, he is advised to keep his reports brief. . . . The novice knows that DC [department of corrections] rules and regulations call for more detailed reports and he points this out to his trainer. The trainer counters this argument by pointing out the many ways in which reports can be misinterpreted by DC officials and by other criminal justice agencies" (McCleary, 1978, p. 159). Full reporting means extra work and perhaps extra trouble for parole officers.

Sometimes the extra trouble is worth it, however, and parole officers report incidents to coerce a parolee, to get rid of a troublesome parolee, or to protect themselves in case of investigations. Since full reporting is the exception rather than the rule, McCleary does not describe any long series of incidents that led to a parole officer's learning to write full reports. Instead, he presents one-shot incidents and quick lessons that are short or truncated time series. For instance, an office supervisor said:

> "There's something final about the written word. A PO can talk himself blue in the face and get no response from the parolee. If he puts it down on paper, though, he gets results. Parolees know that once something gets written down, it's final. I encourage my POs to utilize their reports that way" [McCleary, 1978, p. 140].

This and the following description from a parole officer summarize what may be years of experience. They condense a series of $X_F O$ $X_F O\ X_F O$ into a single report of $X_F O$:

> "One time I had this junkie—a white kid. His mother was causing problems for him. She didn't have any respect for me or for the job I was trying to do. Well, I finally gave him an official warning and mailed a copy to his mother. I never had any trouble with him after that" [McCleary, 1978, p. 139].

Unlike Becker's analysis of the causes of becoming a marijuana user, McCleary's analysis does not specify a sequence of conditions. The five causes of underreporting need not all occur together or in a particular order. They operate singly or in combination. For this reason, I have simply used the X_U to symbolize any combination of causes for underreporting and X_F to symbolize any combination of causes of full reporting. The time-series lines shown above have different lengths, and the observations (either direct or indirect) are not evenly spaced. Some observations or stories about individual parole officers are so brief that they qualify only as one-shot observations rather than actual time series, as in the preceding example, which quotes a parole officer as saying that he never again had trouble after he sent an official warning to a parolee, with a copy mailed to his mother. No one of the lines is as regular and elegant as the archival data described by Campbell in his discussions of time series (Campbell and Stanley, 1966; Cook and Campbell, 1979). This is because McCleary did not "design" his research to take the shape of a time-series design. He did not know what the causes or X's were before he began. He discovered them through analytic induction, so the design is an ex post facto or retrospective creation. Nonetheless, we can decipher the design that emerges, and we can examine the validity of the causal assertions by seeing whether there are plausible rival explanations. The list of rival explanations that follows is taken from Campbell and Stanley (1966) and Cook and Campbell's (1979) lists of threats to internal validity.

- *Selection.* We need not concern ourselves with this because McCleary did not try to compare one group of parole officers with another. He traced the careers of individuals and recorded how they changed, so preexisting selection factors make no difference in his research. Selection is no threat to time-series designs.

- *History*. Is it plausible that some historical quirk was the cause of underreporting? Could something like a change in the police chief, or police department regulations, or mayor, or criminal justice regulations or legislation (for example, in a 1971 decision in the case of *Morrissey* v. *Brewer,* the Supreme Court declared that agencies must protect the due process rights of parolees) explain the underreporting? None of these are plausible explanations, because McCleary tracked the parole officers' behaviors over time and found that each one had to learn to write brief reports. Moreover, their shortened reports did not appear simultaneously, so no single event precipitated a sudden drop in reporting. History does not explain the underreporting.
- *Maturation*. As POs became tired, fatigued, or sophisticated, did they underreport incidents? There was no evidence that fatigue or boredom alone caused underreporting, because POs would immediately give *full* reports when the contingencies changed, no matter how fatigued.
- *Testing*. Did McCleary's observations sensitize the POs and make them change their reporting? If they were aware of what he was observing, they might have done and reported *more* paperwork rather than less, since being observed might have caused them to feel evaluated. Thus, McCleary's presence was not a likely cause of underreporting; more important were the threatened checks by the department of corrections, which led to a flurry of report writing and office attendance. Becker and Geer (1957) compare the demand characteristics of participant observation and isolated interviews and point out that, because the interview is a unique situation, interviews are more likely to produce reactive effects and effects of testing. In the isolation and anonymity of an interview, respondents are free from the constraints of social intercourse. Interviewers hope that respondents will tell the "truth" under such conditions; but respondents are also free to lie or respond to the demand characteristics of the interview (see Rosenthal and Rosnow, 1969). In an ongoing social setting such as a parole office, the daily constraints of the job, the so-

cial group, and the expectations of one's associates are greater than the constraints of being observed, so testing is less likely to influence a respondent's subsequent behavior.

- *Instrumentation.* Did McCleary change his way of perceiving or reporting what was going on? Was the change in the observer rather than the observed? This would be a plausible threat if McCleary had observed one cohort of parole officers, all of whom changed at the same time. Had he observed full reporting at the beginning of this research and underreporting at the end (or vice versa), instrumentation would be a rival explanation because changes in his research procedures could have coincided with the purported changes in the parole officers. Two features of McCleary's research make instrumentation implausible as a rival explanation. The first is that the POs were not a cohort whom McCleary observed from beginning to end. They were an assorted group who had been in the profession for varying lengths of time; POs entered and dropped out during the course of his observations. The second feature is that much of McCleary's data of how POs learned to underreport came not from his direct observations but from retrospective reports by other observers—supervisors and the POs themselves. Both of these features desynchronized changes in the parole officers and changes in the researcher or his instruments, making instrumentation an implausible rival explanation.

- *Regression.* Were the POs observed by McCleary an extreme group of prolific report writers whose behavior could change in only one direction? If anything, the nature of a PO's job attracts people who have another occupation, such as operating a restaurant or studying for a master's degree. They seek parole work because it does not require full-time office attendance, and it is unlikely that they would want to fill their spare time with unnecessary paperwork. They are, therefore, not likely to be compulsive report writers who can only become less so. There is also no indication that those POs whom McCleary observed or heard about were an unusual sample of officers. No one ever described himself or another as being atypical. We can be reasonably confident, therefore, that neither POs as a whole nor those whom McCleary observed represent an extreme sample whose behavior could only regress toward the mean.

• *Mortality*. Do those POs who insist on full reporting drop off the force? This is possible, but McCleary has no evidence that noncompliers left. He does have evidence that an original non-complier changed because of social pressure:

> A new PO in one of the branch offices often wrote lengthy site investigation reports. The other POs in the office disapproved of this, and from time to time would make jokes at the nonconformist's expense. The jokes were meant to be a norm-enforcing mechanism, but as the new PO was not aware of the norm for site investigation reports, the jokes were interpreted as a personal attack: "Whitney called me Sally Social Worker. I don't think that's funny and I don't think he means it as a joke either. He only says it when the other fellows are around. I'm just trying to fit in here. . . . My site investigation reports do have something to do with it but I don't think that it's anybody's business. I'm not saying that anybody has to do it my way. I just happen to think that the site investigation is the most important part of this job. It's a free country, though, and if other people have a different opinion, that's their right. I don't tell them how to do their work and I don't want them telling me how to do mine" [McCleary, 1978, p. 58].

Within a few weeks this PO became "one of the fellows," as a result of both the group's ridicule and his supervisor's refusal to accept long reports.

No participant observation study has ever included an explicit discussion of these threats to internal validity, but this exercise demonstrates how it could be done. In fact, if the researchers themselves were asked to rule out the threats to validity, they could do so even more convincingly, because they have much more data to draw on than appears in the published report. What makes it possible for participant observers to rule out threats to validity in the absence of an explicit design is the richness of the data, the longitudinal observations, and the nonsimultaneity of treatments across persons. The researcher has many pieces of information about each person or incident. Much of this information is gathered without the researcher's foreknowledge of how useful it will be. One of the few rules for participant observation is "write down everything," and it is this which helps the researcher rule out rival explanations and check hypotheses that are formulated after the fact.

Cressey formulated and revised his hypothesis five times before he arrived at his conclusion about the causes of embezzlement. Each time he formulated a new hypothesis, he checked it against not only new interviews but also all of his previously recorded interviews and observations. This ex post facto procedure is a necessary practice in participant observation. It forms the basis for analytic induction and negative case analysis. Negative case analysis requires that the researcher look for disconfirming data in both past and future observations. A single negative case is enough to require the investigator to revise a hypothesis. When there are no more negative cases, the researcher stops revising the hypothesis and says with confidence "This caused that."

This process of revising hypotheses with hindsight which gives the qualitative researcher confidence in his or her findings is the same process that gives a quantitative researcher doubts. It threatens the statistical conclusion validity of findings. Although qualitative researchers do not use statistics and therefore would not discuss statistical conclusion validity, I will examine the procedure from the point of view of a quantitative researcher, because negative case analysis still remains suspect from a hypothetico-deductive standpoint.

Statistical Conclusion Validity and Negative Case Analysis

Cressey (1953) reports in detail how he used negative case analysis to formulate and revise his hypothesis five times: "The first hypothesis . . . was that positions of financial trust are violated when the incumbent has learned in connection with the business or profession in which he is employed that some forms of trust violation are merely technical violations and are not really 'illegal' or 'wrong,' and, on the negative side, that they are not violated if this kind of definition of behavior has not been learned" (p. 27).

He formulated this hypothesis on the basis of other literature and research on white-collar crime. After interviewing only a
few inmates, however, he learned that they knew all along that
embezzling was illegal. He revised his hypothesis, therefore, to
read: "Positions of trust are violated when the incumbent defines
a need for extra funds or extended use of property as an 'emergency' which cannot be met by legal means" (p. 27).

Some embezzlers whom he subsequently interviewed said
theirs had been an emergency, but others said they had taken the
money even without a financial "emergency." Still others said
there had been financial emergencies earlier in their lives, when
they had not embezzled. Both of these groups contradicted the
second hypothesis, so Cressey developed a third: "It shifted the
emphasis from emergency to psychological isolation, stating that
persons become trust violators when they conceive of themselves
as having incurred financial obligations which are . . . nonsocially
sanctionable and which . . . must be satisfied by a private means"
(p. 28). He checked this hypothesis against both subsequent and
previous interviews. These checks revealed that "in a few of them
there was nothing which could be considered as a financial *obligation,* that is, as a debt which had been incurred in the past for
which the person at the present time felt responsible. Also, in
some cases there had been nonsanctionable obligations at a prior
time, and these . . . had not been alleviated by means of trust violations" (p. 28).

Cressey revised his hypothesis again. The fourth version differed by "emphasizing this time not financial obligations . . . but
nonshareable *problems* . . . that is, . . . the subject could be in
financial difficulty not only because of an acknowledged responsibility for past debts, but because of present discordance between
his income and expenditures as well" (p. 29). The fourth revision
accounted for men who had not developed debts but who had
been living above their means and had been afraid to admit this
to their families or friends. It also included some who had been
maintaining separate households without telling their family or
friends. Again, however, there were exceptions: men who said
they had experienced the nonshareable problem for a long time
before they embezzled. "Some stated that they did not violate the
trust at the earlier period because the situation was not in sharp

enough focus to 'break down their ideas of right and wrong' " (p. 30).

This led Cressey to the final revision: "Trusted persons become trust violators when they conceive of themselves as having a financial problem which is nonshareable, are aware that this problem can be secretly resolved by violation of the position of financial trust, and are able to apply to their own conduct in that situation verbalizations which enable them to adjust their conceptions of themselves as users of the entrusted funds or property" (p. 30). Cressey tested this hypothesis against all the data he had gathered, against two hundred cases of embezzlement collected by another researcher, and against additional interviews that he conducted in another penitentiary. He found no negative cases.

Negative case analysis produces the perfect pattern of results shown in Table 1. All four conditions necessary for embezzlement were present in 100 percent of the cases where embezzlement occurred, and one or more conditions were absent whenever embezzlement did not occur. The method ensures a perfect correlation, because the causal hypotheses are continually revised until they fit every case. This is precisely what researchers using the hypothetico-deductive approach consider illegitimate. If the hypotheses are revised to fit the data, the researcher capitalizes on chance, and the probability levels associated with the statistics are meaningless.

The conclusions of participant observation do not depend on comparing a signal-to-noise ratio, as do the conclusions of quantitative research. They are more like the conclusions of some operant conditioning research, which depend not on statistics but on graphic illustrations that demonstrate that a reinforcer works or does not work (Hersen and Barlow, 1976). When we use negative case analysis, we continue until there are no outliers or exceptions to the rule. This is very different from quantitative analysis, which incorporates outliers or exceptions to the rule in the random error term. When we use statistical analysis, we assume there will be error variance. Statistical tests are necessary when the ratio of explained variance to error variance is not obviously great. When the ratio is big enough so that the distributions do not overlap and the difference between treatment and no-treatment conditions can be seen with the naked eye, statistical tests are superfluous. This is the case with qualitative analysis, because there are no outliers, no random error variance. Negative case analysis eliminates all exceptions by revising hypotheses until all the data fit.

No statistical tests are necessary once negative cases are removed, because the data clearly fit the hypotheses. The obvious fit between hypotheses and data can make qualitative research seem either very good or very bad. When the conclusions are so clear that the reader says to him or herself, "Of course, now I see it," the work may appear to be true and insightful. When the conclusions are clear and the reader says, "Of course, it's obvious," the work sometimes appears trivial, because it tells the reader nothing new. The same is true of quantitative research. When the statistical tests are significant and the effects large, the findings may be either very clear and exciting, or very clear, obvious, and uninteresting.

When statistical tests are marginally significant and the effects are small, quantitative work may be interesting but is often unconvincing without additional evidence. Similarly, when the data for qualitative analysis are sparse, and there is not abundant evidence to support the hypothesis, even though there are no negative instances, the conclusions are weak and further evidence is needed.

Therefore, even though inductive analysis violates the assumptions of statistical testing by forming hypotheses to fit the data rather than finding data to test hypotheses, the consequences are not so different. When qualitative research has abundant evidence, it rings true, the conclusions are obvious, and they may be either interesting or trivial. When qualitative research has sparse evidence, it is not persuasive, even though there are no negative cases. Abundant evidence in qualitative research results when one has made many observations and recorded many instances. This is the equivalent of having a large N. The larger the N, the more convincing the conclusions in either case.

Negative case analysis replaces statistical analysis in qualitative research. Both are means to handle error variance. Qualitative analysis uses "errors" to revise the hypothesis; quantitative analysis uses error variance to test the hypothesis, demonstrating how large the treatment effects are compared to the error variance.

Construct Validity and Reliability in Qualitative Research

Construct validity is the appropriate naming of a variable, be it a cause or an effect (Meehl, 1977; Cook and Campbell, 1979). That is, the measurement of a variable must correlate with some other measurement of the same variable, and the theory underlying the variable must be correct. Campbell and Fiske (1959) introduced the multitrait-multimethod matrix as a tool for assessing construct validity. The matrix includes some correlations that should be low in addition to others that should be high as evidence of construct validity. One variable should have low or near zero correlations with another variable which is presumed to be different from the first, even though they are measured by similar methods; and it should have high correlations with other measures of itself, even though the methods of measurement are maximally different. In both cases, the validity rests on correlations between measures which differ in either the method or the trait. Reliability, on the other hand, is the consistency or similarity or replicability of observations. To assess reliability, we make repeated measurements of the same trait with similar or identical methods. Validity and reliability, therefore, represent opposite ends of a continuum of measurement, as shown in Table 2.

Table 2. The Validity-Reliability Continuum

Reliability	*Validity*
Maximally similar ———————————————— methods of measurement	Maximally different methods of measurement

Reliability coefficients indicate how much agreement there is between maximally similar methods of measuring a concept. Validity coefficients indicate how much agreement there is between maximally dissimilar methods of measuring the same construct.

Reliability and validity, broadly construed, are requirements for both quantitative and qualitative research, but the technology for assessing them has been developed primarily in quantitative research on individual differences. Calculating reliability and validity *coefficients* requires quantifying observations. Deciding whether concepts or observations are reliable or valid does not necessarily require quantification, however. Qualitative researchers make such decisions regularly, without the explicit calculations of reliability and validity coefficients. Mills (1951), in his book on white-collar workers, illustrated how qualitative researchers assess the reliability and validity of their observations in the absence of quantified measures. He developed the argument that clerical workers experience "status panic" as a result of the ambiguous position of their work. As a result, they "conceal the nature of their own work, and borrow prestige from the firm or industry, by identifying themselves with such phrases as 'I am with Saks,' or 'I work at Time.' They saved up their salaries and spent them for an evening at expensive places of entertainment, or for a vacation at a costly resort, in order to 'buy a feeling, even if only for a short time, of higher status.' A salesgirl dealing with 'Park Avenue' customers will try to behave with greater dignity and distinction in her off-the-job contacts than the girl who works on 34th Street" (Barton and Lazarsfeld, 1969, p. 242).

These repeated observations of what Mills called "status panic" lie somewhere on the continuum between reliability and validity checks. They are like reliability replications in that they are observations made by the same observer; they are like validity confirmations in that they are different manifestations of the same construct. Mills gathered many instances of status panic, not to demonstrate how frequent it is but to "demonstrate that the variable . . . does exist, that it exists in a number of manifestations, and that some of these manifestations have been observed among white-collar workers of various sorts" (McCall and Simmons, 1969, p. 244).

Repeated observations of a variable or concept in participant observation do not constitute frequency distributions. If something was recorded ten times, we cannot assume that the number 10 represents the frequency with which that event occurs in the population, because a participant observer normally does not select a random sample of people, times, or locations and obtain data from each unit in that sample to ascertain frequencies. Instead, the ten observations demonstrate that the variable exists. When participant observers list repeated instances of an event, the list serves as a reliability check. It shows that the event (and the variable it represents) occurs and that the concept is not based on a chance observation. For instance, in my research on a hypnosis workshop, I argued that the hypnotists rewarded people who behaved like good hypnotic subjects and punished those who resisted. As evidence I listed the following statements made by two different hypnotists at different times during the workshop:

(a) "Sometimes the *very analytic* person won't go along—he thinks to himself 'The hypnotist said I'm going to go to sleep, but hell, I'm not asleep, I'm wide awake, and he's crazy!' "

(b) "The most difficult one is the compulsive. . . . He also has the constant doubt 'Am I or am I not in trance?' Or the terribly rational person who cannot allow himself to go into any fantasy would be terribly difficult."

(c) "There is a myth about hysterics—that they are better subjects. But really, *normal* persons are better subjects."

(d) People who can throw themselves into art, nature, sports—involve themselves in a role and relinquish reality orientation for a time—these are the people who make good hypnotic subjects" [Kidder, 1972, p. 319].

These statements, and others, were made in response to questions from people in the workshop who said they did not feel they had been in a trance. They wondered whether it was "normal" to have difficulty going into a hypnotic trance. Rather than assure them that they were normal, the hypnotists countered by saying that they might be compulsive, very analytic, or terribly rational. Normal people are *better* subjects, said the hypnotists. The messages in these responses were clear to the questioners, all of whom were psychologists, "and given the public nature of the situation, no one could have ignored such rewards and punishments meted out to persons who became hypnotized and others who refused" (Kidder, 1972, p. 319).

The repeated observations of how people were rewarded for behaving like hypnotized subjects and punished for noncompliance can be regarded either as reliability or validity checks or as some combination of the two. The continuum between reliability

and validity shown in Table 2 is both a conceptual and a procedural continuum. Therefore, even in quantitative analysis, the distinction between reliability and validity becomes blurred when the
methods of measurement are neither identical nor maximally dissimilar. Both technically and conceptually, reliability and validity
are even less distinct in qualitative work. Nonetheless, multiple
observations of measurements of the same phenomenon are important in analytic induction, because each observation provides
further evidence for the argument.

    One reason that qualitative researchers are less concerned
with the precise form of measurement—and therefore cannot distinguish between reliability and validity checks—is that they are
concerned more with the content than the form of the observation. What matters is that each additional piece of evidence is *consistent* with the other observations and not that each observation
is *identical.* This is an important difference between quantitative
and qualitative procedures. Reliability in fieldwork lies in an observation's not being contradicted and proved wrong rather than
its being repeated in detail.

    What I have said about reliability and validity in measure-
ment in fieldwork is also true of the reliability and validity of conclusions. They rest on not being contradicted rather than on being
repeated. As a graduate student in social psychology at Northwestern, I occasionally heard Donald Campbell and Howard
Becker disagree over the virtues of exact replication. The problem,
as Campbell defined it, was that field workers are not interested in
replicating results; each would rather make his or her own unique
discovery. And even if one field worker tried to replicate another's
study, the attempt would probably fail, because the second researcher would probably discover something new and pursue that
rather than try faithfully to reproduce the original observations.
Becker considered Campbell's description accurate, but said it
presents no problem because the reliability of fieldwork lies in its
not being contradicted rather than its being repeated in detail. Instead of looking for the same phenomenon, one will do better to
look for new information that may be *consistent* or *inconsistent*
with the first. This is true for reliability across studies as well as
within a single study. Therefore, rather than test-retest reliability
or research-replication reliability, qualitative research calls for

something akin to an internal consistency measure of reliability. What matters in each instance is that there be no negative or inconsistent evidence. Negative case analysis exemplifies the procedure: The researcher "searches for negative or qualifying, as well as for supporting, instances. It has been said that science turns upon negative evidence," writes Strauss (see Strauss, Schatzman, Bucher, Ehrlich, and Sabshin, 1964, p. 21); and that is the rationale for the approach to reliability described above.

External Validity

Conclusions that are externally valid can be replicated across other persons, times, places, and operationalizations of the treatment and effects. The external validity of experiments and quantitative research depends on the researcher's demonstration that the same results can be found in other laboratories, or with another sample of subjects, or with other measures of the same variables. The external validity of qualitative research depends on the researcher's demonstration that similar results occur in other settings. In neither case does the researcher intend to make statements only about the subjects or observations made in that particular study.

Psychological experiments with college sophomores are never intended to be studies of college sophomores; they are intended to be studies about people in general. The same is true of qualitative research. McCleary did not intend his study of parole officers in Chicago to be read as an analysis that is true only of parole officers in Chicago, or even only of parole officers. He wished to make statements about social service agencies in general. He writes in his concluding chapter, "Rule breaking characterizes the official behavior of the social service bureaucrat" (McCleary, 1978, p. 171). Becker's study of marijuana users was not intended to be a study of those particular marijuana users, or even of marijuana users alone. The analysis says something about "the social interpretation of a physical experience which is . . . ambiguous" (Becker, 1963, p. 42). Seldom does a study purport to be about only the people or process examined in that study.

How do qualitative researchers achieve or assess external validity, particularly in the absence of replications? They do so by showing how the process they studied is *similar to* processes that occur in other places and with other people. For instance, my research on how people become hypnotized was a study not only of hypnosis but of the phenomenon Becker observed among marijuana users. I also found that novices had to go through a series of stages to become good hypnotic subjects. The first was learning to behave as though they were hypnotized. The second was redefining that behavior to attribute it to a trance state rather than to a conscious act. The third was redefining "hypnosis" to include the behaviors and feelings the person experienced while behaving hypnotized. The steps in becoming hypnotized are similar to the steps in becoming a marijuana user. The novice first acquires some new behaviors—trancelike behaviors for hypnosis and effective smoking behaviors for marijuana use. Next, the novice must attribute those behaviors to the force of hypnosis rather than to conscious volition. Finally, the novice must change his or her criteria for what can pass as hypnosis, and call even mild states of relaxation hypnosis.

Becoming hypnotized, I argued, is like changing one's attitudes about one's own experiences and behaviors. The change is accomplished through negotiations between the hypnotist and the subject. The hypnotist's first step is to convince the subject that he or she acted like a hypnotized person. The second step is to attribute the behaviors not to compliance or coercion but to the trance state. For instance, when a subject asked: "How do you know if you were in a trance or not? I mean, I know I did some things, but I think they were all under conscious voluntary control," the hypnotist replied: "I think you can tell if someone is in a trance by looking at them . . . the facial expressions. I could walk around the room and tell who wasn't and who was, by how they responded. I thought you were, but maybe you didn't *think* you were." A second hypnotist reinforced this by saying to the same questioner: "You were actually the one that I thought went into trance the quickest" (Kidder, 1972, p. 317).

In addition to pointing out to subjects that they looked as though they had been in a trance, the hypnotists pointed out that the subjects looked and acted that way not because they had been forced but because they were under the influence of the trance. When participants said that although they behaved like hypnotic subjects, they felt they were just "playing the game," the hypnotists pointed out that there were no external constraints that forced them to do so. The responsibility lay with the subjects, who were asked: *"Why* did you feel that you wanted to play the game?"

The exchanges between hypnotists and their subjects were negotiations about who was responsible for the subject's becoming hypnotized. The following are samples of negotiated responsibility:

> *Hypnotist:* Dr. Z tried to help those of you who weren't able to do the arm lift. He said, "For those for whom this was difficult, the arm can get very heavy" [so persons could let their arms go down instead of up].... He pointed out the successes instead of failures and gave other possibilities for achieving success.
>
> *Subject:* You make it sound as if it's the patient's fault instead of yours if he doesn't go into trance....
>
> *Hypnotist:* Well, let me say this. Earlier hypnosis was done in an authoritarian fashion—now it is much more permissive and we conceive of hypnosis as the achievement of the subject, in which the hypnotist helps.... Does this answer your question? [Kidder, 1972, p. 319.]

The hypnotist thus placed the responsibility with the subject. If he behaved like a hypnotized person, it was attributable to him or to his trance, not to external coercion.

The hypnotists negotiated several aspects of reality with their subjects. They negotiated the facts—whether or not someone appeared to have been in a trance; and they negotiated the locus of causality or responsibility for behaving hypnotized. I regard this as a study of negotiated realities and consider it similar to other studies of the social construction of reality (Berger and Luckman, 1967). The process I studied is like the process that Becker (1963) studied in his report on individuals who become marijuana users, and like the process Scheff (1968) studied in his report of therapist's and patient's negotiations and assessments of responsibility. I am less concerned with generalizing to other hypnosis workshops than I am with generalizing to other situations where two or more parties negotiate definitions of what happened and who was responsible.

The external validity of research depends on what the researcher claims to have discovered. In the study of this particular hypnosis workshop, I claim to have discovered something about attitude change and negotiated responsibility more than to have discovered anything about hypnosis in general. It is likely, as the hypnotists themselves pointed out, that other, more "authoritarian" hypnotists would not place responsibility with their subjects, and the process of becoming hypnotized may be quite different. Rather than generalize to other hypnosis workshops which I have not observed, therefore, I wished to generalize to other phenomena which may appear very different on the surface but which share the stages and the negotiations that I described.

By saying this was not a study of hypnosis alone, I am not saying that no one could replicate the research or test it by examining another hypnosis workshop or by examining the same tape recordings that I analyzed. Instead, I contend that the external validity of my findings depends on their similarity to findings from very different settings, such as learning to become a marijuana user or learning to accept someone else's definition of one's ailment. This is not so different from the meaning of external validity for laboratory experiments. Their external validity depends not only on demonstrations that the same results can be obtained in another laboratory with another group of subjects but also on the apparent similarity between the laboratory observations and observations in a nonlaboratory "real-world" setting. With the exception of experiments designed to study the social psychology of the psychology experiment (for example, Orne, 1962, 1969), laboratory research is not meant to be generalizable only to other laboratories. It is designed to be an analogue of social processes outside the laboratory, which may bear no surface similarity to the laboratory procedures.

To demonstrate the external validity of any research, we must actually replicate it, varying either the subjects, the setting, the operational definitions of treatments and effects, or all of these. To assess the external validity of research short of replicating, we work inductively, and we compare the procedures and results of the study in question with the process to which the researcher wishes to generalize. What matters is not their surface similarity but the apparent similarity of their processes, structure, or meaning. Since this comparison depends on inferred rather than obvious similarities, "the question of external validity, like the question of inductive inference, is never completely answerable" (Campbell and Stanley, 1966, p. 5).

Conclusions

I have argued that qualitative research can be not only "deep" and "rich" but also "hard." The same criteria that we apply to quantitative research—criteria for internal, external, and construct validity—can be applied to qualitative research. These criteria are not relevant for all qualitative research, only for that research which contains causal assertions. When qualitative research contains causal assertions, we can discern an implicit quasi-experimental design in the research. Since participant observation is seldom explicitly designed, the timing and frequency of observations rarely match exactly any of the quasi-experimental designs illustrated in the literature. They most nearly resemble multiple time-series designs, however, because they are frequently longitudinal observations of a socialization career. The multiple observations and nonsimultaneous treatments permit the investigator to rule out threats to external validity. Although this has nowhere been done explicitly, I demonstrated how qualitative researchers can and perhaps do rule out rival explanations in arriving at their conclusions about the necessary steps, stages, or conditions for socialization to proceed. These steps or conditions are the alleged causes. They are seldom single causes, and the conclusions do not say "This caused that." They are multistage, multideterministic analyses.

The construct validity and reliability of participant observation depend on the researcher's demonstration that there are multiple instances of a given construct. These multiple observations act like the multiple items or repeated measurements of a quantitative scale—they demonstrate that the construct exists. They demonstrate the reliability of the observations. The methods of making such observations are more varied than identical, so the multiple observations also provide convergent validation. Since procedures for demonstrating reliability and validity differ in degree, and reliability and validity lie on a continuum, I am not violating our understanding of these terms when I say that the multiple observations of a construct may provide evidence of reliability or validity or something in between.

The external validity of qualitative research is ascertainable in the same way as the external validity of quantitative research. I do not claim that qualitative studies outrank quantitative studies in external validity, because naturally occurring events or field settings do not automatically confer high external validity. Field researchers may wish to generalize not to similar field settings but to similar processes in very different settings, as was the case with my study of a hypnosis workshop. The external validity of qualitative research, like that of quantitative research, depends on the underlying rather than the surface similarity between the process studied and the processes the researcher names as analogues.

Qualitative research can be assessed by the same criteria as quantitative research. The logic of internal, external, and construct validity is the same, regardless of whether the researcher uses words or numbers. Good qualitative research, like good quantitative research, is both rich and rigorous.

References

Barton, A. H., and Lazarsfeld, P. F. "Qualitative Support of Theory." In G. J. McCall and J. L. Simmons (Eds.), *Issues in Participant Observation.* Reading, Mass.: Addison-Wesley, 1969.

Becker, H. S. "Becoming a Marihuana User." In *Outsiders.* New York: Free Press, 1963.

Becker, H. S., and Geer, B. "Participant Observation and Interviewing: A Comparison." *Human Organization,* 1957, *16* (3), 28-32.

Berger, P. L., and Luckman, T. *The Social Construction of Reality.* Middlesex, England: Penguin Books, 1967.

Campbell, D. T. "The Mutual Methodological Relevance of Anthropology and Psychology." In F. L. K. Hsu (Ed.), *Psychological Anthropology: Approaches to Culture and Personality.* Homewood, Ill.: Dorsey Press, 1961.

Campbell, D. T. "Qualitative Knowing in Action Research." Kurt

Lewin Award Address presented at 82nd annual meeting of the American Psychological Association, New Orleans, Sept. 1, 1974.

Campbell, D. T. " 'Degrees of Freedom' and the Case Study." *Comparative Political Studies*, 1975, *8*, 178-193.

Campbell, D. T., and Fiske, D. W. "Convergent and Discriminant Validation by the Multitrait-Multimethod Matrix." *Psychological Bulletin*, 1959, *56*, 81-105.

Campbell, D. T., and Stanley, J. C. *Experimental and Quasi-Experimental Designs for Research*. Chicago: Rand McNally, 1966.

Cook, T. D., and Campbell, D. T. *Quasi-experimentation: Design and Analysis Issues for Field Settings*. Chicago: Rand McNally, 1979.

Cressey, D. R. *Other People's Money: A Study in the Social Psychology of Embezzlement*. New York: Free Press, 1953.

Glaser, B. G., and Strauss, A. L. *The Discovery of Grounded Theory*. Chicago: Aldine, 1967.

Hersen, M., and Barlow, D. H. *Single-Case Experimental Designs: Strategies for Studying Behavior Change*. Elmsford, N.Y.: Pergamon Press, 1976.

Kidder, L. H. "On Becoming Hypnotized: How Skeptics Become Convinced: A Case of Attitude Change?" *Journal of Abnormal Psychology*, 1972, *80* (3), 317-322.

Kidder, L. H. "The Inadvertent Creation of a Neocolonial Culture: A Study of Western Sojourners in India." *International Journal of Intercultural Relations*, 1977, *1* (1), 48-60.

Lindesmith, A. R. *Addiction and Opiates*. Chicago: Aldine, 1968.

McCall, G. J., and Simmons, J. L. *Issues in Participant Observation*. Reading, Mass.: Addison-Wesley, 1969.

McCleary, R. "How Parole Officers Use Records." *Social Problems*, 1977, *24* (5), 576-589.

McCleary, R. *Dangerous Men: The Sociology of Parole*. Beverly Hills, Calif.: Sage, 1978.

Meehl, P. E. "Construct Validity in Psychological Tests." In *Psychodiagnosis: Selected Papers*. New York: Norton, 1977.

Mills, C. W. *White Collar*. New York: Oxford University Press, 1951.

Orne, M. T. "On the Social Psychology of the Psychological Experiment: With Particular Reference to Demand Characteristics and Their Implications." *American Psychologist,* 1962, *17,* 776-783.

Orne, M. T. "Demand Characteristics and the Concept of Quasi-Controls." In R. Rosenthal and R. L. Rosnow (Eds.), *Artifact in Behavioral Research.* New York: Academic Press, 1969.

Rosenthal, R., and Rosnow, R. L. (Eds.). *Artifact in Behavioral Research.* New York: Academic Press, 1969.

Scheff, T. J. "Negotiating Reality: Notes on Power in the Assessment of Responsibility." *Social Problems,* 1968, *16,* 3-17.

Sherif, C. W. "Bias in Psychology." In J. A. Sherman and E. T. Beck (Eds.), *The Prism of Sex: Essays in the Sociology of Knowledge.* Madison: University of Wisconsin Press, 1979.

Strauss, A., Schatzman, L., Bucher, R., Ehrlich, D., and Sabshin, M. *Psychiatric Ideologies and Institutions.* New York: Free Press, 1964.

Zelditch, M., Jr. "Some Methodological Problems of Field Studies." *American Journal of Sociology,* 1962, *67,* 566-576.

Znaniecki, F. *The Method of Sociology.* New York: Holt, Rinehart and Winston, 1934.

10

Thomas D. Cook

Dilemmas in Evaluation of Social Programs

This chapter explicates a dilemma that is important for an understanding of past problems encountered in the evaluation of ongoing social programs. My thesis is that nearly all ongoing social programs are extremely variable in the activities implemented at the local project level—so much so that there can be negligible overlap between the activities that occur in different projects funded from the same program. I also contend that outcome-oriented evaluative methods are not sensitive to such variability and

Note: This chapter was partially funded by the Food and Nutrition Service, U.S. Department of Agriculture, Contract # 53-3198-9-26. Grateful thanks are extended to Laura Leviton, who will co-author with me a response

depend on having preselected for study a very restricted set of apparently homogeneous activities. Given the number and heterogeneity of local projects in most programs, I am skeptical about the desirability of focusing program evaluation on the program as the unit to be analyzed.

Fortunately, some sources of variability *within* a program represent already implemented approaches to solving a designated social problem. I suggest that evaluation might be used to assess whether some successful practices can be selected from among manipulable, policy-relevant activities that occur within a program and that stand a reasonable chance of being implemented elsewhere. I focus on transferable local projects, rather than on transferable elements within projects, as the unit to be analyzed. However, I freely recognize that in some circumstances project elements are more important than whole projects, that sometimes project elements are all that can be analyzed, and that for some stakeholders in evaluation (for instance, project managers) project elements are nearly always more important than whole projects. Whatever the preferred unit of transfer to other sites, I contend that the evaluation of ongoing social programs threatens to be an enterprise built around sophisticated procedures for selecting possible solutions to problems (for example, the social experiment or structural equation modeling), but with the selection often coming from a curtailed distribution of the possible solutions that can be found in the typical, heterogeneous social program.

Defining Terms

In the United States governments design social policies to further goals that they and many others consider important—goals such as the promotion of economic growth, universal literacy, or safe driving. In the social domain national or state-level *programs* are developed for bringing about policy goals. Several programs often are targeted at the same apparent problem. Thus, if universal literacy and numeracy are goals, Title I is developed, *Sesame*

to the dilemma presented here and who helped recast and document some of the first ten propositions. The incisive editorial input of Marilynn Brewer, Barry Collins, and Donald Fiske helped considerably in focusing the argument.

Street and *The Electric Company* are funded, and Follow Through is designed. If the revitalization of cities is a goal, Community Development Block Grants are let, and General Revenue Sharing becomes a reality. I could go on and on with examples. The point is that we shall refer to programs as agency attempts to coordinate activities in order to realize policy goals.

The activities that most directly shape the behavior of potential program beneficiaries usually occur at a local level. Though bureaucrats set the objectives for Title I in education and write regulations designed to realize these objectives, ultimately it is local officials who decide which children should receive Title I services and who then manage the services and teach the children. In nearly all programs, the manner and context in which services are provided reflect local concerns, and considerable heterogeneity exists from site to site in which mix of services is provided. Such heterogeneity suggests the need to invoke a concept like *local project* in order to designate each of the different local organizations that is supported from a particular "program." In essence, this casts the program as a funding umbrella more than as a blueprint for highly specific actions that are to impinge directly on beneficiaries.

I realize that a few federal programs do not have local projects in any important sense. Social security is an obvious instance, as are other programs that directly transfer cash from governmental agencies to individuals (Hargrove, 1980). However, such programs are rare when compared to service delivery programs. I am also aware that the word *program* is often used very broadly—even to designate what I have called a local project. Indeed, I think that many practitioners and clients refer to the services they provide or receive as coming from a local program—say, the Head Start program in Evanston. It is unfortunate that the word *program* is often used to refer to both the control agency initiative and the local realization of that initiative; for, as I hope to point out, such confusion obscures important lessons and complicates communication. Finally, I am conscious that many local organizations funded from a national program are themselves internally heterogeneous. For instance, the national program of General Revenue Sharing is deliberately designed so that local municipalities have maximal freedom, subject to certain accountability constraints, to spend their funds as they want. Thus, different sets of activities emerge between municipalities. But perhaps just as important is that each city does many different things with its funds, and what it does can shift from year to year. To say that local projects differ from each other even though they are funded from the same program in no way minimizes the problem that each local project is itself a complex mix of often shifting activities. Though the activities of local projects are typically more homogeneous than those of national or state-level programs, they are far from being homogeneous in any absolute sense.

In the case of programs that deliberately foster variability both within and between the organizations receiving funds, my inclination is to designate as the local project that set of presumably coordinated activities that is aimed at directly touching potential beneficiaries for the purposes of fulfilling a specific goal. In the context of General Revenue Sharing, for example, this would mean the street repair project in Evanston rather than all of the many different types of activities financed through General Revenue Sharing funds. The major distinction is that *programs* are administrative entities that concentrate on disbursing funds and regulations, while *projects* are organizational vehicles which directly provide services to the ultimate target audience of beneficiaries. Programs are part of the context of projects; but the two can be differentiated.

I have repeatedly referred to *ongoing programs*—meaning, at the federal level, programs approved by Congress and already implemented by agencies of the executive branch. Such programs should be distinguished from demonstration projects, where the aim is to develop and test a model for the delivery of a particular set of services. Ongoing projects should also be distinguished from other attempts to test plans for possible programs, as with so-called "social experiments" (Riecken and Boruch, 1974). In neither of these cases are already implemented programs at issue— programs that have developed powerful supporting constituencies and have probably evolved many significant local variants on how services are provided.

What is *evaluation*? This is not the place to attempt a formal definition. Instead, I want briefly to list the types of questions that evaluators try to answer, for a study aimed at any subset from among these question types would qualify in my eyes as an evaluation. One type of question concerns the demography of services, and issues here relate to such matters as the number and nature of the clientele and service providers of a program or the extent to which target groups are among the clients. A second type of question involves implementation. Among other things, it is necessary to describe the nature, frequency, and quality of the delivered services and to describe the context in which services are received. A third type of question involves effectiveness, and here the major question is how clients who receive services are affected by them in both expected and unexpected ways. A fourth type of question is about impact, and for me this involves questions about how higher-order aggregates (such as families, neighborhoods, and towns) are affected by the services. Such questions are important since many programs and projects strive for impact and since many local projects do not try to reach all the persons in the aggregate. The fifth set of questions refers to causal process, and for both the effectiveness and impact domains the crucial question is: *Why* did the services affect the particular outcome measures in the manner they did. Knowledge of causal processes is useful for replicating findings and for modifying services so as to increase their efficiency. A sixth set of questions concerns financial costs—the total cost of a program or project; the cost per client per year; and the cost-effectiveness of different ways of achieving a particular result. I am suggesting that any study aimed at answering any question from among these question types counts for me as an evaluation study.

Features of Ongoing Social Programs and Local Projects

To explicate some important features of most ongoing social programs, I shall discuss a series of propositions about programs, particularly stressing for each proposition its implications for evaluation policy. The purpose of the explication is to try to locate leverage points where evaluative information might play a significant role in influencing substantive policy about ongoing social programs. Since programs consist of local projects in which services are delivered, project evaluations lie at the heart of program evaluation. I shall therefore also attempt to explicate the crucial features of local projects.

Proposition 1. Ongoing programs rarely die. Federal programs are rarely funded unless one or more politically powerful constituencies support them; and, once implemented, they typically develop additional support. Thus, after revenue sharing began, many cities funded Washington offices, apparently to protect their new funds (Yin, 1980). Since programs have such political ties, it is nearly impossible to terminate them. Certainly, to our knowledge, it is currently impossible for negative evaluation results to lead to the termination of programs. Consequently, the evaluation of social programs cannot be justified on grounds that the resulting information is used to decide the ultimate fate of programs.

Proposition 2. While program budgets are sometimes marginally affected by evaluations, typically they are not. At appropriations time, congressional committees routinely question administrators about the consequences of their programs, and administrators often back up their requests for funds with information derived from evaluations. The data are doubtless often used to justify decisions that have already been made on other grounds, which I consider an entirely legitimate form of usage. In some instances, closer examination may reveal that the data played a more active role in determining decisions; for example, Upward Bound may have received more funds from Congress than the administration requested because of evaluation, and a negative evaluation of nonprofit organizations that helped with school desegregation probably contributed to the reduction of their budget by two thirds (Leviton and Boruch, 1980). Moreover, Rein and White (1977) have argued that the growth of Head Start funds was slowed down by negative evaluation results.

Yet despite these probable counterinstances, the scholarly literature suggests that evaluations are rarely used in such instrumental fashion (for a review, see Cook and Leviton, in press). Instead, evaluations tend not to be used at all or to be used in more diffuse ways that influence the thinking of decision makers without affecting specific funding decisions. Weiss (1977) refers therefore to social science knowledge as fulfilling more of an "enlightenment" than a decision-forming function.

Propositions 1 and 2 suggest that it is difficult to defend the position that the proper business of program evaluation is the evaluation *of* ongoing programs. But, as the next propositions suggest, the proper business of program evaluation may be evaluation *within* ongoing programs.

Proposition 3. Most social programs consist of local projects, and there is a regular turnover of such projects. In most programs, new projects are continually beginning and old projects are being revamped. One reason for this is that some ongoing programs grow, and this growth is typically manifested as the funding of new projects at new sites. Also, top-level personnel change at the local level, because of retirement, promotion, and the like. Such changes are often associated with new directors bringing in new theories or new ways of fulfilling old functions, especially when the explicit motivation for change is to bring in "new blood." Even when change is not initially expected from a new director, she can nonetheless feel pressure to differentiate what she does from what her predecessors did, making her seek out changes to make in the project. Established project directors also want change for its own sake sometimes, and sometimes they feel pressured into promoting a major change effort because evidence has mounted to indicate that change is needed.

It is not yet clear how often new projects are set up or how often project philosophies and fundamental practices change in ongoing projects. The viability of considering the project as the unit of major focus in the evaluation of ongoing programs depends on the frequency of the spontaneous (or easily induced) turnover of projects. Certainly the program itself does not seem to offer much potential as a source of leverage.

Proposition 4. The implementation of federal programs is heterogeneous at the project level. Local governments control the proximal delivery of services in most cases, irrespective of whether a program is funded federally or more locally (Danielson, Hershey, and Banes, 1977). The form of these services reflects local needs and preferences (Pressman, 1975; Nathan, 1979). Danielson, Hershey, and Banes describe this fact of political life: "State and local governments play a significant role in the nation's affairs because of the nature of the American federal government. Unlike regional and local governmental units in most other nations, the states and localities are not merely instruments for the implementation of national policies. Instead, the American federal system devolves real power to its components. State and local governments

make meaningful policy choice, allocate public resources, and resolve conflict without the involvement of the national government" (p. 4).

A consequence of decentralized authority for programs is
that they are usually heterogeneous in the form in which they
reach the public. Some programs, such as revenue sharing, were
created with this heterogeneity in mind. Others, such as compensatory education, were not planned to be heterogeneous, but time
has revealed that they are (Berman and McLaughlin, 1978; Hawkridge, Chalupsky, and Roberts, 1968; National Institute of Education, 1977).

Much of the heterogeneity arises because of local differences in service providers and clients and because of local peculiarities in the political and administrative context of the project.
However, projects also differ because they take different paths in
attempting to reach the ultimate objectives of the program that
funds them. These different paths are embodied ("operational-
ized") as activities aimed at lower-order goals, which are typically
set by local project personnel themselves in the hope that reaching
the lower-order project goal will eventually help reach the ultimate
program goals (Suchman, 1967). For example, the ultimate goal of
the Community Block Development Grants is the alleviation of
distressed urban areas (Nathan, 1979). Project personnel may try
to reach this ultimate goal by setting their own goals for community improvement, such as housing rehabilitation, new housing, or
public works of many different types. These goals (which Suchman calls "intermediate") can themselves be met in many different ways. Thus, to select only one of them, housing rehabilitation
may involve, for example, owner rehabilitation, contractor rehabilitation, gutting of inner structures, instituting minor repairs, training consumers in rehabilitation, or offering loans and matching
grants. Goals at this last level Suchman calls "immediate," and it is
clear that immediate goals can vary enormously from project to
project, even though all of the projects are part of the same program and purport to be pursuing the same ultimate objectives.

*Proposition 5. For most ongoing social programs little sys-
tematic documentation of project activities exists.* Interest in how
federal money is spent seems to be growing. Monitoring requirements were attached to the Comprehensive Employment and

Training Act (CETA) (Van Horn, 1979) to Law Enforcement Assistance Administration (LEAA) Block Grants (Lavrakas and Normoyle, 1980), and even to the most decentralized program, General Revenue Sharing (Van Horn, 1979). Even in such sectors as health and education, which are by their nature decentralized (Banta and Bauman, 1976; Farrar, DeSanctis, and Cohen, 1979), concern over cost in federal health programs has fostered a need for central control, which is expressed in regulation and legislation (Enos and Sultan, 1977).

Federal monitoring and enforcement efforts are necessarily limited, since containing the size of the federal bureaucracy is a political issue on which both parties agree (Cook and Buccino, 1979), communication within bureaucracies is impeded by the selective passage of information (Downs, 1967), and the political will to use federal muscle against local forces is often lacking. Consequently, federal bureaucrats are usually chronically underinformed about the programs they administer, and local bureaucrats know this and use that knowledge to suit their own needs (Blau, 1955; Pressman and Wildavsky, 1973). Nathan (1979, p. 11) makes the point: "Try to get information about what is happening under any one of your favorite federal grant-in-aid programs. The Feds—in many cases—simply do not know. Not only do they lack sophisticated control mechanisms with which to implement grandiose grant plans, they simply do not know what is being done. . . . They are likely to have relatively little information about specific programmatic uses of these funds and even less information about the effects and effectiveness of the dollars spent."

An important implication is that decision makers would be helped if evaluations simply described the nature and range of project-level activities that take place within a social program. Traditional accountability goals would thereby be served, and feedback about project-level activities would considerably help federal and state managers, who are often asked to describe their program in detail. Indeed, some evidence exists at the national level that descriptive information about the services and social groups on which funds are actually spent can directly lead to changes in regulations (at least in education; see Hill, 1980) and to changes in agency policies (at least in mental health; see Cook and Shadish, in press). Although such information is not of a causal nature, some descriptive information about projects is probably a precondition for the responsible assessment of causal consequences. Without it, one does not know what is being evaluated at the local level; and at the more global program level, knowledge of projects is necessary if evaluators are somehow to choose which of the local projects they should study in detail.

Proposition 6. Not all sources of program variability are equally important for evaluation purposes. Fortunately, it is not the business of program evaluation to try to "explain" all the sources of variation in major program or project outcomes. Donald Campbell and I have argued elsewhere (Cook and Campbell, 1979) that a primacy exists in applied social research on discovering the consequences of those sources of variability that can be manipulated at will, that are—or promise to be—reasonable in cost, and that can probably be transferred with success to other sites. The emphasis, then, is on policy-relevant, transferable manipulanda, as opposed to structural aspects that cannot be varied at will. In many cases, variables that cannot be changed include the clientele, the personality of project developers and their often enthusiastic "true believer" staffs, and many of the political and administrative factors within which the program or project operates.

Even though a concentration on manipulanda reduces the range of variables to which evaluators' attention needs to be directed, a large number of real and potential manipulanda nonetheless exist in most programs and in most projects funded from a program. At the program level, the manipulanda include ways of phrasing regulations, of monitoring for compliance with regulations, and of coordinating evaluation activities. At the project level, potential manipulanda include ways of recruiting clients, of structuring the services that are delivered, and of conducting and reporting evaluative information. As a result, evaluation policymakers, or evaluators themselves if policymakers are not explicit enough, have to try to determine which of all these possible manipulanda they should concentrate on. Obviously, some can be eliminated from consideration quite easily, perhaps because they are too costly, require personnel who are not available, or entail material that is not readily available. But even after some manipulanda are excluded on grounds of irrelevancy, many remain, and a decision has to be made about which to concentrate on.

We shall see later that many different viewpoints exist on this issue. Some persons stress that most decisions about evaluative questions should be in the hands of project directors; others stress that the search should be for successful and transferable project elements, such as ways of implementing a management information system, or of recruiting new clients, or of structuring intake services. My stress is embarrassingly more grandiose. It is aimed at discovering the types of projects that can be transferred, with suitable modifications to accommodate local needs, to sites that are just starting up or to sites where the need is felt for fundamental changes in the project activities. Thus, I am prepared to make the tentative assumption that the variability of greatest utility for evaluation policy is variability in the theory behind different classes of projects, as opposed to variability in the specific activities that occur within projects.

Proposition 7. A tentative and corrigible causal theory of each project can be inferred. A tentative causal theory of project functioning can be inferred from any combination of the following: the written rationale for a project, which is often contained in funding and interim proposals and is based in some part on existing substantive theory; discussions with key project personnel, who may have developed their own rationales for relating project activities to immediate and intermediate goals; and direct observation in the project setting of the activities of service providers and clients. An elaborated causal theory should do more than suggest the role that each project activity is supposed to play in the presumed causal path to intermediate goals. Particularly when it is based on direct observation, the theory should also suggest forces that might facilitate or hinder the presumed causal flow. The latter forces are particularly important as variables that specify the contingencies on which a causal relationship is thought to depend. I am prepared to assume, therefore, that from a variety of sources each evaluator can develop his or her own working understanding of the contingency causal theory that buttresses, however implicitly, the design of a project and of the activities it promotes.

Proposition 8. Project types can be inferred from overlapping theories and activities. Some projects within a program may have substantially overlapping theories and major activities. These permit the projects to be considered as belonging to the same type. For example, several home health-care projects share a team approach, with the activities of individual team members being comparable to a greater or lesser extent across projects (Trager, 1973) and with the exact nature of important similarities and differences being presumably amenable to behavioral description. Without such description one might be able to learn which projects are supposed to have overlapping theories and activities; but one could not be sure whether the causal theory of project process has been modified in practice or whether the activities designed to "operationalize" the theory are deficient and fail to implement what was planned.

Examples may help clarify what a "project type" is. Primarily because of the costs of long-term hospitalization, interest has increased in ways of keeping elderly or disabled persons out of hospitals. One way, emphasizing home-help services, is built around the following theory: Elderly or disabled persons can be made aware of home-help services and will request them. Moreover, health personnel can visit the home and do any combination of four activities: (1) They can clean the house and ensure that the home remains physically fit for the elderly or disabled person to remain in; (2) they can offer a companionship that prevents the deterioration of social skills; (3) they can offer counseling and modeling in areas of self-management (such as nutrition and personal grooming) that make the person more independent; and (4) they can detect early symptoms of physical problems, which they then bring to the attention of medical authorities. All the above factors are presumed to reduce the number of subsequent days of hospitalization for the elderly person. Various contingencies are specified by some commentators as being crucial for the above theory to work, including an administrative tie to a hospital or the formal coordination of the home-help services with other services, primarily medical.

Home-help services do not constitute the only type of project designed to keep the elderly out of hospitals. Another, predicated on the viability of adult day care, has the following postulates: (1) The elderly can be made aware of, and will accept, adult day-care services. (2) Transportation is easily available to centers on a continuing basis. (3) Adult day-care services provide companionship, a good meal, and the chance to detect gross physical and mental problems. The last factors are all independently supposed to reduce hospitalization. Once again, contingency conditions of a crucial nature have been postulated, dealing mainly with the quality of services and their ability to sustain the motivation of clients to return for adult day care.

For both the home-help and day-care services, one can think of projects that appear on paper to be similar but that have little overlap in the activities that actually take place. For instance, the persons providing home-help services may differ considerably in whether they perform some of the functions listed previously, in whether they behave in ways that actually promote goals they are supposed to be reaching, and in whether they provide additional services for which they do not bill. While it would be useful to consider such factors as "heterogeneous irrelevancies" within a project type, some of them may be absolutely crucial for success, however success is defined. It follows, therefore, that evaluators should strive to sample a wide range of well-implemented, conceptually different instances within a presumed project type; and they should not expect all examples of a type to be successful even if some are.

Additional complexities arise in considering project types with more elaborated theories of causal process than the ones I have just mentioned. Consider compensatory education. Here, nearly all the programs and projects are aimed at increasing children's achievement and academic self-concept, mostly on the assumption that there is a relationship of mutual causation between the two. However, some projects are initially aimed more at stimulating academic achievement than academic self-concept, others have the opposite emphasis, and yet others are initially aimed about equally at stimulating both academic achievement and academic self-competence. Within these emphases, various subtypes might be differentiated. For instance, some projects with an initial emphasis on achievement relate the child's learning problem to the low motivation of middle-class teachers to teach poorer children, and so they use computers or programmed learning materials in ways that reduce the teacher's role. Other projects stress the need for children simply to spend more time in instruction with teachers, while others stress modifications in instructional practices—for example, more drilling or the use of new curriculum.

Doubtless within any one of these approaches, finer subcategories could be differentiated; for instance, several curricular approaches to teaching seventh-grade English might exist.

At first glance, the compensatory education example seems to reflect a domain with a particularly large number of potential project types. However, the number of types decreases once certain crucial decisions are made. Perhaps the major one concerns the extent to which ends, process theories, or modes of delivery are stressed in defining project types. In the compensatory education context, I mean by ends whether achievement is stressed over academic self-concept; by process theories I mean whether behavior modification principles are preferred to other principles, such as those based on rote memory or on active discovery; and by modes of delivery I mean whether the consumer of evaluations is interested in learning how teachers compare to computers or programmed texts or in how standard English differs from some vernacular as the medium of instruction. My preference is clearly for definition in terms of theory, though I acknowledge that theories are often related to particular ends and methods. Thus, behavior modification principles are usually targeted more to fostering achievement and the use of programmed learning texts than to fostering a child's academic self-concept or the use of discovery methods of learning. In any evaluation the concepts stressed in a working definition of types will depend on the nature of the projects implemented and the use to which the evaluative information is to be put. Consequently, I do not want to be dogmatic about preferring a definition of project types in terms of process theories, but I do want to be dogmatic about the difficulty of developing simple definitions of project types.

Proposition 9. A finite number of project types exist for any program. On an inductive basis, I assume that, for any group of program-funded projects, the range of process theories is limited. Even though the number of project types can sometimes be large, the number of methods used by practitioners is probably small and is restricted by training, experience, and the limitations of existing theories and human inventiveness. If the number of types was not generally small, one would find clinicians inventing their own therapies, police their own criminal justice systems, and teachers their own reading materials. While this can happen, most practitioners rely on existing practices and arrangements, though with their own modifications in some areas of their work. I am treating such modifications as irrelevancies, across which evaluators would like to generalize in their search for transferable solutions to problems.

To illustrate some of the basic project types, consider the case of home health care, where the causal theories seem to be based on (1) adult day care in the home or in centers where social services are supplied; (2) a strictly medical model, such as that of the visiting nurse associations; and (3) a model based strictly on voluntary organizations. Combinations of all these models are, of course, also possible. If one looks to individual psychotherapy, only a limited number of theories of the path from particular sets of therapist activities to curing schizophrenia are postulated. And I have been led to believe that in urban housing policy there are only a few theories of how to build public housing so as to have satisfied residents and other citizens.

Proposition 10. Some project types are likely to be associated with a number of successful project exemplars. In most programs multiple projects can be categorized as belonging, to varying degrees, to a particular type. The success of a type can be inferred when several of its exemplars are themselves evaluated as "successful." The evaluator cannot expect all project-level exemplars of a type to be successful. This is because (1) some project-specific procedures for accomplishing a particular subgoal are not effective, though alternative procedures for achieving the same end that are implemented elsewhere may be; (2) some activities that are generally effective do not happen to be implemented well in particular projects; and (3) the idiosyncratic nature of client and practitioner groups at some sites means that effects are not found there that are found with more modal client and practitioner groups. I do not want to be specific about the number or percentage of projects that have to be considered "successful" for the project type to be considered so. Obviously, one's confidence increases with the percentage, but only if the successful projects are not distinguished from the unsuccessful ones in sharing irrelevant biases that operate in the same direction.

Proposition 11. Knowledge of successful project types facilitates the transfer of project-level activities to other sites. In propositions 6 through 10, I have briefly described a rationale for considering the project as an important leverage point in evaluation and have tentatively suggested that projects can be grouped into categories about which knowledge is possible. I have, in essence, enjoined evaluators to look for a concept of project type that increases the probability of finding replicated solutions across projects that share a common theory of the path from initial activities to outcome. The projects in question do not have to have equivalent activities. Indeed, they may well have quite different ways of reaching the same subgoal. But they should have similar functional subgoals.

The purpose of the analysis is to suggest that knowledge about project types is particularly useful because it has to be based on findings that have been replicated across the multiple heterogeneous irrelevancies that are inevitably associated with each exemplar of a project type. Critically interpreted knowledge about a solution that was multiply obtained in the past holds more promise for successful transfer to other sites in the future than does knowledge about the effects of a single project. However, knowledge of project types does not guarantee successful transfer to particular sites elsewhere. It merely makes such transfer more likely.

Propositions About the Conduct of Evaluation Research

The propositions that follow are derived from the painfully acquired experiences of many practitioners and commentators on evaluation. They reflect some of the major mistakes made in evaluation since about 1965, as well as some suggestions about how evaluation might be improved.

Proposition 12. Multiple constituencies have a stake in any evaluation, and their interests should be represented in the work. In the past, narrow evaluative questions have been asked that reflect the concerns of only one of the major groups with a stake in the evaluation. Typically, these have been the concerns of federal officials. It is obvious that their interests in evaluation are not necessarily isomorphic with those of other stakeholders. In education, for example, federal officials have typically wanted to know how a program affected achievement; school superintendants have wanted to know about financial accountability; teachers have wanted to know how they could improve their practices; and designers of educational materials have wanted to know how they could improve their products. It is not totally clear what parents have tended to want to know. There is typically some overlap in the questions each constituency would like to have answered; but even among the overlapping questions priorities can differ from one stakeholder group to another. The major implication of this proposition is that evaluators will have to work with multiple indicators of "success" that permit each group to feel it has at least some stake in the evaluation.

Proposition 13. For each project the theory of the process from inputs to outcomes should be explicated, and the explication should influence the design and the measurement framework. Projects with supposed common aims differ considerably in the assumptions that are made (not always so explicitly) about how some designated effect can be brought about. Sometimes, laying out the preliminary theory of the process from manipulanda to outcome will be sufficient to indicate the inadvisability of evaluation, especially when crucial links in the causal chain are "known" to be wrong. At other times, analysis of the presumed causal links and of factors that might condition their operation will suggest constructs worth measuring that might otherwise have been overlooked.

Proposition 14. Evaluation plans should be general guidelines and not rigid prescriptions. Murphy's Law teaches evaluators to expect the unexpected and to anticipate that it will be bad. Consequently, evaluators have to think of themselves as learners and be prepared to modify evaluation plans during the course of their work. Indeed, evaluation plans should include analyses of all the anticipatable forces that might vitiate the plans and of any reasonable steps that can be taken to prevent such forces from operating. The plans should also include the necessary measurement and other design features that will permit fallback analyses of outcome should the desired plan not be fully implemented. To these ends, a system of field monitoring is needed.

Proposition 15. Evaluations increase in difficulty of implementation and ambiguity of results as the number of presumed causal agents increases. In theory there is no problem with conducting planned variation experiments with, let us say, twenty different treatments. In practice the problem is immense. First, the projects implementing the different treatments would disagree about the appropriateness and/or importance of each outcome measured. Second, logistical difficulties are compounded in the case of multiple treatments, and it is not easy to achieve comparable conditions and measurement across the groups. Third, the quality of implementation is likely to differ from project to project and from project type to project type; hence, differences in results can be attributed not to intrinsic differences but to seeming irrelevancies in how well the treatments were delivered. Fourth, some treatments aimed at a particular problem are oriented toward fairly immediate results, while others look for slower-operating changes achieved by more circuitous psychological and sociological routes, which may, however, "anchor" the desired outcomes better; as a result, research which has to do justice to multiple sites and has to be coordinated across multiple sites is likely to produce lower-quality information than research which is more circumscribed and permits easier logistics, greater local anthropological wisdom, closer monitoring, and more flexibility during data collection.

Proposition 16. The single evaluative study is inevitably flawed in multiple ways. Like science in general, evaluation presumably achieves higher levels of dependability through multiple, heterogeneous replication. The unit of progress is, as it were, not the single study but the review of a literature in which irrelevancies are made heterogeneous. Fortunately, multiple replication is now increasingly found in evaluation. One needs only note the four Negative Income Tax Experiments, the seven *Sesame Street* evaluations, the hundreds of studies on school desegregation and prison rehabilitation. Moreover, Glass's (1978) work on meta-analysis has largely used evaluative data, and the mandate to evaluate (as with Title I and other programs) may well accelerate the growth of academic activity in the meta-analysis domain. Yet the reality remains that heterogeneous replication at different time points is still the exception rather than the rule.

Proposition 17. Findings of "no difference" (or small effect size) are common in evaluation and are often particularly uninformative. In evaluations with an outcome or impact focus, it is frequently concluded that "no statistically significant difference" was found or that effect sizes were so small as to be of no practical value. Many of the problems associated with such negative conclusions about covariation are generic to all knowledge processes. Particularly difficult in the evaluation context, however, is the uncertainty such conclusions typically leave about why no effects were found. Was it because the underlying theory is deficient in major ways? Was it that one small step in the theory is wrong but it is otherwise correct? Was it that the project and its underlying theory were not implemented well at some stage? Was it that the evaluation design was not sensitive to detecting unbiased effects or could not estimate them with a precision that gives confidence? The ubiquity of disappointing findings and their often uninformative nature suggest, first, that evaluators would do well to avoid the "black box" trap by devoting resources to evaluation planning and to measuring and analyzing process data; and second, that they would also do well to erect a foundation of heterogeneous and replicated no-difference findings before claiming that it would be reasonable to act *as though* the null hypothesis were true.

The Apparent Dilemma and Current Approaches
to Dealing with It

The foregoing propositions lead to two conclusions about the quality of evaluation studies. One follows from the desirability of local knowledge, of on-site monitoring capabilities, of being oriented to a few causal agents, and of involving the interests of local constituencies. In other words, site-intensive evaluations with multiple indicators of success are superior to evaluations with more perfunctory site-level knowledge. The second conclusion, that replicated evaluations are superior to single-shot studies, follows from the paucity of high-quality substantive knowledge in many fields and from the evaluator's consequent need to be educated through immersing himself in evaluative events. It also follows from the need to make heterogeneous the conceptual and practical irrelevancies that inevitably accompany single studies, just as they accompany single measures (Campbell and Fiske, 1959). Unless unrealistically high expenditures on evaluation are magically possible, *the model of evaluation suggested by the prop-*

ositions and the conclusions derived from them is a model of ex-tended investigation in great depth at a small number of sites.

Yet program evaluators who take seriously the task of eval-uating federal or state programs are faced with many, many sites and what appears at first to be a bewildering complexity of opera-tional theories and activities across the population of sites. The dilemma they face is how to select the restricted number of sites in which it is feasible to conduct evaluations of high quality when pressures exist to sample many sites because (1) clients expect ex-tensive generalizability, (2) there is usually considerable variability in the quality with which any set of services is implemented, (3) several project types can usually be inferred, and (4) operational details differ across projects with comparable theories. Evaluation methods for intensive site-level study are not well suited to an ini-tial need to make sense of the considerable heterogeneity in the typical social program.

My tentative answer to the dilemma is that somehow the evaluator needs to select, as examplars of project types, projects that are policy relevant because they are affordable, potentially transferable, and give some promise of being successful. Arriving at such a selection point will not be easy, since project activities have to be described and project types inferred; where many types exist, criteria of an empirical or nonempirical nature have to be de-veloped for sampling among them; where many projects exem-plifying a type are available, criteria have to be developed for sam-pling among them; and then the resulting smaller sample of projects has to be evaluated in intensive detail. This is not the place to lay out a theory of how these tasks can be accomplished program-matically. Such a theory is discussed elsewhere (Cook and Leviton, in press). However, this is the place to discuss whether the dilem-ma is as I have tried to outline it; and whether, if it is a dilemma, it is already being dealt with adequately.

Denial of Need for Information About Generalized Project Types. I have suggested that a dilemma exists between the hetero-geneity in theories and activities that characterize the projects within a typical social program and the desirability of small inten-sive evaluations conducted at the project level. Not everyone would accept such a formulation of the problem. I would deduce from the philosophy of evaluation of the National Institute of

Mental Health (NIMH) with respect to community mental health centers (Cook and Shadish, in press) that for this heterogeneous program, NIMH officials see the most important issues of evaluation policy relating not to questions about the effectiveness of local centers but to questions about improving record keeping and meeting any information needs that individual center directors might have. The emphasis on responsiveness to local evaluation questions reflects the belief that results are more likely to be used when questions are local than when they come from funding sources. (However, most center directors do not seem to pose questions that evaluators can answer, or else they ask simple questions about managerial processes. Of minimal importance are questions about solutions to major problems, and certain topics—for instance, issues concerning the community mental health center's overall effectiveness, or the effectiveness of different methods of delivering services or of different types of service providers—are rarely broached.)

Another analysis of evaluation policy which minimizes the need to discover project types stresses the desirability of identifying successful and potentially transferable methods of improving practices within projects; for example, methods of structuring intake or of designing and operating management information systems (see Chelimski, 1979). In this analysis, outcome research could be conducted, but with the project element as the unit of interest and not the global project. (When Cronbach, 1978, claims that all evaluation should be "formative," I believe that he is referring to the identification of ways to improve project elements.)

Much can be said for basing evaluation on the identification of successful project elements. First, the approach stresses useful immediate feedback to evaluatees; second, generalized knowledge about project elements is possible from syntheses of studies; and third, the approach avoids the problems inherent in defining and measuring project types. Its major drawback, from my point of view, is that one can improve a practice within projects whose underlying theory is so incorrect, or whose overall implementation is so flawed, that improving the practice will not improve the final product.

Haphazard Accumulation and Haphazard Synthesis. Most agencies do not have clearly defined philosophies of evaluation.

Instead, local researchers who want to evaluate, or who are mandated to do so, can largely do what they want. In such a haphazard system, evaluative studies of projects within a program accumulate until someone wants to synthesize them to abstract more generalizable information. In a sense, this is what happened with psychotherapy research when Smith and Glass (1977) conducted their meta-analysis. Delving into a literature composed of haphazard, small, locally controlled studies, they attempted to synthesize the research by creating project types (that is, schools of psychotherapy) and then analyzing the data so as to discover which types were more successful than others. Is haphazard accumulation a good model for gaining information about project types?

One fundamental problem with meta-analytic efforts as they are currently practiced is that the same source of bias can run through all the studies. Director (1979) has claimed that this is the case with evaluations of job training programs, and Campbell and Erlebacher (1970) have claimed the same for most evaluations of compensatory education programs. That is why sophisticated meta-analysts like to break out of their data the subset of studies where, for instance, random assignment to treatments was carried out. This subset of studies then serves as the validation criterion for the remaining studies of lower technical quality. Alternatively, an attempt could be made to analyze separately studies where the bias is presumed to be in one direction *versus* another. Attention is then directed to findings that replicate *despite* presumed biases in each direction. A second fundamental problem lies in the categorization of individual projects into project types. In the past, this has been done on the basis of how the treatments were labeled by investigators, as opposed to how the treatments were described empirically. In psychotherapy, for example, it is not inconceivable that similar activities regularly occur despite different therapeutic labels being attached to these activities.

Mandated Local Evaluations Planned for National Synthesis. When locally controlled evaluation is mandated, multiple project-level evaluations will be done. The dilemma I have discussed may not seem to operate in this instance, since no sampling of projects is called for; the data from all the program-funded projects simply come into a central location for synthesizing. Is this a good model for gaining dependable knowledge about project types?

The likelihood of fruitful synthesis is increased if enough control is exercised over local evaluators to compel them to use a restricted set of measures and experimental designs, as with the very heterogeneous Title I program in education. The local evaluators of that program had to use either their own testing instruments or nationally standardized achievement tests with one of three experimental designs: a randomized experiment, a regression-discontinuity design, or a design where the performance of Title I children could be related to norms. The vast majority of school districts chose the last design, probably for reasons of simplicity of execution and because statistical regression increased the likelihood of making their project look successful (Trochim, 1980). In any event, the decision to restrict the experimental design alternatives was made in the expectation that generalized information about successful projects was more likely to result than would have been the case if school districts did whatever they wanted in order to comply with the mandate to evaluate. So for Title I there was local evaluation in the context of national designs specifying a restricted choice in how to evaluate.

This strategy is dependent on the stronger designs' being implemented in reasonable fashion in a reasonable number of school districts. More important, from my point of view, it also depends on having reasonable descriptive data on the nature of the treatment implemented. This is a difficult issue in the Title I context, because funds for Title I students are mingled with funds from other sources and because some of the funds are spent on children who are not eligible for Title I. But it is not necessarily a problem for the strategy as a whole, since appropriate data on project activities can be collected. Indeed, if one is to use any attempts to synthesize local evaluations of local projects in order to reach generalizable conclusions about project types, some descriptive data on the theory and activities of sites is a necessity, albeit often an expensive one.

A more major problem with the strategy, from my point of view, is that it is not cost efficient. One need not study all the school districts receiving Title I funds in order to draw confident generalizations about successful types of projects or successful project practices. In the Title I case, outcome evaluation was mandated, and so relevant data would anyway be collected from nearly all the eligible schools. But for more voluntary contexts of evaluation, the Title I example would be wasteful of evaluation resources. One would instead want to target resources to conduct intensive local evaluations of process and outcome at a more restricted set of projects representing different theories of how Title I funds should be allocated for educational services.

Dissemination Review Panels. In a situation like that currently operated by the Department of Education, the evaluator is spared a major role in detecting successful project types for purposes of transfer to new sites. The directors of local education projects are invited to submit details of their project and its evaluation to a review panel, which considers all the information. The panel then recommends that some of these projects be adopted elsewhere because they appear to be both relevant and successful. Plans are sometimes even made to help potential adopters financially. This is a system designed to detect and transfer successful projects, and it is driven at the initial nomination stage by self-selection. As a result, the bewildering initial array of project-level diversity is reduced.

A difficulty with this approach concerns the primacy it attaches to self-nomination. My guess is that such a nomination procedure will result in self-enhancing replies by developers and their staffs, whose commitment to their project is so strong that it inadvertently influences the evaluations, making them appear more positive than is warranted. Also, many projects will not be nominated, or will be rejected for transfer elsewhere, because of technical problems in the evaluation that might not have arisen if the evaluation had been conducted by persons with different sorts of experience or professional skill. But, most important of all, decisions about transfer are usually made on the basis of a single evaluation of a single project. In my opinion, this may not turn out to be a particularly sound basis for *successful* transfer, though I realize that the panel's practical wisdom and theoretical expertise are decision-relevant inputs over and above project descriptions and evaluations.

Evaluation, Experiments, and Evolution

The dilemma explicated here has much in common with Cronbach's (1978) discussion of bandwidth. He stresses that a question that is not yet precisely targeted or whose answer demands that a research net be widely cast, perhaps because more populations or causes or time points are at issue, usually receives a less adequate answer than a more targeted question or a question pertaining to fewer and more circumscribed entities. Perhaps the major problem with the evaluation of ongoing social programs is that a considerable amount of apparent noise initially confronts the honest evaluator, and this needs to be investigated in a general way before clearer signals can be discerned. However, no valid methods are currently available for moving quickly and with certainty from noise to clearer signals that are worth further investigation. It is our sad predicament at this time that, with the knowledge about social programs that is usually available, we are much more overwhelmed by the initial appearance of noise than we are pointed toward important signals worthy of intensive evaluation.

A second analogy may help elucidate the dilemma. In evolutionary theory the most crucial constructs are variation, selection, and retention. One premise of this chapter has been that local projects provide an important source of policy-relevant variations within programs; a second has been that evaluation is supposed to provide a mechanism of selection among these projects. Further, I have assumed that, for a project which is successful in one setting to be viable in another site, we need evaluative information about something more generalizable than a single project. However, even with knowledge about project types, we should not expect perfect transfer from tested project sites to novel project sites. This is because each novel site operates in a unique ecology with unique criteria for selection. In the very distant future, essentialist theories of causation (Cook and Campbell, 1979) that permit perfect prediction and transfer may be developed for some project types. But no one should expect such theories at this time, if ever.

As I understand evolutionary theory, variations will be selected that increase adaptation to the local environment, but they need not be the "optimal" adaptations. Variations that might have had an even better fit may not be tried if earlier variations were "satisfactory." "Satisficing," not "optimizing," is the criterion. For the evaluation of ongoing social programs, therefore, an important rhetorical stance should be that project types identified as "successful" need further scrutiny and development and should not rest on their laurels. More important, it suggests that projects and project types that have not been evaluated should not be treated as though they were unsuccessful. Such a tendency might occur once successful projects or project types are publicly acknowledged, and it is especially likely to be troublesome if reimbursement for services comes to be linked to demonstrations of effectiveness. There is great utility to the stance that the unevaluated—or the poorly evaluated—should be presumed to satisfice.

The present fumblings toward a theory of the evaluation of ongoing social programs does not assume that significant sources of satisficing variability are *necessarily* present in a heterogeneous program and are only waiting to be uncovered. Because no guarantee can be made that successful sources of variability exist, the present argument implies that the identification of successful ongoing project types should not preclude research and development activities designed to develop new project types. The evolutionary analogy that buttresses this chapter predicts that failures of adaptation occur when not enough sources of novel variability are generated quickly enough. To increase the sources of variation, and to make sure that truly novel sources are included, research and development activities probably should take place within ongoing programs. Indeed, a major function of demonstrations (and their offspring, social experiments) is, as I see it, the generating and evaluating of *novel* sources of variability that are not now part of an ongoing program. (Another function is to provide trials of existing sources of variability whose effectiveness appears likely from past studies that nonetheless were flawed in ways that might plausibly account for the obtained results.)

There is little debate among evaluation theorists about experimental methods being the preferred means of selecting successful projects, project types, or project elements. Debate concerns two other issues related to experimentation. The first has to do with its feasibility, a matter that Donald Campbell and I have addressed elsewhere (Cook and Campbell, 1979, ch. 8). The second concerns possible unintended side effects of experimentation, including restrictions to external validity that sometimes follow from strategies designed to make experiments feasible. The third issue concerns the timing of experimentation in evaluation. I can make a case for experiments coming early in an evaluation sequence when demonstrations, carefully considered policy options for the future, or circumscribed project elements are being investigated. But even here, I would want to assume that the evaluator has immersed herself in both the relevant substantive theory and in the experience gained by earlier evaluators of similar entities.

I find it more difficult to make a case for experimentation coming early in a research sequence when heterogeneous ongoing programs are being evaluated and comparisons are made at the project level. As Provus (1971) has pointed out, evaluating projects requires substantial prior knowledge about the number and types of projects in a program, what goes on in the projects, which ones represent high-quality implementation, and which types should be chosen for study. With ongoing social programs and a focus on project-level comparisons, the problem to be avoided with experimentation is prematurity. Since so few treatments can be assessed in so few contexts, a great deal depends on which instances of which project types one evaluates. Rushing into this decision means that a sophisticated tool for selecting among a restricted set of alternatives is used (that is, the experiment) when the variant or variants being tested have not been chosen at a high level of consciousness from a well-documented array with many heterogeneous alternatives.

To pursue the evolutionary analogy further, it may not be advisable to evaluate whichever projects come blindly to hand. While nature has made an eloquent case for "blind" variation, we would not do well in program evaluation to understand "blind" in terms of "projects which somehow happen to be selected for evaluation"; or "projects that developers recommend"; or "projects that seem most amenable to experimentation"; or "projects that seem to be the most developed"; or "projects that program managers think are effective"; or whatever. We need theories that take us from the bewildering array of heterogeneous projects, about which we typically have little reliable knowledge, to the several instances we can effectively study of each of a few project-level manipulanda.

No discussion of an analogy with evolutionary theory is complete without reference to the need for mechanisms by which satisficing mutations are retained. In the present context I shall understand "retain" to mean "widely adopted in the community," with the community being program funders at federal and state levels and project directors at more local levels. To encompass the retention function, a theory of the social transfer and adoption of projects is required. Presumably, the theory would include reference to the felt need for change; who endorses a particular project as "successful"; project costs; the ease with which project activities can be manipulated; and the extent to which success depends on factors related to the setting, the historical moment, and the population of clients and service providers.

Setting and population factors are particularly important since project activities will be transformed and "reinvented" at new sites, and even those that are not greatly transformed will be implemented in a somewhat new context that can significantly affect transfer. As with species in evolutionary theory, project types have a greater capacity for survival in novel environments the less precisely they are engineered to the unique features of specific local contexts. This is why it is so important to note that inferences about project types depend on having replicated results across projects that inevitably differ from each other in a myriad of theoretically irrelevant ways. Unlike knowledge of individual projects, knowledge of project types informs us that transfer to other sites has already been demonstrated. If I may simultaneously use the technical languages of both evolution and knowledge transfer, knowledge of project types gives us some assurance that we do not have a product (a variant) that is so precisely developed (adapted) that it is only effective in a specific, and perhaps infrequently occurring, setting (ecological niche).

References

Banta, H. D., and Bauman, P. "Health Services Research and Health Policy." *Journal of Community Health*, 1976, *2*, 121-132.

Berman, P., and McLaughlin, M. W. *Federal Programs Supporting Educational Change.* Vol. 8: *Implementing and Sustaining Innovations.* Santa Monica, Calif.: Rand, 1977.

Blau, P. M. *The Dynamics of Bureaucracy.* Chicago: University of Chicago Press, 1955.

Campbell, D. T., and Erlebacher, A. E. "How Regression Artifacts in Quasi-experimental Evaluations Can Mistakenly Make Compensatory Education Look Harmful." In J. Hellmuth (Ed.), *Compensatory Education: A National Debate.* Vol. 3: *Disadvantaged Child.* New York: Brunner/Mazel, 1970.

Campbell, D. T., and Fiske, D. W. "Convergent and Discriminant Validation by the Multitrait-Multimethod Matrix." *Psychological Bulletin,* 1959, *56,* 81-105.

Chelimski, E. "Evaluating Broad-Aim and Large-Scale Social Demonstration Programs." In F. Zweig (Ed.), *Evaluation in Legislation.* Beverly Hills, Calif.: Sage, 1979.

Cook, T. D., and Buccino, A. "The Social Scientist as a Provider of Consulting Services to the Federal Government." In J. Platt and J. Wicks (Eds.), *The Psychological Consultant.* New York: Grune & Stratton, 1979.

Cook, T. D., and Campbell, D. T. *Quasi-Experimentation: Design and Analysis Issues for Field Settings.* Chicago: Rand McNally, 1979.

Cook, T. D., and Leviton, L. C. "Towards a General Theory for Evaluating Ongoing and Proposed Programs and Projects." In G. Lindzey and E. Aronson (Eds.), *The Handbook of Social Psychology.* Reading, Mass.: Addison-Wesley, in press.

Cook, T. D., and Shadish, W. R. "Evaluating the Two-Percent Evaluation Requirement in Community Mental Health Centers." In G. Stahler and W. R. Tash (Eds.), *Innovative Approaches to Mental Health Evaluation.* New York: Academic Press, in press.

Cronbach, L. J. "Designing Educational Evaluations." Unpublished manuscript, School of Education, Stanford University, 1978.

Danielson, M. N., Hershey, A. M., and Banes, J. M. *One Nation, So Many Governments: A Ford Foundation Report.* Lexington, Mass.: Lexington Books, 1977.

Director, S. M. "Underadjustment Bias in the Evaluation of Manpower Training Programs." *Evaluation Quarterly,* 1979, *3,* 190-218.

Downs, A. *Inside Bureaucracy.* Boston: Little, Brown, 1967.

Enos, D. D., and Sultan, P. *The Sociology of Health Care.* New York: Praeger, 1977.

Farrar, E., DeSanctis, J., and Cohen, D. K. *Views from Below: Implementation Research in Education.* Cambridge, Mass.: Huron Institute, 1979.

Glass, G. V. "Integrating Findings: The Meta-Analysis of Research." *Review of Research in Education,* 1978, *5,* 351-379.

Hargrove, E. W. "A Theory of Implementation." Unpublished manuscript, Institute for Policy Analysis, Vanderbilt University, 1980.

Hawkridge, D. G., Chalupsky, A. B., and Roberts, A. O. H. *A Study of Selected Exemplary Programs for the Education of Disadvantaged Children.* Palo Alto, Calif.: American Institutes for Research, 1968.

Hill, P. "Evaluating Education Programs for Federal Policy Makers: Lessons from the NIE Compensatory Education Study." In J. Pincus (Ed.), *Educational Evaluation in the Public Policy Setting.* Santa Monica, Calif.: Rand, 1980.

Lavrakas, P. J., and Normoyle, J. *Evaluation and Its Utilization in Criminal Justice.* Evanston, Ill.: Center for Urban Affairs, Northwestern University, 1980.

Leviton, L. C., and Boruch, R. F. "Case Studies of Utilization." In R. F. Boruch and D. S. Cordray (Eds.), *An Appraisal of Educational Program Evaluations: Federal, State and Local Agencies.* Evanston, Ill.: Northwestern University, 1980.

Leviton, L. C., and Hughes, E. S. X. "A Review and Synthesis of Research on the Utilization of Evaluations." *Evaluation Review,* in press.

Nathan, R. P. "Federal Grants-in-Aid: How Are They Working in 1979?" Paper presented for conference on "The City in Stress," sponsored by the Center for Urban Policy Research, Rutgers University, March 9, 1979.

National Institute of Education. *Compensatory Education Services.* Washington, D.C.: National Institute of Education, 1977.

Pressman, J. L. "Political Implications of the New Federalism." In R. P. Inman, M. McGuire, W. E. Oates, J. L. Pressman, and R. D. Reischauer (Eds.), *Financing the New Federalism: Revenue Sharing, Conditional Grants, and Taxation.* Baltimore: Johns Hopkins University Press, 1975.

Pressman, J. L., and Wildavsky, A. *Implementation.* Berkeley: University of California Press, 1973.

Provus, M. *Discrepancy Evaluation.* Berkeley, Calif.: McCutchan, 1971.

Rein, M., and White, S. H. "Can Policy Research Help Policy?" *The Public Interest,* 1977, *49,* 119-136.

Riecken, H. W., and Boruch, R. F. (Eds.). *Social Experimentation: A Method for Planning and Evaluating Social Innovations.* New York: Academic Press, 1974.

Smith, M. L., and Glass, G. V. "Meta-Analysis of Psychotherapy Outcome Studies." *American Psychologist,* 1977, *32,* 752-760.

Suchman, A. E. *Evaluative Research.* New York: Russell Sage Foundation, 1967.

Trager, B. *Homemaker–Home Health Aide Services in the United States.* Washington, D.C.: U.S. Department of Health, Education and Welfare, 1973.

Trochim, W. "Regression-Discontinuity and the Evaluation of the Title I Program in Education." Unpublished doctoral dissertation, Northwestern University, 1980.

Van Horn, C. E. *Policy Implementation in the Federal System.* Lexington, Mass.: Heath, 1979.

Weiss, C. H. "Research for Policy's Sake: The Enlightenment Function of Social Research." *Policy Analysis,* 1977, *3,* 531-545.

Yin, R. K. "Creeping Federalism: The Federal Impact on the Structure and Function of Local Government." In N. Glickman (Ed.), *Urban Impacts of Federal Policies.* Baltimore: Johns Hopkins University Press, 1980.

11

David A. Kenny

Interpersonal Perception: A Multivariate Round Robin Analysis

Donald Campbell has studied how individuals and scientists know the world. Although many of his illustrations have focused on visual perception, he has always been interested in shared social knowledge. To Campbell knowledge has two seemingly contradictory sides: it is both veridical and errorful. Perceivers know the world but are subject to many biases. In science our tests usually measure the implicit construct, but they are also subject to numer-

Note: This research was supported in part by National Science Foundation Grant BNS-7913820. I would like to thank Lawrence La Voie, Charles M. Judd, Joseph Burleson, Charles A. Lowe, and the editors of this volume for their many helpful comments.

ous sources of error. Also, our research designs aid us in answering many questions, but the results of research are always subject to alternative explanations. Knowing, then, represents a mix of truth and error.

Essential to Campbell's analysis of knowledge is the duality of subject and object. An object can be known by the subject, but the subject still makes various mistakes. In person perception this duality of subject and object becomes a reciprocal duality, since the object can also perceive the subject. Consider two persons named Peter and Paul interacting. The behavior of Peter is a stimulus or an object which Paul experiences subjectively. Paul's behavior then becomes an object for Peter to experience. In social interaction the participants are both subject and object, stimulus and response, actor and partner. The marvel of social interaction is that there is considerable overlap in the two persons' perceptions of each other, even though they perceive different objects—namely, each other.

Shared experience, a fact of social life, is not well documented by social science in general or social psychology in particular. We detach subject and object: we examine people as perceivers and carefully document their biases and presuppositions, or we carefully study how particular social objects, such as males and females, are perceived. We now have a good idea how information is processed and filtered, and how certain types of persons are perceived, but we rarely capture the full *social* reality that we share.

Campbell has focused on this reciprocal duality in his research in person perception. Campbell, Miller, Lubetsky, and O'Connell (1964) had members of nineteen living groups rate all the other members of their group on twenty-seven personality traits. Thus, each person served as stimulus or ratee, since he or she was rated by the others; and each person served as a subject or rater, since he or she rated the others. The major focus of the study was an attempt to measure and analyze the components of social projection. Campbell and his associates computed a *reputational* score, which was the mean on a given trait for a given ratee across all raters; and a *rater bias* score, which was the mean across all ratees for a given rater. Their measure of projection was the correlation between the reputational score and the rater bias score. For the trait "hostile," a negative correlation would imply that

persons who are seen by others as hostile see others as being unhostile. A positive correlation would indicate that hostile persons
see others as being hostile.

The purpose of the present chapter is to revise and extend
the model of analysis developed by Campbell and his associates.
The first part of the chapter will reexamine their measure of projection in light of a new statistical model; in the second part a
multivariate generalization of that model will be presented.

Components of Social Perception

The design that Campbell and his associates employed is
called the *round robin design*. Table 1 presents an example of the
design. The rows of the table represent the subject, judge, or rater.

Table 1. Example of Round Robin Design of Ratings of Friendliness

Rater	Ratee					Mean
	Mary	Helen	Linda	Jane	Deirdre	
Mary	—	8	4	7	9	7.0
Helen	6	—	4	9	7	6.5
Linda	5	7	—	8	8	7.0
Jane	4	7	3	—	6	5.0
Deirdre	7	6	5	8	—	6.5
Mean	5.5	7.0	4.0	8.0	7.5	6.4

The columns which are ordered as the rows represent the object,
target, or ratee. Thus, for example, Mary's rating of Jane on the
trait "friendliness" is 7, while Jane's rating of Mary on the same
trait is 4.

Recently, Warner, Kenny, and Stoto (1979) have developed
a new model and method of analysis for the round robin design.
The model is very similar to the classical two-way analysis of variance model:

$$X_{ijk} = \mu + \alpha_i + \beta_j + \gamma_{ij} + \epsilon_{ijk}$$

For a rating study of friendliness, the term μ represents the average rating of friendliness across all raters and ratees. The row effect, or rater effect (α_i), represents the degree to which a given
rater sees others as friendly. The column effect, or ratee effect

(β_j), represents the average friendliness of a given ratee; that is, the degree to which he or she is seen as friendly. The interaction term, γ_{ij}, represents the degree to which a specific rating cannot be explained by the row or column effect. For instance, if Mary thinks that Helen is much more friendly than she is seen by others and more friendly than Mary sees others, then the interaction term is positive. This term, then, represents the interaction of rater and ratee. The final term, ϵ_{ijk}, simply represents noise or measurement error.

The round robin design captures Campbell's notions concerning social perception. First, each participant in the study serves as both a subject and an object. Mary is a stimulus for Jane, but Mary also serves as a rater. Second, the ratings contain a mixture of truth and bias. The row effect can be thought of as a bias: the tendency to use either the upper or lower part of the scale— or, as it is called in psychometrics, the leniency effect (Guilford, 1954). The column effect measures the degree to which someone is friendly and can be considered the truth or, at least, the social truth.

There are two major types of information that result from the Warner, Kenny, and Stoto analysis. The first type is variance. It is possible to partition the variance of ratings into four components: variance due to raters (σ_α^2), variance due to ratees (σ_β^2), variance due to the rater-by-ratee interaction (σ_γ^2), and random error (σ_ϵ^2). We can then judge how much of the variance of ratings is due to who is doing the rating, who is being rated, and who is rating whom. The second type of information is correlation. The row and column effects may well be correlated if raters who see others as friendly are seen by others as unfriendly. The interaction effects may also be correlated. If May thinks Helen is friendly, Helen might well see Mary as friendly. These two correlations will be considered in much more detail later in the chapter.

Campbell has always viewed himself more as a meta-methodologist than as a statistical methodologist. He has seen his role as that of shaping methods to problems, while exact statistical mechanics have usually been left to his more statistically minded colleagues. Campbell has always recognized that statistical analysis might elaborate and refine and, in many cases, modify his intuitions. Consistent with Campbell's own emphasis on multiple modes of knowing, the convergence of intuitive and analytic approaches achieves a more complete picture than can be attained through either mode alone.

As was stated above, Campbell, Miller, Lubetsky, and O'Connell (1964) attempted to measure projection by a round robin design. They had members of nineteen groups (such as fraternities and sororities) rate all other members of the groups on twenty-seven personality traits, as is illustrated in Table 1. Their major analysis consisted of computing the means averaged across ratees (the row means in Table 1) and correlating them with the means computed across raters (the column means in Table 1). The correlation between the two was taken to be the measure of projection. I will ignore their complex double-standardization procedure, since it did not significantly alter their major results. Their major conclusion was that for a number of traits the reputational score was *negatively* correlated with the rater bias score. They interpreted this result as evidence of *contrast projection*: a person high on the trait attributes less of it to others than does someone with a lower rating on that trait. Although for a number of traits the contrast projection correlation was significantly negative, most correlations were only slightly negative, the median being $-.04$. Raised here are three concerns about the Campbell analysis. These concerns, derived from the Warner, Kenny, and Stoto formal analysis, have to do with statistical artifacts, level of generalization, and conceptual interpretation.

As stated above, Campbell and his associates used the correlation between the reputational score, or ratee effect, and the rater effect as their measure of projection. The Warner, Kenny, and Stoto analytic results demonstrate that, in the absence of any real correlation between ratee's reputation and rater's bias, for n raters these means still correlate between $-1/(n-1)$ and -1! (These values were obtained from the expected mean cross-products in Table 6 of the Warner, Kenny, and Stoto paper.) For Campbell and his associates, this negative bias would be at least $-.043$, since there are on the average twenty-four members in each group. This statistical problem with the analysis by Campbell and his associates calls into question their small contrast projection effect.

How are we to understand the analytically derived negative

correlation? Two independent and rather subtle factors underlie this artifact. The first may be called the *single-variable problem.* Usually two variables are used when correlations are computed. When observations of the same variable are used to compute a correlation, bias can result. Suppose one has a pool of observations and one randomly selects two observations (X_1 and X_2). If one takes each observation and subtracts the same mean from each and then forms $(X_1 - \overline{X})(X_2 - \overline{X})$, this value tends to be negative. Given that X_1 is above the mean, we know that the sum of the other observations must be below the mean, making the tendency for $X_2 - \overline{X}$ to be negative. The bias is created by the components of the cross-product sharing the same sample mean. (If there are only two observations, the bias is clear. For instance, if $X_1 = 2$ and $X_2 = 4$, then $(X_1 - \overline{X})(X_2 - \overline{X})$ equals -1. The product must be minus, since it must equal $-(X_1 - X_2)^2/4$.) Since the sum of cross-product deviations is the numerator of a correlation, the single-variable problem produces a negative correlation. The problem arises for Campbell and his associates because the means computed from a single variable are correlated. The single-variable problem refers to correlations between pairs of values taken from the same pool of observations.

The second source of artifact might be called the *missing-diagonal artifact.* Consider a simple round robin model that has only a rater effect and a ratee effect; thus, the rating of Jane by Mary is a function of Mary's average rating of others (Mary's rater effect) and the degree to which Jane is seen to be high on the trait (Jane's ratee effect). However, the simple column mean for Jane equals not only Jane's ratee effect but also the average of the *other* rater effects, since Jane does not rate herself (the missing diagonal). So if Jane has a higher rater effect, her average rating given by others will be low vis-à-vis the averages for the other raters, since she rates them but not herself. This bias is also present in the frequent observation that baseball's good pitching teams are usually the good hitting teams also. One reason they are good hitting teams is that they never have to face their own good pitching.

Formally, the simple row mean measures the row effect, minus the column effect times a constant. Consider the example in Table 2. Three persons—Jimmy, Teddy, and Jerry—rate each other. An idealized model of only ratee effects is assumed. The

Table 2. Illustration of the Missing-Diagonal Bias

| Rater | Ratee | | | Mean |
	Jimmy (−2)	Teddy (4)	Jerry (−2)	
Jimmy		4 + 10 = 14	−2 + 10 = 8	11
Teddy	−2 + 10 = 8		−2 + 10 = 8	8
Jerry	−2 + 10 = 8	4 + 10 = 14		11
Mean	8	14	8	10

ratee effects are −2, 4, and −2, respectively. The grand mean is set to 10. Note that although there are no row effects, the row means still vary. Teddy's row mean is lower because he does not rate himself, and he has a larger ratee effect than the others. On the other hand, Jerry and Jimmy's row mean is higher than Teddy's because they do not rate themselves, and they have a lower ratee effect than Teddy. In general, the row mean measures the row effect, minus the column effect times a constant. The constant is minus one over one less than the number of raters, or −1/2 in this case. So when the row mean is correlated with the column mean, a bias results, since the row mean is minus one half the column effect. The resulting correlation is a perfect −1. The column means are similarly biased estimates when there are row effects. These biased row and column means, which are created by the failure to rate self, produce a negative bias in the correlation.

Both the single-variable problem and the missing-diagonal problem result in a negative bias in the correlation between ratings given and received, which would explain the contrast projection results of Campbell and his associates. In fact, it would appear that they more than explain the negative correlations, and the artifact-free correlations are probably positive. It would be instructive to reanalyze their data using the Warner, Kenny, and Stoto method, since this method is not subject to either the single-variable or the missing-diagonal problem.

There is also a level-of-generalization problem in Campbell and his associates' definition of projection. Their definition focuses on the correlation between a rater's bias and a ratee's reputation; the unit of analysis is the rater or equivalently the ratee. To illustrate this level-of-analysis problem, imagine a counseling psychologist who is interested in relationships between people. Returning to Table 1, our counseling psychologist would correlate the observations as in Table 3. Here the unit of analysis is the dyad,

Table 3. Data in Table 1 Organized by Dyad

Dyad	Data	
Mary-Helen	8	6
Mary-Linda	4	5
Mary-Jane	7	4
Mary-Deirdre	9	7
Helen-Linda	4	7
Helen-Jane	9	7
Helen-Deirdre	7	6
Linda-Jane	8	3
Linda-Deirdre	8	5
Jane-Deirdre	6	8

not the rater. There are then two different correlations: one computed across persons and the other across rater-ratee pairs. This second correlation would determine whether they both saw each other as friendly or unfriendly.

The Warner, Kenny, and Stoto model allows for two different correlations. First is the correlation between row and column *effects* (not means); they designate this correlation as rho_1. The second correlation is that between γ_{ij} and γ_{ji}, where γ refers to the rater-ratee interaction; they designate this correlation as rho_2. For Campbell and his associates, rho_1 is the projection effect, while rho_2 measures something quite different. Rho_2 assesses the correlation computed across dyads after rater and ratee effects are removed. It would then measure whether persons in a relationship share a mutual view of each other. For example, if the trait is "hostile" and rho_2 is positive, then if person A is seen by person B as being hostile, controlling for A's rater effect and B's ratee effect, B will see A as more hostile, controlling for B's rater effect and A's ratee effect. We can view rho_2 as the partial correlation between A's rating of B with B's rating of A controlling for row and column effects.

The Campbell measure of projection, which is the correlation of the row mean with the column mean, does not estimate rho_1; instead, it estimates a mixture of rho_1 and rho_2 —since the row mean measures not only the row or rater effect but also the average ratee-rater interaction effect for that rater. The column mean also contains the average rater-ratee interaction effect for that ratee. Thus, when the row and column means are correlated, they partially measure rho_2, which is the correlation between interaction effects. (In a similar fashion, the correlation computed across rater-ratee combination estimates a mixture of rho_1 and rho_2.)

Since Campbell and his associates' measure of projection is confounded with a correlation computed across dyads, it does not necessarily measure an individual-level process. So a finding that hostile persons see others as unhostile may only reflect the fact that if one member of a dyad is seen as hostile by the other member, the first member sees his or her partner as unhostile. The negative correlation may be true only for dyads and not for persons.

The third problem in the Campbell analysis concerns the interpretation of the measure of projection. Polley (1979), using the Warner procedure, estimates rho_1 for the trait of friendliness to be .286. This correlation is not subject to the problems of statistics or level of generalization that were raised above. Does it indicate that friendly persons project friendliness to others? Consider a plausible explanation for the correlation that is an alternative to projection. Friendly people cause social interactions to be amicable. So when friendly persons have to rate a partner in a social interaction, they may well misattribute their good feelings about the interaction to the other person rather than to themselves. If there is such a two-stage process of friendliness causing friendly interactions and friendly interactions causing friendly attributions, then the previously described positive correlation will arise. The result, as was pointed out to me by my colleague Bill Wilson, is that the correlation does not measure the intrapsychic fantasy process of the person, which is what projection is supposed to be, but is only a misattribution. Rho_1 can be interpreted as either measuring projection or a misattribution.

Multivariate Generalization

The procedures for the round robin design of Warner, Kenny, and Stoto (1979) can be applied to Campbell and his associates' (1964) analysis of contrast and similarity projection.

However, Campbell also considered complementary projection; for example, do anxious persons see others as hostile? To measure such effects, a *multivariate* round robin analysis is required, since we seek to correlate the rater effects for the trait "hostile" with the ratee effects for the trait "anxious." The Warner methods are univariate and cannot handle the multivariate case.

This section extends the round robin analysis to include the multivariate case. Instead of presenting only the necessary mathematical derivations, we will find it useful to situate our discussion in the problem area of interpersonal knowing. In particular, we will leave the topic of projection and move on to the two variables of attraction (how much does Peter like Henry?) and perceived attraction (how much does Peter think Henry likes him?). There are, then, two parallel purposes of this section. The first is the extension of round robin analysis to the multivariate case. The second is a rethinking of the aspects and levels of social perception.

Tagiuri, Bruner, and Blake (1958) have distinguished three major aspects of social perception: mutuality, congruence, and accuracy. Let us consider each in terms of attraction and perceived attraction. *Mutuality* refers to reciprocity of attraction. If Peter likes Henry, does Henry like Peter? *Congruence* concerns perceived reciprocity. If Peter likes Henry, does he think Henry likes him? *Accuracy* refers to the veridicality of perceptions. If Peter thinks Henry likes him, does Henry like him? Mutuality, congruence, and accuracy are obviously crucial to any meaningful theory of social perception. McLeod and Chaffee (1973) have systematically reviewed the three aspects and have pointed out that numerous theorists have given different names to them. Furthermore, the three aspects, while conceptually independent, need not be so empirically. If there is both mutuality and congruence, then accuracy must follow.

Although these three aspects pervade social life, they have not received equal attention in social science. Mutuality has received some limited attention. Most research on it has been in the area of interpersonal attraction. For instance, researchers have investigated whether similar persons (persons who have a mutual view of social reality) are attracted to each other.

Since congruence is the only *intra*personal aspect of social perception, it can be studied in single individuals. Congruence focuses on an individual's implicit theories of relationships and has received a disproportionate amount of study. The measures of congruence tend to be larger in size and hence more replicable, while mutuality and accuracy results are often subtle, to say the least.

Accuracy has now become almost a taboo area of research in person perception. The study of individual differences in accuracy was an important topic during the post-World War II years of social psychology. It was an especially important variable because clinical psychologists had theorized that accuracy—or, as they preferred to call it, empathy—brought about positive therapeutic outcomes in psychotherapy. However, a devastating critique by Cronbach (1955) of this literature brought accuracy research to a standstill (Campbell also played an important role in this critique). Researchers who did apply the methods suggested by Cronbach usually obtained disappointing results. Thus, we no longer see much research on accuracy.

Most of the orientation toward the three aspects has been on whether variation in them explains certain outcomes. For instance, Laing, Phillipson, and Lee (1966) compared happily married versus distressed couples in mutuality, congruence, and accuracy. Researchers have also studied how mutuality (similarity) affects everything from happiness to attributions. Most of the pre-Cronbach focus on accuracy was on individual differences in accuracy. Thus, the orientation in investigating the three aspects has been "differential." Do differences in mutuality, congruence, or accuracy make a difference? Whether these differences are created experimentally or vary naturally, the interest has been in their impact on various outcomes. An alternate orientation is simply to measure the degree of mutuality, congruence, and accuracy across persons. This has been done for congruence. Do persons assume reciprocity, or do they assume similarity? We can describe this orientation as global.

Global assessments of mutuality, congruence, or accuracy usually do not include a wide range of social objects. For instance, for most measures of reciprocity, subjects interact with only a single confederate. Such a procedure means that the measure of reciprocity refers to that particular confederate alone. It is obviously more desirable to be able to generalize beyond the particular stimulus persons used in the experiment.

A multivariate round robin analysis can be used to obtain global measures of mutuality, congruence, and accuracy. Moreover, such measures refer not to the specific persons studied but to the population of such persons. Finally, a multivariate round robin analysis can be used to measure the three aspects at two levels of analysis: at the individual level and at the group level. The remainder of this chapter details such a multivariate model.

Table 4 contains a hypothetical round robin design in which Peter, Henry, David, and John state their attraction toward each

Table 4. Attraction of a Rater Toward a Ratee

| | Ratee | | | |
Rater	Peter	Henry	David	John
Peter		7	6	9
Henry	9		10	4
David	6	9		9
John	5	7	8	

other. For instance, Peter's attraction toward Henry is 7, while Henry's toward Peter is 9. A similar table could be made for perceived attraction. For illustrative purposes we will consider Peter and Henry. The round robin equation for how much Peter likes Henry on a given day is:

Peter's attraction toward Henry on day k (A_{ijk})
= average attraction across raters and ratees (m)
+ Peter's level of giving liking (a_i)
+ Henry's level of receiving liking (b_j)
+ Peter's relative attraction toward Henry (g_{ij})
+ noise (e_{ijk})

It is assumed that attraction equals the overall level of attraction; the degree to which Peter likes people in general, which will be called his level of *giving liking*; the degree to which Henry is liked by others, which will be called his level of *receiving liking*; the relative attraction of Peter toward Henry; and noise. The relative attraction term measures attraction at the level of the dyad, unconfounded with individual differences. That is, it measures how much Peter likes Henry after we adjust for how likable Henry is and how much liking Peter gives.

Consider now a measure of perceived attraction: how much Peter thinks Henry likes Peter. The model for perceived attraction is:

Peter's perception of Henry's attraction toward him on day k (P_{ijk})
    = average perceived attraction across raters and ratees (n)
        + Peter's perception of his level of receiving liking (c_i)
        + Henry's perceived level of giving liking (d_j)
        + relative perceived attraction of Peter toward Henry (h_{ij})
        + noise (f_{ijk})

The terms in this model are analogous to those in the previous one. Perceived attraction is assumed to equal the degree to which Peter believes he receives liking, the degree to which Henry is believed by others to be a liker, relative perceived attraction of Peter toward Henry, and noise. Let us denote A_{PH} as how much Peter likes Henry, A_{HP} as how much Henry likes Peter, P_{PH} as how much Peter thinks Henry likes him, and P_{HP} as how much Henry thinks Peter likes him.

To measure *mutuality*, one examines the equations for A_{PH} and A_{HP}:

$$A_{PH} = m + a_P + b_H + g_{PH}$$

$$A_{HP} = m + a_H + b_P + g_{HP}$$

The correlations between a and b measure mutuality at the individual level; they assess the degree to which persons who give liking receive liking—that is, are likers liked? The correlation between g_{HP} and g_{PH} measures mutuality at the dyadic level. These correlations can be measured by the univariate round robin analysis of Warner, Kenny, and Stoto and have been extensively discussed by Kenny and Nasby (1980). Thus, there are two mutuality correlations. One refers to individuals and the other to relationships.

To measure *congruence*, one examines the equations for A_{PH} and P_{PH}:

$$A_{PH} = m + a_P + b_H + g_{PH}$$

$$P_{PH} = n + c_P + d_H + h_{PH}$$

There are two individual level correlations. The first (between a

and c) measures the degree to which people who give liking think they receive liking, and the second (between b and d) measures the degree to which people who receive liking are thought to give liking. The correlation between g_{PH} and h_{PH} measures congruence or perceived reciprocity at the dyadic level.

To measure *accuracy*, one examines the equations for A_{PH} and P_{HP}:

$$A_{PH} = m + a_P + b_H + g_{PH}$$

$$P_{HP} = n + c_H + d_P + h_{HP}$$

Again, there are two individual-level correlations. The first, between b and c, measures the degree to which those who receive liking perceive that they do. The second, between a and d, measures the degree to which those who give liking are perceived by others to do so. The correlation between g_{PH} and h_{HP} measures accuracy at the level of the dyad.

Thus, a multivariate round robin analysis allows us in a very natural way to measure mutuality, congruence, and accuracy in dyadic relationships. These global measures of the three aspects can be partitioned into individual and dyadic correlations. At the individual level, there is one mutuality correlation and two congruence and accuracy correlations. At the dyadic level, there is one correlation for each aspect.

There is a formal equivalence between these measures of accuracy and Cronbach's (1955) four measures of accuracy. He considered accuracy for a rater (his term is "judge") measured across a set of traits and ratees (which he calls "targets"). The analysis here measures accuracy for a trait across a set of raters and ratees. If one interchanges trait and rater, then the translation of each of Cronbach's terms into the term here is (1) elevation: the difference between average attraction (m) and average perceived attraction (n); (2) differential elevation: the correlation of a person's average level of receiving liking (b) with his perception of how likable he is (c); (3) stereotype accuracy: the correlation of a person's average level of giving liking (a) with his average level of being perceived as a liker (d); and (4) differential accuracy: the correlation of person A's relative attraction toward B (g_{AB}) with the relative perception of B of A's attraction toward B (h_{BA}).

The major improvement of the measures presented here is that both raters and ratees are random factors. The 1955 Cronbach analysis treated both target and trait as fixed, thus limiting generalization to the specific targets and traits sampled. Also, the interchanging of rater and trait may prove to be more than a statistical nicety. Researchers who pursued the Cronbach measures of accuracy have generally obtained disappointing results. They find that raters do not vary much in their differential accuracy. However, when one examines accuracy within a trait across raters, global accuracy may prove to be quite high.

Estimation. The procedure to estimate the correlations from the multivariate round robin analysis is as follows: The two measures of mutuality of attraction can be estimated by the univariate procedures given by Warner, Kenny, and Stoto. The remaining six correlations require a multivariate approach. These correlations to be estimated are:

(1) a with c
(2) a with d
(3) b with c
(4) b with d
(5) g_{ij} with h_{ij}
(6) g_{ij} with h_{ji}

Since the variances of the variables can be obtained by the univariate method, only the covariances need be estimated. One can posit two other covariances:

(7) e_{ijk} with f_{ijk}
(8) e_{ijk} with f_{jik}

which are analogous to rho$_3$ in Warner, Kenny, and Stoto. Thus, there are eight unknown covariances that must be estimated.

In Table 5 are eight mean cross-products which can be estimated from data. (It has been posited that there are n subjects and r occasions. The prime notation means a reversal of the first two subscripts.) These cross-products were chosen to correspond to naive estimates of the eight unknown covariances. As shown in the Appendix, one can use these eight known mean cross-products to estimate the eight unknown covariances. It can be shown that the results in the Appendix subsume the derivations in Warner, Kenny,

Table 5. Round Robin Mean Cross-Products

Row by row:	$$\dfrac{rn \sum\limits_{i} (\bar{A}_{i..} - \bar{A}_{...})(\bar{P}_{i..} - \bar{P}_{...})}{n - 1}$$
Row by column:	$$\dfrac{rn \sum\limits_{i} (\bar{A}_{i..} - \bar{A}_{...})(\bar{P}_{.i.} - \bar{P}_{...})}{n - 1}$$
Column by row:	$$\dfrac{rn \sum\limits_{i} (\bar{A}_{.i.} - \bar{A})(P_{i..} - \bar{P}_{...})}{n - 1}$$
Column by column:	$$\dfrac{rn \sum\limits_{i} (\bar{A}_{.i.} - \bar{A}_{...})(\bar{P}_{.i.} - \bar{P}_{...})}{n - 1}$$
Cell by cell:	$$\dfrac{r \sum\limits_{i} \sum\limits_{\substack{j \\ i \neq j}} (\bar{A}_{ij.} - \bar{A}_{...})(\bar{P}_{ij.} - \bar{P}_{...})}{n(n - 1) - 1}$$
Cell by cell':	$$\dfrac{r \sum\limits_{i} \sum\limits_{\substack{j \\ i \neq j}} (\bar{A}_{ij.} - \bar{A}_{...})(\bar{P}_{ji.} - \bar{P}_{...})}{n(n - 1) - 1}$$
Error by error:	$$\dfrac{\sum\limits_{k} \sum\limits_{i} \sum\limits_{\substack{j \\ i \neq j}} (A_{ijk} - \bar{A}_{ij.})(P_{ijk} - \bar{P}_{ij.})}{n(n - 1)(r - 1)}$$
Error by error':	$$\dfrac{\sum\limits_{k} \sum\limits_{i} \sum\limits_{\substack{j \\ i \neq j}} (A_{ijk} - \bar{A}_{ij.})(P_{jik} - \bar{P}_{ji.})}{n(n - 1)(r - 1)}$$

and Stoto. In other words, this chapter presents the multivariate generalization of the univariate methods of Warner, Kenny, and Stoto.

Example. The data set analyzed here is that of Curry and Emerson (1970), who attempted to replicate Newcomb's (1961) classic roommate study. There are six 8-person clusters for a total of 48 individuals and 336 dyads. Dormitory students previously unacquainted stated their attraction toward and perceived attraction from their fellow cluster members on 100-point scales at weeks one, two, four, six, and eight. Four of the six clusters were all male, and the other two all female. (For additional details concerning the data collection, one should consult Curry and Emerson.) The multivariate round robin analyses were performed on each of the six clusters separately, and the results were averaged across clusters. (Standard errors of each estimate within each cluster were computed by the jackknife techniques described in Warner, Kenny, and Stoto. The standard errors were then pooled, assuming homogeneity, across clusters for significance test purposes.)

Table 6 presents the basic results at the individual level. The −.210 correlation indicates no mutuality at the individual level.

Table 6. Mutuality, Congruence, and Accuracy at the Individual Level

	Giving Liking (a)	Receiving Liking (b)
Receiving liking (b)	−.210	1.000
Perceived amount of liking received (c)	.919[a]	.119
Perceived amount of liking given (d)	−.307	.981[a]

[a] $p < 0.05$

That is, likers are not liked; in fact, they are slightly disliked; however, the correlation is not statistically significant. The negative sign may indicate a trend that unpopular persons are attempting to buy friendship by liking others.

Congruence at the individual level is very strong, approaching a perfect 1.000 correlation. Persons who are liked by others are thought by others to return the liking (.981), and persons who like others tend to think that others return the liking (.919). The latter correlation could be simply measuring a leniency effect. If each person used a different part of the scale for both attraction and perceived attraction, this latter correlation would be substantial. However, since the .981 correlation cannot be explained by a leniency effect, it seems more plausible to interpret the .919 correlation as not being due to a leniency effect.

Accuracy at the individual level is nonexistent. This result must hold for statistical reasons, since mutuality is essentially zero and congruence is virtually perfect. Individuals who receive liking do not realize that they do (.119), and persons who give liking are not recognized by others (−.307). The latter result is not very surprising, while the former may seem surprising. For opposite-sex attraction, persons presumably realize how attractive they are. Why not so in same-sex dyads? First, physical attractiveness is a major determinant for opposite-sex attraction, while its effect for same-sex dyads is substantially less. Also, in same-sex interaction, it is difficult to obtain feedback about the degree to which one is attractive. It should be recalled that the subjects were acquainted with each other for only eight weeks.

As can be observed in Table 7, at the dyadic level there is more texture in the results. Mutuality or reciprocity of attraction

Table 7. Mutuality, Congruence, and Accuracy at the Dyadic Level

	Relative Attraction of A toward B (g_{AB})
Relative attraction of B toward A (g_{BA})	.617
Perceived relative attraction of A toward B (h_{AB})	.909
Perceived relative attraction of B toward A (h_{BA})	.532

Note: $p < 0.05$

is substantial (.617, a result previously reported by Kenny and Nasby, 1980). After we subtract the effects of giving and receiving liking, persons in a relationship return liking with liking. This substantial correlation stands in sharp contrast to the individual-level correlation of −.210. To mix together both the individual- and dyadic-level correlations is to obscure the different result at the two levels.

The result for congruence at the dyadic level parallels the individual-level results. Dyadic congruence is .909. One way of interpreting the results is as follows: If Paul likes Henry more than he likes Peter, and Henry and Peter are equally liked by others, then Paul will think that Henry likes him more than Peter does, assuming that other people think Peter and Henry give the same amount of liking. Dyadic congruence is a mouthful! As Laing, Phillipson, and Lee (1966, p. 21) have said, "The theoretical problem constantly facing us is that we find it easier to think of each person in a dyad separately or one at a time, rather than together."

Accuracy at the dyadic level is moderately high and statistically significant (.532). This result is a necessary consequence of the mutuality and congruence results. Since mutuality at the dyadic level is .617 and congruence is .909, accuracy should be around .500. The subjects, then, are no more accurate than would be expected, given mutuality and congruence.

Had accuracy been computed by a simple correlation between attraction and perceived attraction across occasions, it would be only .227. Thus, the multivariate round robin analysis shows a moderately high level of accuracy at the dyadic level, even though it is entirely explainable in terms of mutuality and congruence.

Conclusion

A major theme in social psychology is the interplay between the individual and the situation. Behavior is a function of both the individual's dispositions and the situation. One important fact, which is easy to lose sight of, is that much of what we call situation is another person. For social behavior the stimulus is another person. A complete theory of personality must elucidate individual differences that refer not only to the consistent behaviors of the person but also to the behaviors that the person elicits from others. A complete description of the neurotic personality, for instance, should elaborate not only the behavior of such persons but also the behavior that neurotics consistently elicit from others.

But the behavior of persons in social interaction is not just a function of the personalities of the participants. The dyad is much more than the sum of its parts. The term that best embodies this notion is *relationship*. Two people form a relationship with each other that is unique to them.

We have, then, the basic model for studying dyads. If Mary and Jane are interacting, Mary's behavior and perceptions are a function of Mary, Jane, and the unique relationship between Mary and Jane. The round robin design represents one way to estimate the parameters of this model. My associates and I are currently developing simpler designs, which can also estimate the parameters of the model. The model is then not a "round robin" but, rather, a model of social relations. This *Social Relations Model* is being expanded in a number of ways. First, we are allowing for the effects of personality variables and experimental manipulations. Second, we are considering an additional level of analysis besides the individual and the dyad: the group. Third, we are exploring designs and models for triads; that is, forming all possible triads. Thus, the Social Relations Model represents a very general orientation to the study of interpersonal behavior. Its application in this chapter to projection and to the relationship between attraction and perceived attraction demonstrates both the model's versatility and its potential.

Appendix

Given the eight known mean cross-product terms in Table 5, we can derive their expected values by applying the following two theorems:

Theorem 1. Let X and Y be random variables whose covariance is σ_{XY}, and let the covariance of X_i (i odd) with Y_{i+1} and X_i (i even) with Y_{i-1} be σ_{XY}'. For a sample of n observations (n even) and

$$\overline{X} = \frac{\sum\limits_{i=1}^{n} X_i}{n} \qquad \overline{Y} = \frac{\sum\limits_{i=1}^{n} Y_i}{n}$$

it follows that

$$E\,(X_i - \overline{X})(Y_i - \overline{Y}) = \sigma_{XY}\left(\frac{n-1}{n}\right) - \sigma_{XY}'\left(\frac{1}{n}\right)$$

Theorem 2. Let X and Y be two random variables such that $E(X_{ij}Y_{ij}) = \sigma_{XY}$ and $E(X_{ij}Y_{ji}) = \sigma_{XY}'$, and define

$$\overline{X} = \frac{\sum\limits_{\substack{i \\ i \neq j}}^{n}\sum\limits_{j}^{n} X_{ij}}{n(n-1)} \qquad \overline{Y} = \frac{\sum\limits_{\substack{i \\ i \neq j}}^{n}\sum\limits_{j}^{n} Y_{ij}}{n(n-1)} \qquad \overline{X}_i = \frac{\sum\limits_{\substack{j=1 \\ j \neq i}}^{n} X_{ij}}{n} \qquad \overline{Y}_i = \frac{\sum\limits_{\substack{j=1 \\ j \neq i}}^{n} Y_{ij}}{n}$$

Then

$$E\,(\overline{X}_i - \overline{X})(\overline{Y}_i - \overline{Y}) = \sigma_{XY}\left(\frac{1}{n}\right) - \sigma_{XY}'\left(\frac{1}{n(n-1)}\right)$$

Table 8 presents the set of coefficients for each mean cross-product term. So, for instance, the mean cross-product for Row by Row equals nr times the true Row by Row covariance, minus $nr/(n-1)$ times the Row by Column covariance, and so on.

References

Campbell, D. T., Miller, N., Lubetsky, J., and O'Connell, E. J. "Varieties of Projection in Trait Attribution." *Psychological Monographs,* 1964, *78* (entire issue 592).

Cronbach, L. H. "Processes Affecting Scores on 'Understanding of Others' and 'Assumed Similarity.' " *Psychological Bulletin,* 1955, *52,* 177-193.

Curry, T. J., and Emerson, R. M. "Balance Theory: A Theory of Interpersonal Attraction?" *Sociometry,* 1970, *33,* 216-238.

Guilford, J. P. *Psychometric Methods.* New York: McGraw-Hill, 1954.

Kenny, D. A., and Nasby, W. "Splitting the Reciprocity Correlation." *Journal of Personality and Social Psychology,* 1980, *38,* 249-256.

Laing, R. D., Phillipson, H., and Lee, A. R. *Interpersonal Perception.* New York: Harper & Row, 1966.

McLeod, J. M., and Chaffee, S. H. "Interpersonal Approaches to Communication Research." *American Behavioral Scientist,* 1973, *16,* 469-500.

Newcomb, T. M. *The Acquaintance Process.* New York: Holt, Rinehart and Winston, 1961.

Polley, R. B. "Both Sides of the Mirror: Small Groups and Subjectivity." Unpublished doctoral dissertation, Harvard University, 1979.

Tagiuri, R., Bruner, J. S., and Blake, R. R. "On the Relation Between Feelings and Perception of Feelings Among Members of Small Groups." In E. Maccoby and others, *Readings in Social Psychology.* New York: Holt, Rinehart and Winston, 1958.

Warner, R. M., Kenny, D. A., and Stoto, M. "A New Round Robin Analysis of Variance for Social Interaction Data." *Journal of Personality and Social Psychology,* 1979, *37,* 1742-1757.

Table 8. Expected Mean Cross-Product Table

Mean Cross-Product	Covariance Component							
	Row by row	Row by col	Col by row	Col by col	Cell by cell	Cell by cell'	Err by err	Err by err'
Row by row	nr	$-nr/q$	$-nr/q$	nr/q^2	nr/q	$-nr/q^2$	n/q	$-n/q^2$
Row by column	$-nr/q$	nr	nr/q^2	$-nr/q$	$-nr/q^2$	nr/q	$-n/q^2$	n/q
Column by row	$-nr/q$	nr/q^2	nr	$-nr/q$	$-nr/q^2$	nr/q	$-n/q^2$	n/q
Column by column	nr/q^2	$-nr/q$	$-nr/q$	nr	nr/q	$-nr/q^2$	n/q	$-n/q^2$
Cell by cell	rq^2/p	$-rq/p$	$-rq/p$	rq^2/p	r	$-r/p$	1	$-1/p$
Cell by cell'	$-rq/p$	rq^2/p	rq^2/p	$-rq/p$	$-r/p$	r	$-1/p$	1
Error by error	0	0	0	0	0	0	1	0
Error by error'	0	0	0	0	0	0	0	1

Note: $n - 1 = q$; $n(n - 1) - 1 = p$.

PART III

In the realm of psychological and social events, the issue of whether a particular phenomenon or causal relationship exists "out there" or only "in the mind of the beholder" is frequently a matter of considerable controversy. More than in the physical domain, questions of validity and objectivity are pervasive, and the social science researcher requires a full armament of sophisticated methodologies to make any finding convincing. In a number of different substantive areas, Donald Campbell has provided novel methodological and conceptual tools to disentangle the observed from the observer.

Substantive Applications

Composite Dispositions

Campbell's own research papers on a number of heterogeneous substantive problems have analyzed the joint contribution of stimulus characteristics and perceiver judgment biases in determining what is observed. This approach is illustrated by Campbell's study of interpersonal perception (Campbell, Miller, Lubetsky, and O'Connell, 1964; see Kenny chapter in the preceding section of this volume). Other research projects have analyzed contrast effects in social perception—particularly, the tendency of perceivers

311

to exaggerate the differences between stimuli that are categorized into separate categories or groups (Campbell, 1956). Campbell has applied this enhancement-of-contrast effect to the analysis of intergroup stereotyping (1967), to the study of ingroup bias, and to ethnocentrism more generally (LeVine and Campbell, 1972; Brewer and Campbell, 1976).

A common theme running through Campbell's treatment of these diverse judgmental phenomena is the idea that such a judgment is a composite response reflecting the individual's compilation and integration of multiple sources of information from past and present experience. The most accurate knowledge, according to this analysis, is a composite formed by a combination, triangulation, and integration of information garnered from maximally different (heterogeneous, divergent) sources of information. In Campbell's teaching, the slogan "heterogeneity of irrelevancies" is continually stressed. The best knowledge will have been garnered from several modes of knowing. Thus, the composite analysis is both descriptive and prescriptive.

Campbell's own illustration of the various modes of acquiring behavioral dispositions (1961, 1963) demonstrates the application of this conceptual theme to a substantive problem and also illustrates the use of a concrete example to make an abstract point:

> I am reminded of the multiple-unit maze for humans which Prof. Warner Brown used to have at Berkeley—a trough maze with 3-inch-high pathway edges through which one shuffled. Socialized human beings acquiring behavioral dispositions in this trough maze . . . will provide the concrete illustrations as we examine the several ways in which these behavioral dispositions can be acquired.
>
> 1. *Blind trial and error—locomotor exploration.* Shuffling blindfolded and with ears covered, the individual could learn Warner Brown's maze without benefit of distance receptors. . . . If the person starts with no initial behavioral dispositions, blind trial and error is the only method of approach. . . .
>
> 2. *Perception.* Were one not blinded in Warner Brown's maze, one acquired the appropriate behavioral disposition so rapidly that the process eluded

study. The fact that perception may be substituted for blind trial and error is so ubiquitous as to go unrecognized. . . . Since a major purpose of the argument is to establish the equivalence of these modes of acquisition in leading to the same behavioral coordination or disposition, let us specify that the subject be allowed to see the maze, explore it visually, and then run the maze blindfolded, as does the person learning the maze by blind trial and error.

3. *Perceptual observation of another person's responses.* Behavioral dispositions may also be learned by observation of another organism's responses to stimuli. This is the traditional social science mode of behavioral-disposition acquisition, known as imitation of responses. . . . Let us suppose that we permit an observer to watch a person who already has learned the maze run it repeatedly. And let us suppose that the observer can see the choice-point stimuli but not the clues for distinguishing blind alleys and main paths. By memorizing the responses and then imitating them, the observer might acquire a behavioral disposition externally identical to that which others had acquired by perception of the maze or by blind trial and error.

4. *Perceptual observation of the outcomes of another's explorations.* The previous mode referred to the observation of another only after he had already acquired an asymptotic behavioral disposition; it represented a learning of responses. The present mode refers to observing another only during his initial exploration of the environment; it refers to learning about the environment and its rewardingness rather than about responses.

5. *Verbal instruction about responses to stimuli.* . . . [It] is obvious that adaptive behavior dispositions can be acquired by verbal instruction. We can instruct a person as to what response to make at the first and second choice points, etc., so that he can run the maze blindfolded, perfectly the first time, with overt behavior which is not distinguishable from that of other modes of acquisition.

6. *Verbal instruction about the characteristics*

of objects. For purposes of a balance to be made clear
later, it should be emphasized that behavioral dispositions can be verbally induced by means of descriptions
of the objects with no mention of responses. Instead
of saying "Turn right," one can say "The left path is
blocked, the right one open."
. . . Not only are the results of these different
modes of dispositional acquisition comparable, they
also combine in ways that so far seem to be additive,
or at least monotonically cumulative, or decremental,
as the case may be [Campbell, 1963, pp. 107-110].

Campbell has applied this notion of composite response dispositions to the study of social conformity (1961; Jacobs and
Campbell, 1961), to the formation and change of social attitudes
(1963; Miller and Campbell, 1959), to the role of learning and cultural factors in visual perception (Segall, Campbell, and Herskovits, 1966), and to the analysis of the subjective experience of hedonic satisfaction (Brickman and Campbell, 1971).

Overview of Chapters

The five chapters in this final section of the volume all reflect Campbell's concern with the interplay among philosophical
issues of knowing, methodology, substantive knowledge, and social application. The first two chapters, by William Crano and
Marilynn Brewer, provide a logical transition from the chapters of
the preceding section, demonstrating how comparative, cross-
cultural data can be used in the spirit of triangulation (as suggested
in the earlier chapter by LeVine) to assess the validity of a particular social science construct. A similar use of cross-cultural comparison is illustrated in the succeeding chapter, by Marshall Segall,
who reviews available evidence on the role of experience in various
aspects of visual and social perception. The composite disposition
(or multiple modes of knowing) theme emerges in this context in
Segall's discussion of biological evolution and learning as adaptive
mechanisms and as separate contributions to the perception of
visual illusions.
The next two chapters illustrate the relevance of hypothetical realism, construct validity, and methodological triangulation to
research in applied areas of clinical psychology and social policy.

The chapter by Barry Collins recapitulates the themes from Campbell's own work (see Chapter One of this volume) in the context of an assessment of the syndrome of "hyperactivity," suggesting (as did Kenny in the preceding section) that the "reality" of a psychological entity may reside not entirely within an individual but, in part, in the interaction between situation and individual. Finally, Norman Miller illustrates how theoretical preconceptions—particularly in policy-relevant domains—can exert a conservative influence on the processing and interpretation of relevant data until competing perspectives are applied to the same settings.

Consistent with the theme that runs throughout this volume, the chapters in this section confirm the relevance of theories of knowledge to the methodology of social science and to the development of the social scientist as competent knower.

References

Brewer, M. B., and Campbell, D. T. *Ethnocentrism and Intergroup Attitudes: East African Evidence.* New York: Halsted Press, 1976.

Brickman, P., and Campbell, D. T. "Hedonic Relativism and Planning the Good Society." In M. H. Appley (Ed.), *Adaptation-Level Theory: A Symposium.* New York: Academic Press, 1971.

Campbell, D. T. "Enhancement of Contrast as Composite Habit." *Journal of Abnormal and Social Psychology,* 1956, *53,* 350-355.

Campbell, D. T. "Conformity in Psychology's Theories of Acquired Behavioral Dispositions." In I. A. Berg and B. M. Bass (Eds.), *Conformity and Deviation.* New York: Harper & Row, 1961.

Campbell, D. T. "Social Attitudes and Other Acquired Behavioral Dispositions." In S. Koch (Ed.), *Psychology: A Study of a Science.* Vol. 6: *Investigations of Man as Socius.* New York: McGraw-Hill, 1963.

Campbell, D. T. "Stereotypes and the Perception of Group Differences." *American Psychologist,* 1967, *22,* 817-829.

Campbell, D. T. "On the Conflicts Between Biological and Social Evolution and Between Psychology and Moral Tradition." *American Psychologist,* 1975, *30,* 1103-1126.

Campbell, D. T., Miller, N., Lubetsky, J., and O'Connell, E. J.

"Varieties of Projection in Trait Attribution." *Psychological Monographs,* 1964, *78* (entire issue 592).

Jacobs, R. C., and Campbell, D. T. "The Perpetuation of an Arbitrary Tradition Through Several Generations of Laboratory Microculture." *Journal of Abnormal and Social Psychology,* 1961, *62,* 649-658.

LeVine, R. A., and Campbell, D. T. *Ethnocentrism: Theories of Conflict, Ethnic Attitudes, and Group Behavior.* New York: Wiley, 1972.

Miller, N., and Campbell, D. T. "Recency and Primacy in Persuasion as a Function of the Timing of Speeches and Measurements." *Journal of Abnormal and Social Psychology,* 1959, *59,* 1-9.

Segall, M. H., Campbell, D. T., and Herskovits, M. J. *The Influence of Culture on Visual Perception.* Indianapolis: Bobbs-Merrill, 1966.

12

William D. Crano

Triangulation and Cross-Cultural Research

To attain a more precise and complete understanding of human behavior, social psychologists long have been encouraged to adopt the "physical science model" in the practice and refinement of their craft. The methodological rigor and general scientific orientation of the physicist are put forward as characteristics to be emulated; and, on the whole, our field has profited from this emphasis. Paradoxically, however, in our adaptation of the physicists' model to the study of social beings, we typically have adopted an orientation of Newtonian characteristics, whose causal, deterministic features were rejected by the majority of physical scientists more than fifty years ago. In one sense, the social sciences' deterministic orientation is appropriate, since it parallels that characteristic of

317

the physical sciences at a comparable level of scientific development. If the historical progression of our field follows that of physics, then it is conceivable that the continued refinement of our basic research technique will occasion a revision in our fundamental view of the structure of social reality, comparable to that which occurred in the early years of this century with the development of the "new physics."

Probabilistic and Deterministic Views in Social Science

The parallel development of our fields is noteworthy because it provides possible clues concerning the directions that social science might traverse in its gradual evolution. As in physics, the abandonment (at least philosophically) of an overliteral operationalism, an operationalism "by definitional fiat" (Webb and others, 1966, p. 3), witnessed the accumulation of reliable, well-grounded information at a rate unprecedented in previous generations in social science. Such an accumulation has proceeded unabated in our field, with differential rates of advance in its various subdisciplines a function of the alacrity with which the scientific method was embraced. The social sciences have yet to arrive at a widespread consensus on a theoretical position that subsumes the mass of information that we have accumulated, the methodology appropriate for the issues at hand, or even common agreement on what these issues are; while preparadigmatic in the extreme (see Kuhn, 1962), however, the general epistemological orientation of social science is causal. Despite our lip service to a probabilistic universe, and the long-standing admonitions of some of our science's greatest minds (for example, Tolman and Brunswik, 1935), our closed-system linear model statistics, our basic experimental research methods, our theories, and our fundamental conceptualizations of the forces that intervene in social affairs all presume a causal, deterministic universe. The typical plaint "If only our measures were better, we would be better able to understand the causal pattern that exists in this interaction" betrays an ignorance or a conscious rejection of the forces of uncertainty that Heisenberg demonstrated more than a half century ago. There can be no doubt that we need better measures, but with them an effect opposite that often presumed is most likely to occur; a more developed methodological armamentarium would likely have the effect of demonstrating ever more compellingly the indeterminacy of the basic stuff of our science.

In physics the shift from the causal, deterministic, and ultimately comprehensible world of Newton and Einstein to the probabilistic, indeterminate reality of Bohr and Heisenberg was based on the development and refinement of measurement technique and the concomitant reduction of error variance that such a refinement allowed. The indeterminists found that the regularities observed in Newton's universe were in large part the biased outcomes of improper or insensitive measuring devices. With more sophisticated techniques and theories of measurement, for example, Heisenberg demonstrated that the nearer one approached an accurate measure of either the position or the momentum of a subatomic particle, the less accurate, of necessity, was the assessment of the other component.

The relevance of this observation for the social sciences is critical, for if measurement impinges on the measured at fundamental physical levels, might it not also influence the outcome of the analysis of social organisms by social observers? While such speculation is premature, the forces of uncertainty that impinge in the analysis of the behavior of subatomic particles might operate, too, in the analysis of social interaction. The recognition of a "social uncertainty principle," should such be warranted, would not negate the possibility of arriving at a set of general laws in the social sciences; such laws, however, would operate as actuarial principles, summarizing the operations and outcomes of the conglomerate social system, rather than as predictors of the specific behaviors of the individual actors comprising the system.

Is uncertainty to be the lot of the social scientist? Will forces similar to those postulated by Heisenberg also be found to impinge on the fundamental data of our field? If the developmental progression of the social sciences is destined to parallel that of the physical sciences, then it is critical that we move toward a resolution of these questions. A review of current practices in our field, however, indicates that such a resolution will not be soon forthcoming, for only rarely are concerns of this sort even noted.

Admittedly, some progress has been made, and methodological development has occurred in this field over the years. Indeed, the rapidity of the evolution of the social sciences, from a general philosophical orientation to a more rigorous scientific system, is noteworthy. Fundamental to this progress was the gradual refinement of measurement techniques and the concomitant development of a general philosophy of measurement that accepted the present inevitability of instrument or observational error. Such an orientation, rather than assuming the perfection of our measurement techniques (an unrealistic assumption at best), compelled social scientists to recognize the unreliability of their observations and to attempt to account for such imperfections through the establishment of more sophisticated and comprehensive designs in their research.

Triangulation

Central to this general methodological orientation is the concept of triangulation, the principal focus of this chapter. In its most simple aspect, triangulation is little more than a reasonable methodological rule of thumb calling for the application of multiple, heterogeneous measures, all of which relate in some specified way to the theoretical construct of central interest. Through the use of multiple indicators, all of which "home in" or triangulate on the critical variable, we hope to identify the common features that our measures and construct share.

In ordinary usage in the social sciences, triangulation is taken to mean little more than the use of multiple indicators in the assessment of a hypothetical construct. As will be noted, however, this simple and reasonable methodological point of view carries with it a host of implications regarding the proper conduct of social research, the effects of imperfection or unreliability of measurement operations on the development of theory, and the manner in which our field might gradually attain the status of the more developed sciences.

In a tribute to Donald Campbell, a review of triangulation is quite appropriate, since this methodological emphasis is fundamental to all of the more complex technical contributions he has made over the years. In this discussion, therefore, I shall describe the central features of the concept, show its relevance to some of Campbell's major methodological innovations, and illustrate the utility of the "triangular" orientation in a secondary analysis of cross-cultural archival data.

Methodological Triangulation. In theory, if not in practice, triangulation has been adopted widely as a fundamental of social science research methodology. At the same time, the implications of this orientation often have been misunderstood. Triangulation of measures follows from the fact that no single measure is, or can be, a perfect indicator of the hypothetical construct of interest. The acceptance of this position is important; it emphasizes the necessity for a continual technical refinement of our measurement methodology, which, in turn, will allow for the development of theoretical positions of ever increasing validity.

Campbell's recognition of the imperfection of our common techniques of measurement, and the implications of such imperfections for the construction of a science of human behavior, is long standing, and this fact helps to explain his stress on the necessity of triangulation. Consider the following observation:

> As I have documented in more detail elsewhere (Campbell, 1953, 1963b), when stimulus is defined as a single sense-receptor activation, and response as muscle contraction, almost no stimulus-receptor consistencies are in fact found, particularly in higher animals. Just as any one retinal cell activation may be a part of a wide range of differently significant percepts, so too a given muscle contraction (or larger leg movement) can be a part of many different acts. These specifiable particles are inevitably equivocal. Their utility (or the utility of lever-press records) in regard to a learning theory is not as operational definitions of theoretical terms, but as corrigible but nonetheless useful confirmations of hypothesized entities and states, such as habits or meanings or perceptions, that are only indirectly confirmed and that are thus relegated to that limbo of uncertainty shared by the entities and processes of physics, chemistry, and astronomy.... Direct knowing, nonpresumptive knowing, is not to be our lot [Campbell, 1969a, pp. 49, 52].

If, as Campbell observed, the interpretation of any single measure is "inevitably equivocal" (because of the multiplicity of factors that might have affected it), then the use of multiple indicators is clearly indicated as a common methodological practice. The rationale for this suggested methodological orientation is based on widely accepted beliefs regarding the imperfection of measurement processes in our science. We know, for example, that no single measure can capture completely the meaning of a theoretical construct. Each indicator that we employ will possess extraneous features that are not in common with the construct it is designed to identify. How does triangulation help alleviate this problem? By encouraging the scientist to employ many indicators, all of which share some common feature with the core construct but which also have nonoverlapping "unique" (or extraneous) aspects. We admit that none of the measures that we use are perfect; all will miss the mark to some extent, but they are chosen so as to miss it in different ways. The resulting intersection of measures helps to identify the boundaries of the construct under study. The pattern of overlap of the indicators employed provides the best picture (though often an ephemeral one) of the core conceptualization.

A good example of the use of triangulation in social science is provided in the area of attitude measurement. When attempting to assess attitudes, we usually are not content to employ a single item to represent a person's feelings about an issue. Why? Because we acknowledge that any single item can be susceptible to a host of extraneous influences that might have little, if anything, to do with the concept under consideration. The use of multiple items, triangulated on the attitude of interest, however, helps to overcome this difficulty. Theoretically, each item is thought to consist of a combination of "true score" and "error." True score represents the reliable, consistent relationship that exists within the item set. Measurement errors across items are thought to be uncorrelated; that is, though the imperfection of any single indicator is accepted, it is assumed that, over all items, these imperfections will not cumulate or fall into a consistent pattern. The resulting overlap of measures (the true score relationship) thus provides the best available indication of the construct.

A Physical Analogue. To elaborate further the concept of triangulation, let us consider a physical analogue: Suppose that we were to attempt to pinpoint the origin of a radio signal. To do so, we would require for our task two mobile, directional antennae, which we would move through space until each had fixed strongly on the signal. The direction of the signal relative to each unit would then be recorded, and the intersection of the projections that extended from each monitor would identify the wave source in two-dimensional space.

It is instructive to translate this example to the realm of the social sciences. To begin with, we employed only two monitors (radio detection units) in our example. This was somewhat risky because, if the monitors were prone to even small amounts of measurement error, our identification of the source of the signal would have missed the mark. Obviously, accuracy could have been improved substantially if more monitors had been employed. The distortions introduced through our somewhat faulty monitors would have tended to cancel if numerous units had been used; the resulting intersection of numerous projections would have provided a better estimate of the location of the signal than that derived from the intersection of only two projections. In the social sciences, the use of multiple, triangulated indicators is likewise the almost invariably preferred measurement alternative. As suggested in the physical analogue, it is almost inevitable that, within certain constraints (noted below), the more measures employed, the more precise the identification of the construct of central relevance. This observation does not imply an operationalism run amok but suggests, rather, a reasoned recognition of the frailty of our measurement processes.

This is not to suggest that the multioperational approach will address all the difficulties that afflict the scientific enterprise in our field. While the use of multiple, triangulated measures is clearly indicated, even with perfectly reliable and accurate measures—a sought-for end point of all of our measurement processes—there will always remain the unresolvable question of validity, in the epistemological sense. Even under conditions of "perfect" measurement, we must still contend with the possibility of an ultimately indeterministic social universe.

In continuing our emphasis on a multiple indicators or triangular approach, we must note that not any measure, chosen haphazardly, will enhance the "fit" between measurement operations and the theoretical construct. Returning to our example, it is clear that the closer the antennae to the origin of the signal, the more likely a pinpoint identification, since minor angular distortions over short distances would not result in major errors of specification. Clearly, measures must be chosen judiciously, so that they bear some obvious theoretical or empirically established relationship to the construct of interest.

Perhaps not so obvious is the necessity for some "distance" between the identifiers. In the physical analogue, the angular projections of our monitors would have overlapped or paralleled one another had little or no distance separated the signal detection units. Much the same difficulty exists when two identical measures are employed in the social sciences. If our measures differ only in error variance (that is, if they provide identical true score estimates), then they cannot triangulate on the construct of interest. An isomorphism of indicators rules out the possibility of triangulation. The point here is that the relationship between indicators must be known. Whether the identifiers are convergent or divergent makes little difference in terms of the quality of the triangulation. But we must know their relationship. We cannot triangulate measures, even very accurate ones, when the distance between them is zero.

Methodological Outgrowths of the Triangular Approach. The physical analogue will, I hope, bring to mind some of Campbell's classic methodological contributions: his insistence on a multiple operational orientation in the social sciences; his invention, with Fiske, of the multitrait-multimethod matrix; his observations regarding the limitations on the appropriate use of partial and multiple correlational approaches in hypothesis testing (see Campbell and Fiske, 1959; Brewer, Campbell, and Crano, 1970); and his invention of the cross-lagged panel correlational technique (Campbell, 1963a; see also Pelz and Andrews, 1964).

In this book so much has been said already of Campbell's multiple operationalism that we need not recapitulate this aspect of his thinking. The point that might be reemphasized in all of this, however, is that the multioperational approach implies much more than a mere piling on of instruments. Triangulated measures whose relationship is known or given (theoretically) are implied in Campbell's multiple operationalism. This necessity is not always recognized in the mundane practice of our science.

Recent developments in the use and extension of the multi-trait-multimethod matrix approach suggest a renewed appreciation of the role of triangulation in our science. This approach, which calls for the intercorrelation of a number of similar and dissimilar traits, each of which is measured in a number of different ways, allows for a partitioning of the resulting matrix of covariation into validity and error components. The correlation of traits that are theoretically similar, across diverse methods of measurement, represents an instance of a methodological triangulation that provides evidence for the validity of the construct of interest (as does the divergence between theoretically dissimilar traits or indicators). Between-trait similarities that exist solely as a function of a shared measurement operation also can be identified and distinguished from theoretically relevant dependencies. The importance and the implications of the multitrait-multimethod matrix are now recognized. The "rules of evidence" that Campbell and Fiske suggested be employed in the assessment of such matrices, for example, have been mathematized (Boruch, Larkin, Wolins, and MacKinney, 1970; Kavanaugh, MacKinney, and Wolins, 1971; Stanley, 1961), and the basic methodology extended, in the special case of longitudinal panel data, to provide a more precise estimate of the convergent and discriminant validity contours in the data matrix (Mellon and Crano, 1977).

Another methodological innovation that stems directly from a consideration of some of the central features of triangulation was presented in Campbell's discussion of the use and misuse of the partial correlational approach in the testing of hypotheses (Brewer, Campbell, and Crano, 1970). In the radio wave example, it was argued that the use of two identical monitors, which differed only in their differential susceptibility to distortion or error, would prove relatively uninformative. Similarly, it was argued, the use of essentially isomorphic measures—which, because they have nonoverlapping error patterns, are thought to tap different aspects of the phenomenon of central interest—also will probably produce misleading outcomes. This point, which is quite obvious when presented in the abstract, is a source of major confusion in practice. The paper by Brewer, Campbell, and Crano emphasizes this point and presents a number of examples from the literature that suggest a misunderstanding of this fundamental idea. The principal message in this research concerns the necessity for the triangulation of multiple measures, separated not in physical distance, as in the radio analogue, but in operational distance between the identifiers. Though observed in practice, the "triangulation of isomorphisms" is a logical nonsequitur. It is unfortunate that the obviousness of this truism, so marked in physical settings, is so invidiously murky in social science situations.

As a final generative extension of the concept of methodological triangulation, it seems reasonable to consider the cross-lagged panel correlational (CLPC) technique. While the link between triangulation and this method might not be immediately obvious, it is interesting to speculate on the formative relationship between this orientation and the development of the cross-lagged panel approach, one of Campbell's most intriguing methodological innovations (see Campbell, 1963a; Crano, Kenny, and Campbell, 1972; Kenny, 1973, 1975; Rozelle and Campbell, 1969).

The CLPC technique, which combines the logic of the experimental method with the descriptive richness of the multivariate correlational approach, developed as a result of Campbell's dissatisfaction with traditional methods of data collection and the concomitant effect of these methods on the development of theory. In the CLPC technique, the temporal order of repeated panel measurements is employed to generate inferences regarding the functional relationships that exist among a set of measures. In most typical applications of this technique of late, the method of investigation involves the combination of reasonably large numbers of panel variables over multiple time periods (see, for example, Crano, 1977; Crano and Mellon, 1978). As such, the technique implicitly propounds a methodological triangulation, in that the pattern of correlational discontinuities among a series of heterogeneous measures provides the basis for the interpretation of results. A consideration of the whole, rather than of minor topographical features of the whole, is a necessity for proper interpretation in such applications.

Theoretical Triangulation. It would be a mistake to confine our consideration of triangulation to a strictly methodological format, for an understanding of much of Campbell's work on the construction of theory is tied to an appreciation of his view of the role of what might be called "theoretical triangulation." In the development of models of social behavior, Campbell has suggested that the most promising approach to the construction of plausible theory involves a matching of the contours of the data field with an appropriately curved or malleable theoretical template—a pattern-matching process, through which the consistencies between observations and theoretical expectations become increasingly obvious with each successive iteration. In his classic "Pattern Matching as an Essential in Distal Knowing" chapter, Campbell was quite explicit on this point:

> It has long been a common property among logical positivisms to describe scientific theory as an internally consistent formal logic (analytically valid) which becomes empirical (gains synthetic truth) when various terms are interpreted in a data language. A variant of this general model is [proposed] here. The formal theory becomes one "pattern," and against this pattern the various bodies of data are matched. ... These empirical observations provide the other pattern, but somewhat asymetrically. The data are not required to have an analytical coherence among themselves, and indeed cannot really be assembled as a total except upon the skeleton of theory. In addition, the imperfection or error of the process is ascribed to the data pattern, for any theory-data set regarded currently as "true" ... the error in matching theory to data is allocated to the imperfect representation of the theoretical concepts by the data series. ... In practice, no theory that has been judged useful in the past of a science is ever rejected simply upon the basis of its inadequacy of fit to data. Instead, it is only rejected when there is an alternative that fits better to replace it [Campbell, 1966, pp. 97-99].

Here we have the kernel of Campbell's theoretical triangulation. He suggests a triangulation of data pattern with theory pattern, which—given a continuous series of adjustments, a continuing modification of the theoretical template—leads to a more precise understanding of the nature of the social reality. We despair of arriving at a final truth in such a system, because the match of theory with data never can be perfect. But variations in degree of perfection are certainly possible in this scheme, and approaching the ideal is far from an unreasonable preoccupation. In a less ethnocentric social science, the overlap of subdisciplines would be expected to accelerate the template adjustment process (Campbell, 1969b). In thinking back to the radio wave analogue, we find that our example is not sufficient for the conceptualization of theoretical triangulation, for now we are concerned with the triangulation of planes rather than of lines. In pattern matching, we attempt to locate a phenomenon in an n-dimensional (theoretical), as opposed to two-dimensional (methodological), space.

The general principle of theoretical triangulation does not demand that the overlap of patterns of data and theory be all-encompassing; indeed, we presume that such a fit would prove ultimately unreachable. A theoretical projection that covers only a small portion of the data field, however, is not an unreasonable starting point in today's social science. (How many times in the depths of the existential dither that only a student of experimental social psychology could experience were we encouraged by Campbell's aphorism "A psychology of the college sophomore is better than no psychology at all!")

An Application of Triangulation in Cross-Cultural Research

While it is true that a general methodological orientation consistent with the concept of triangulation can enhance the potential validity of any social inquiry, this observation is perhaps nowhere as compelling as in the study of the most fundamental of social institutions, the family. A good deal of the controversy that surrounds the social psychological analysis of the family can be traced to the (sometimes unavoidable) failure of researchers to frame their inquiries in a form consistent with the general multi-operational orientation that has been the focus of discussion throughout this chapter. Neglect of the research principles that can be inferred from a consideration of triangulation often results in the postulation of theoretical positions of apparently wide generality which, in reality, represent more the failure of a measure than the accurate depiction of existing relationships.

Because this area provides a compelling argument for the adoption of a triangular orientation, the remainder of this chapter will be devoted to a discussion of research on factors that influence the allocation of roles, and role complementarity, in the family. In this section, the plausibility of Parsons and Bales' (1955) widely acknowledged theory of family role allocation will be discussed in light of our recent analyses.

The Ethnographic Atlas. In this research two data sets are employed. The first, Murdock's (1967) Ethnographic Atlas, contains nominally coded information regarding the social, cultural, political, and economic practices of more than 800 societies. These nominal classifications were recently recoded into ordinal form by Smith and Crano (1977a), and this revised form of the Atlas was employed in the research to be discussed. The second archive is the Standard Cross-Cultural Sample (Murdock and White, 1969), a set of 186 societies, selected from the larger Atlas, on which more extensive amounts of information are available.

These archives are the outcome of an almost Herculean feat on the part of Murdock and his colleagues to catalogue all the known ethnographic information on all societies studied over time, to establish a standard set of variables or categories along which each society is referenced (information on a common set of sixty-five traits was included in Murdock's Atlas), and to do so for as many different societies (even those no longer extant) as possible. That Murdock actually was able to produce such an archive is extremely impressive.

This research, employing Murdock's archives, is based on the observations of the hundreds of anthropologists and ethnographers whose work is summarized in the Atlas. As such, it can be viewed in light of the principles of methodological triangulation that are presented throughout this chapter. Let us consider each of the ethnographers who contributed to the Atlas as a different "method" (of measurement), and each society investigated as a different "trait." If this view is adopted, we can conceptualize our data set as a special form of an incomplete multitrait-multimethod matrix, a methodological outgrowth of the general principle of methodological triangulation. In this incomplete multitrait-multimethod matrix, multiple measures (that is, ethnographers) are typically "applied" to each trait (that is, a given society might be investigated by many researchers); but since not all measures are applied to all traits, the method-by-trait matrix is incomplete. The application of multiple, heterogeneous measures—whose individual data-biasing features would not be expected to cumulate—in a host of varied observational situations conforms to the general emphasis on triangulation that has been advocated throughout this chapter. While not a perfect example of methodological triangulation, the Atlas does appear to provide many of the advantages that such an approach to research involves.

Criticism of the Atlas. Not surprisingly, perhaps, there are a number of critics of Murdock's quantitative approach and of other attempts at quantification in cultural anthropology (for example, the Human Relations Area Files). The general objection voiced by such critics is that the data gathered by the many ethnographers whose labors are summarized in the Atlas might be biased; that is, since the contributing ethnographers might have been working under a series of powerful theoretical presuppositions, their descriptions of the social reality cannot be trusted. This "ethnographer bias" argument is similar to the contention that the results of a laboratory experiment cannot be trusted because the central dependent measure was biased. However, such a criticism is not plausible in situations that conform to the general guidelines of methodological triangulation discussed earlier in this chapter. By providing multiple, heterogeneous measures, all focused on the same construct, event, or trait (as in the Atlas), we circumvent the difficulties that a more restrictive, single-indicator approach inevitably raises.

In light of this observation, it appears that the critics of Murdock's quantitative approach fundamentally are misinformed. Their failure to understand the twin principles of methodological and theoretical triangulation, along with their neglect of the advantages of a "heterogeneity of irrelevancies," are at best incorrect and at worst antiscientific. Of course, the ethnographies on which the Atlas was based might well have been biased by the theoretical presuppositions of the ethnographers who created them. But to rule out all such research because the data represent a mixture of truth and error (or bias) is to reject the entire social science enterprise. One might just as well rule out all interview or questionnaire data.

As was indicated, we welcome a heterogeneity of irrelevancies in archives of this type. It is not surprising, that is, to find less

than perfectly reliable indicators (or ethnographers); but we assume that, while all reports of all cultures are in some ways errorful, the errors observed across societies are not related—that is, in
the absence of evidence to the contrary, they are assumed independent. While it is not likely that any ethnographer will reach
perfection of reportage, it is even less likely that all ethnographers
will err in the same way. Such misses would not be expected to
cumulate systematically; thus, bias would not be expected. Surely
the factors that influenced Fra Junipero Serra's ethnography were
different from those that affect Bronislaw Malinowski's. To dismiss ethnographic evidence in favor of that collected only by the
"right" informant is not only elitist and antiscientific but self-
defeating as well. In short, we would argue that the regularities observed under highly error-laden circumstances are quite likely to
be valid indicators of the underlying reality.

The critics of the ethnographic archival approach err in focusing their attention on individual ethnographers or their works.
When applied in the context of archival cross-cultural research,
such critiques rarely can prove compelling, because almost always
they focus on the consideration of (random) error instead of the
more crucial issue of systematic bias. It is the total pattern of evidence, all of whose components might be quite imperfect, that is
at issue, rather than the validity of any single ethnography (or set
of ethnographies) considered in isolation. A demonstration of the
imperfection of an indicator, or of entire sets of indicators, should
not prove surprising, nor should it become the sought-for end
point of the methodological critique. Rather than the mere destruction of the given, a plausible replacement must be provided if
scientific advance is to occur. In a discussion of the scientific and
social responsibilities of the methodologist, Crano and Brewer
(1973, p. 319) made this point explicit:

> While research studies in social psychology are
> like fingerprints in that no two are exactly alike, they
> all do share one common feature—their imperfection.
> A point that has been made repeatedly . . . is that the
> results of any research study considered in isolation
> are always subject to multiple interpretation. To say,
> then, that the outcome of any particular study has

failed to "prove" a theoretical point is to state the obvious, and clearly should not constitute the limits of research criticism. The burden of the methodologist is to identify those alternative interpretations that have the status of general principles with applications beyond a particular research instance. It is not sufficient that the methodologist be able to suggest alternative explanations for given research findings, but rather that he provide alternatives of demonstable plausibility. The methodologist, in other words, is subject to the same rules of scientific credibility as the theorist.

The critics of the archival approach in cross-cultural research also appear to misunderstand a very important aspect of theoretical triangulation; namely, the stress on the plurality of information sources necessary to arrive at a persuasive social science regularity. If the cross-cultural archive were the only source employed in the development of a science of human behavior, then the criticism of such an approach would be much more compelling. But that is far from the case. The information generated in such studies is subject to the same processes of verification and falsification as those of any other social finding. And in those cases involving a failure of fit, the same winnowing processes between expectation and observation are, or can be, employed.

This is not to suggest that further improvements in the development of the archival data base are not needed, or that the methods of analysis applied to such sources of information are perfect, or even sufficient. My criticism is aimed at those who despair of the application of the scientific method to cultural anthropological data; it is not meant to devalue the legitimate and necessary developments currently observed in archival cross-cultural research methodology. For example, Naroll's work in this area remains a shining example of a reasoned and constructive approach to the maximization of the signal-to-noise ratio of ethnographic data (see Naroll, 1962, 1965, 1976; Naroll and Cohen, 1970; Naroll and Naroll, 1973). Efforts of this type clearly are to be nurtured and encouraged.

Substantive Applications

Having described the use of the cross-cultural archive as an example of the reasonable extension of the principle of triangulation, we can now consider some recent substantive investigations of family structure that have made use of such archives. The underlying theme that runs throughout this work (see Aronoff and Crano, 1975; Crano and Aronoff, 1978; Smith and Crano, 1977a, 1977b) is our conceptualization of each of the ethnographies subsumed in the Atlas as a measure, though imperfect, of the social reality experienced in the societies under study. As such, the principles of triangulation discussed in this chapter can come into play. Let us, then, examine the utility of triangulation in a consideration of family role structure.

Role Allocation. From the earliest days of our discipline, the principle of role complementarity has played a central part in the social psychological analysis of the family (Durkheim, 1893; Parsons and Bales, 1955). Following Durkheim, the sociological analysis of the family pictured this group as an economic unit, a central aspect of which is concerned with the division of labor in the tasks of production. The division of labor in the fundamental subsistence activities of the society came to be seen as a basic and invariant feature of the family, and was thought to arise in the service of the maximal utilization of labor through task specialization. This reasoning was developed in Murdock's (1937) research, which suggested a segregation of task activities by sex. In Murdock's study, hunting, fishing, mining, and herding were found to fall almost exclusively in the domain of the male, while burden bearing, mat weaving, cooking, and other such tasks were assumed, almost without exception, by women. In his later application of these results, Murdock (1949) employed his findings to support a more inclusive biosocial principle of role differentiation. As males possessed greater strength, Murdock hypothesized that they generally could be expected to undertake the more strenuous tasks of subsistence production, while women, constrained by the twin physiological functions of pregnancy and nursing, could be expected to concentrate their activities on the more "simple" tasks that could be performed near the home. These biosocial observations formed the core of Murdock's view of task specialization in the family.

Later theoretical developments generally were consistent in form and substance with the logic of this approach. For example, in dealing with what had become by then a vast store of empirical observations, Parsons and Bales (1955) attempted to locate the study of the family within the perspective of the small group, and thus bring to bear the extensive insights of this area of research to the relatively formless scientific literature on the family. In his study of small groups, Bales found that two contrasting behavior patterns regularly accounted for a good deal of the variation observed and that, typically, these contrasting patterns resided in different persons. The "task specialist" role was assumed by persons who directed the group's behaviors toward the successful completion of the activity at hand, while the "socioemotional specialist" was seen typically as more concerned with expressive relationships and thus more involved in the maintenance of the integrity of the group. In their discussion of the family as a specific exemplar of a more general social system, Parsons and Bales (1955, p. 151) translated their task-versus-socioemotional-specialist distinction directly: "Considered as a social system, the marriage relationship is clearly a differentiated system. . . . [The] more instrumental role in the subsystem is taken by the husband, the more expressive by the wife. . . . [The] husband has the primary adaptive responsibilities, relative to the outside situation . . . whereas the wife is primarily the giver of love."

In a further extension of these observations, Zelditch (1955), a contributor to the Parsons and Bales (1955) volume, developed a theoretical proposition that integrated the insights of the small-group researchers with the ethnographic-based observations of Murdock (1937, 1949). In this extension, which ultimately was to assume the status of a cross-cultural universal of family role structure, differentiation into instrumental and expressive specialist roles was postulated as necessary for the integrity of the family. In addition, it was hypothesized that these roles could not be assumed consistently by the same person; thus, the optimal structure, in Zelditch's view, was characterized by the segregation of task-related and expressive activities.

In the family, the allocation of roles was determined by a

specific function of the female that could not be assumed by the male; namely, the nursing of infants (note that Murdock's allocation principle was based on sex-related variations in physical strength). Nursing, in Zelditch's theory, established the mother as a source of security, comfort, and stability. Because of the special emotional relationship that developed as a result of this activity, the woman was thought to become the family member most likely to specialize in the socioemotional well-being of the family unit. The male, almost by default in this scheme, was allocated the role of production or task specialist.

Reflection on the principles of methodological and theoretical triangulation indicates clearly the riskiness of Murdock's and (later) Zelditch's postulates, both of which attempt to build a theory of social structure on the foundation of a single variable. The existence of even minor biases or unreliabilities in these indicators would prove extremely troublesome, given each scheme's almost complete dependence on a single variable. The discovery of universal regularities in family structure and role allocation truly is a compelling quest; but, in retrospect, an analysis based only on sex-linked variations in physical strength or mammary functioning seems almost predestined to fail.

In the absence of a large and complete cross-cultural data set focused on the allocation of roles, and information regarding the relative contribution of such roles to the maintenance of the social unit, the acceptance of such a universalist position is more a matter of palate than of scientific logic or empirically based observation. Such a deficiency was alleviated with the publication of Murdock's (1967) Atlas, which provided information on the relative dependence of each of more than eight hundred societies on five central subsistence activities, and on the relative contribution within societies, by sex, to each of these activities.

With this archive we can move beyond the single-variable foundation on which Murdock (1937, 1949) built his earlier theory of role specialization, and on which the later theoretical structures of Parsons and Bales and of Zelditch were constructed. As noted, in these earlier attempts Murdock based his propositions on the fact that certain economic tasks were allocated almost exclusively to males, and others were allocated almost exclusively to females. However, he was unable to determine the relative importance of these exclusionary activities to the physical well-being of the society, and this information is critical if one is to construct a functionalist theory of universal sex-linked role allocation rules.

Instrumental Specialization. In the Atlas Murdock (1967) and his colleagues used the available ethnographic evidence to estimate the degree to which each society was dependent on the central subsistence traits of gathering, hunting, fishing, animal husbandry, and agriculture. In addition, information was provided on the extent to which males and females contributed, independently, to each of these central subsistence activities. By a simple process of multiplying a given society's dependence on a particular subsistence trait by the extent to which females contributed to the activity, and summing over all five central subsistence categories, we can obtain a rough but telling measure of the extent to which women are involved in task-oriented production activities. If males truly assume the burden of the task specialist—that is, if the "more instrumental role in the subsystem is taken by the husband," who has "the primary adaptive responsibilities, relative to the outside situation" (Parsons and Bales, 1955, p. 151)—then the lion's share of subsistence contributions over the mass of preliterate societies would be provided by the males.

An analysis of this type was undertaken by Aronoff and Crano (1975) and provided almost no support for the universalist hypothesis of Parsons and Bales. In this research a series of analyses inspecting sex-linked subsistence contributions over six major culture regions of the globe all demonstrated an appreciable female involvement in the instrumental subsistence activities of their societies, both within and across culture regions. So great were women's contributions that if they were removed from the subsistence larders of their respective societies, and not replaced, starvation almost inevitably would have resulted.

In terms of data quality, and of the possibility of a systematic bias in the archive, it is important to note that no theory of social structure has ever postulated the centrality of female subsistence contributions. As such, ethnographic observations biased by one or another theoretical preconception would have militated against the observed results rather than in their favor. Thus, if bias exists in the analysis, it probably operates to cause an underestimate of the true extent of women's instrumental contributions.

It is not reasonable to assume that the findings presented here are a result of some major variations in sex-linked contributions within culture areas. As Aronoff and Crano (1975, p. 18) demonstrated, in fully 45 percent of all societies investigated, women's subsistence contributions accounted for more than 40 percent of the foodstuffs of their respective groups. Although these data do not test for possible power differentials associated with the instrumental role (see Zelditch, 1955; Stephens, 1963), it is obvious that, in terms of sheer magnitude of contribution, the role of women in task-oriented activities long has been underestimated.

Another feature suggesting the validity of these results is the relative consistency of findings across widely diverse geographical areas. In some instances in ethnographic research, a good deal of the information concerning the relationships among a set of variables, over a restricted geographical region, is the product of the research of one particular "school" of anthropology. In such a case the theoretical biases of the school might well influence the outcome of the analysis. In the present instance, however, the relative consistency of females' subsistence contributions to their respective societies, observed over all areas of the globe, would cast severe doubt on the plausibility of such an alternative interpretation. In this case a triangulation of results over a large number of observations (societies), employing a large number of measures (ethnographers) and culture traits, lends greater confidence to the interpretation of the findings.

The summary implications of Aronoff and Crano's (1975) results suggest that the biosocial differences on which Zelditch based his role allocation system did not shape the definition of expressive and instrumental specialization within societies consistently. The results provide no support for the proposition that males generally assumed the role of instrumental leader or that females uniformly adopted the socioemotional specialist role—though they undoubtedly assumed the task of nursing.

Expressive Specialization. Although there appears to be little evidence of sex-linked role allocation or complementarity in terms of basic subsistence contributions, it is important to determine whether there exists some degree of sex-linked expressive specialization in the special realm of childrearing, an endeavor that long has been thought in our field to fall within the domain of the female. This issue is of obvious relevance to the general theoretical thrust of Zelditch and of Parsons and Bales, since both approaches propose that women are the primary expressive specialists across societies and, further, that there exists some complementarity of instrumental and expressive role contributions within societies.

In a study that examined the absolute extent to which males and females contributed to caretaking responsibilities (see Crano and Aronoff, 1978, for a description of the exact operations involved), we find a considerable female investment in their infants. For example, in more than 90 percent of the societies for which data (from the Standard Cross-Cultural Sample) were available, mothers were viewed as the principal caretakers of the infant. While the distribution of fathers' involvement scores was not nearly so skewed, there was evidence nonetheless of considerable father-infant interaction; in 32 percent of all societies, fathers maintained a close or frequent relationship with the infant. In addition, there is some evidence of complementarity within the expressive role, as the correlation between mothers' and fathers' caretaking involvement with their infants was significantly, and negatively, correlated ($r = -.29, p < .05$). That is, the more intense the mother's involvement on behalf of the infant, the less involved the father. While the strength of this relationship is far from overwhelming, its direction is as expected.

As the offspring reach early childhood (one year to four or five years of age), however, the apparent caretaker consistencies observed in infancy begin to evaporate. First, the intense level of mothers' involvement with caretaker activities is strongly attenuated. For example, in none of the societies was the young child's (versus the infant's) care assumed exclusively by the mother. Further, in nearly 32 percent of the societies sampled, the majority of the young child's time was spent away from the mother.

Fathers' contributions to the caretaking of their young children appeared to be somewhat similar in extent to that which they exhibited on behalf of their infants. We found a very strong correlation between fathers' caretaker involvements on behalf of their infants and young children ($r = .76, p < .01$). Apparently, the fathers studied in Barry and Paxon's (1971) archive were relatively consistent in their caretaking behavior, at least across the infancy and early childhood of their offspring. Interestingly, evidence for complementarity of parental caretaker investments, somewhat apparent in the infant analysis, was totally absent in the early childhood comparison. The correlation between fathers' and mothers' expressive involvements with their young children ($r = .00$) was as weak as any relationship ever reported in the literature.

Instrumental and Expressive Complementarity. The final analysis to be reported here combined the results of the study of expressive specialization with the instrumentally oriented findings of Aronoff and Crano (1975), discussed earlier. In this analysis Crano and Aronoff (1978, pp. 469-470) attempted to determine the degree of complementarity between expressive and instrumental contributions. This issue represents an important test for the Zelditch and the Parsons/Bales hypotheses, as complementarity of instrumental and expressive roles is a central expectation in both theoretical statements. In our analysis, however, we found no evidence whatsoever for complementarity of roles. The correlations between males' and females' expressive and instrumental involvements accounted for practically no variance. While it is not possible to disprove a theory on the basis of a null result, an entire matrix of null findings can prove somewhat persuasive.

Taken as a whole, the data of our two investigations seem reasonably compelling. There is almost no evidence supportive of a complementarity effect, either within a specific role (across sex) or between roles. The modest association between fathers' and mothers' expressive commitments in the infancy of their offspring is simply not a powerful enough base on which to build a universal principle of family structure. The universal role allocation rules, so plausible at first glance, now appear unfounded. We come to these conclusions somewhat reluctantly but without surprise. The universal regularities sought did not exist—perhaps they could not exist. If such fundamental processes as the visual perception of the Müller-Lyer illusion can be altered by the social world of the observer (Segall, Campbell, and Herskovits, 1966), how much more must the social fabric affect the allocation of expressive and instrumental specializations, and their mutual interdependence?

In the exploration of this issue, Aronoff and I were fortunate in having available in usable summary form the insights of a host of ethnographers, whose toil over the past two centuries has made possible the development of one of our science's most valuable repositories of information on human behavior. Considering each of the available ethnographies as an imperfect projection of the underlying social reality enables a triangulation of indicators that results in a much more accurate map of the reality than could be derived from a single-indicator approach. The more widespread application of a triangular orientation in social science will minimize the measurement error that afflicts us currently. As I have argued throughout this chapter, the end result of this process will not, indeed cannot, result in the discovery of a causal, deterministic universe. Rather, the widespread adoption of such a research orientation will result in a more precise picture of the probabilistic uncertainties with which we must forever contend. The ultimate promise of methodological and theoretical triangulation is not truth in an absolute sense but an accurate picture of the limits of our science.

Epilogue

In his later years, Einstein is said to have admitted to James Franck, "I can, if the worst comes to the worst, still realize that the Good Lord may have created a world in which there are no natural laws. In short, a chaos. But that there should be statistical laws with definite solutions, i.e., laws which compel the Good Lord to throw the dice in each individual case, I find highly disagreeable" (quoted in Seelig, 1956, p. 209). While it can be foolhardy to argue with a giant, we might evaluate Einstein's Deity in a very different light. Far from the perverse crapshooter pictured in Einstein's mind, we might envision instead a God who plays dice with the world as a more complex and interesting, if somewhat impish, God than One who would set up an ultimately knowable, and thus ultimately boring, deterministic universe. We do not despair of the indeterminism of our social world; indeed, we might well revel in it. As scientists, such a vision clearly defines our role, and it is a more interesting role than that possible under conditions of eventual certainty. Rather than predicting with unerring accuracy the outcome of each throw of the dice, we have as our task at this stage of our development the determination of the number of sides of each die, and the number of dice being used in the game.

References

Aronoff, J., and Crano, W. D. "A Reexamination of the Cross-Cultural Principles of Task Segregation and Sex Role Differentiation in the Family." *American Sociological Review*, 1975, *40*, 12-20.

Barry, H., III, and Paxon, L. M. "Infancy and Early Childhood: Cross-Cultural Codes." *Ethnology*, 1971, *10*, 466-508.

Boruch, R. F., Larkin, J. D., Wolins, L., and MacKinney, A. C. "Alternative Methods of Analysis: Multitrait-Multimethod Data." *Educational and Psychological Measurement*, 1970, *30*, 547-574.

Brewer, M. B., Campbell, D. T., and Crano, W. D. "Testing a Single-Factor Model as an Alternative to the Misuse of Partial Correlations in Hypothesis-Testing Research." *Sociometry*, 1970, *33*, 1-11.

Campbell, D. T. "Operational Delineation of 'What Is Learned' via the Transportation Experiment." *Psychological Review*, 1953, *61*, 167-174.

Campbell, D. T. "From Description to Experimentation: Interpreting Trends as Quasi-Experiments." In C. W. Harris (Ed.), *Problems in Measuring Change*. Madison: University of Wisconsin Press, 1963a.

Campbell, D. T. "Social Attitudes and Other Acquired Behavioral Dispositions." In S. Koch (Ed.), *Psychology: A Study of a Science*. New York: McGraw-Hill, 1963b.

Campbell, D. T. "Pattern Matching as an Essential in Distal Knowing." In K. R. Hammond (Ed.), *The Psychology of Egon Brunswik*. New York: Holt, Rinehart and Winston, 1966.

Campbell, D. T. "A Phenomenology of the Other One: Corrigible, Hypothetical, and Critical." In T. Mischel (Ed.), *Human Action: Conceptual and Empirical Issues*. New York: Academic Press, 1969a.

Campbell, D. T. "Ethnocentrism of Disciplines and the Fish-Scale Model of Omniscience." In M. Sherif and C. W. Sherif (Eds.), *Interdisciplinary Relationships in the Social Sciences*. Chicago: Aldine, 1969b.

Campbell, D. T., and Fiske, D. W. "Convergent and Discriminant Validation by the Multitrait-Multimethod Matrix." *Psychological Bulletin*, 1959, *56*, 81-105.

Campbell, D. T., and Stanley, J. C. "Experimental and Quasi-Experimental Designs for Research on Teaching." In N. L. Gage (Ed.), *Handbook of Research on Teaching.* Chicago: Rand McNally, 1963.

Crano, W. D. "What Do Infant Mental Tests Test? A Cross-Lagged Panel Analysis of Selected Data from the Berkeley Growth Study." *Child Development,* 1977, *48,* 144-151.

Crano, W. D., and Aronoff, J. "A Cross-Cultural Study of Expressive and Instrumental Role Complementarity in the Family." *American Sociological Review,* 1978, *43,* 463-471.

Crano, W. D., and Brewer, M. B. *Principles of Research in Social Psychology.* New York: McGraw-Hill, 1973.

Crano, W. D., Kenny, D. A., and Campbell, D. T. "Does Intelligence Cause Achievement? A Cross-Lagged Panel Analysis." *Journal of Educational Psychology,* 1972, *63,* 258-275.

Crano, W. D., and Mellon, P. M. "Causal Influence of Teachers' Expectations on Children's Academic Performance: A Cross-Lagged Panel Analysis." *Journal of Educational Psychology,* 1978, *70,* 39-49.

Durkheim, E. *The Division of Labor in Society.* (G. Simpson, Trans.) New York: Free Press, 1933. (Originally published 1893.)

Kavanaugh, M. J., MacKinney, A. C., and Wolins, L. "Issues in Managerial Performance: Multitrait-Multimethod Analyses of Ratings." *Psychological Bulletin,* 1971, *75,* 34-49.

Kenny, D. A. "Cross-Lagged and Synchronous Common Factors in Panel Data." In A. S. Goldberger and O. D. Duncan (Eds.), *Structural Equation Models in the Social Sciences.* New York: Seminar Press, 1973.

Kenny, D. A. "Cross-Lagged Panel Correlation: A Test for Spuriousness." *Psychological Bulletin,* 1975, *82,* 887-903.

Kuhn, T. S. *The Structure of Scientific Revolutions.* Chicago: University of Chicago Press, 1962.

LeVine, R. A., and Campbell, D. T. *Ethnocentrism: Theories of Conflict, Ethnic Attitudes, and Group Behavior.* New York: Wiley, 1972.

Mellon, P. M., and Crano, W. D. "An Extension and Application of the Multitrait-Multimethod Matrix Technique." *Journal of Educational Psychology,* 1977, *69,* 716-723.

Murdock, G. P. "Comparative Data on Division of Labor by Sex." *Social Forces,* 1937, *15,* 551-553.

Murdock, G. P. *Social Structure.* New York: Macmillan, 1949.

Murdock, G. P. "Ethnographic Atlas: A Summary." *Ethnology,* 1967, *6,* 109-236.

Murdock, G. P., and White, D. R. "Standard Cross-Cultural Sample." *Ethnology,* 1969, *8,* 329-369.

Naroll, R. *Data Quality Control.* New York: Free Press, 1962.

Naroll, R. "Galton's Problem. The Logic of Cross-Cultural Analysis." *Social Research,* 1965, *32,* 428-451.

Naroll, R. "Galton's Problem and HRAFLIB." *Behaviour Science Research,* 1976, *11,* 123-148.

Naroll, R., and Cohen, R. (Eds.). *A Handbook of Method in Cultural Anthropology.* New York: Natural History Press/Doubleday, 1970.

Naroll, R., and Naroll, F. *Main Currents in Cultural Anthropology.* New York: Appleton-Century-Crofts, 1973.

Parsons, T., and Bales, R. F. (Eds.). *Family, Socialization, and Interaction Process.* New York: Free Press, 1955.

Pelz, D. C., and Andrews, F. M. "Detecting Causal Priorities in Panel Study Data." *American Sociological Review,* 1964, *29,* 836-848.

Rozelle, R. M., and Campbell, D. T. "More Plausible Rival Hypotheses in the Cross-Lagged Panel Correlation Technique." *Psychological Bulletin,* 1969, *71,* 74-80.

Seelig, C. *Albert Einstein. A Documentary Biography.* (M. Savill, Trans.) London: Staples Press, 1956.

Segall, M. H., Campbell, D. T., and Herskovits, M. J. *The Influence of Culture on Visual Perception.* Indianapolis: Bobbs-Merrill, 1966.

Smith, F. J., and Crano, W. D. "Cultural Dimensions Reconsidered: Global and Regional Analyses of the Ethnographic Atlas." *American Anthropologist,* 1977a, *79,* 364-387.

Smith, F. J., and Crano, W. D. "Patterns of Cultural Diffusion: Analyses of Trait Associations Across Societies by Content and Geographical Proximity." *Behaviour Science Research,* 1977b, *12,* 145-167.

Stanley, J. C. "Analysis of Unreplicated Three-Way Classifications, with Applications to Rater Bias and Trait Independence." *Psychometrika,* 1961, *26,* 205-219.

Stephens, W. N. *The Family in Cross-Cultural Perspective.* New York: Holt, Rinehart and Winston, 1963.

Tolman, E. C., and Brunswik, E. "The Causal Texture of the Environment." *Psychological Review,* 1935, *42,* 43-77.

Webb, E. J., Campbell, D. T., Schwartz, R. D., and Sechrest, L. B. *Unobtrusive Measures: Nonreactive Research in the Social Sciences.* Chicago: Rand McNally, 1966.

Zelditch, M., Jr. "Role Differentiation in the Nuclear Family: A Comparative Study." In T. Parsons and R. F. Bales (Eds.), *Family, Socialization, and Interaction Process.* New York: Free Press, 1955.

13

Marilynn B. Brewer

Ethnocentrism and Its Role in Interpersonal Trust

Among the universal features of social structure that have been postulated in the social sciences is that of "ethnocentrism" or "ingroup bias." As described by Sumner (1906, pp. 12-13), ethnocentric values, attitudes, and behavior are the inevitable consequence of the formation of distinct social groups:

> A differentiation arises between ourselves, the we-group, or ingroup, and everybody else, or the others-groups, outgroups. The insiders in a we-group are in a relation of peace, order, law, government, and industry, to each other. Their relation to all outsiders, or others-groups, is one of war and plunder. . . .

345

Sentiments are produced to correspond. Loyalty to the group, sacrifice for it, hatred and contempt for outsiders, brotherhood within, warlikeness without— all grow together, common products of the same situation.

Ethnocentrism is the technical name for this view of things in which one's own group is the center of everything, and all others are scaled and rated with reference to it. Folkways correspond to it to cover both the inner and outer relation. Each group nourishes its own pride and vanity, boasts itself superior, exalts its own divinities, and looks with contempt on outsiders. . . . For our present purpose the most important fact is that ethnocentrism leads a people to exaggerate and intensify everything in their own folkways which is peculiar and which differentiates them from others. It therefore strengthens the folkways.

In these paragraphs Sumner provides a perspective on intergroup relations with three major facets: (1) that ethnocentrism is a syndrome involving mutually reinforcing interactions among attitudinal, ideological, and behavioral mechanisms that promote ingroup integration and outgroup hostility; (2) that this syndrome is a universal concomitant of the formation and differentiation of social groups; and (3) that it is functionally related to intergroup conflict and competition, both in that conflict is prerequisite to the formation of ingroup-outgroup distinctions and that the resulting sentiments provide the conditions that sustain that conflict.

Following a comprehensive theoretical analysis of ethnocentrism in intergroup relations (LeVine and Campbell, 1972), social psychologist Donald Campbell and anthropologist Robert LeVine collaborated on a large-scale cross-cultural project designed to bring the concept of methodological triangulation to bear on this particular social science construct. For a ten-year period, with funding from the Carnegie Corporation of New York, the collection and analyses of data from the Cross-Cultural Study of Ethnocentrism (CCSE) were undertaken to assess the universality of ethnocentrism by examining the correlates of ingroup-outgroup perceptions, attitudes, and behavior in multiple cultural settings. Stated in general terms, the operating hypothesis derived from

Sumner's formulation was that "all dimensions of group difference will show a positive correlation with each other, will represent a general factor, and will scale as a unidimensional scale" (Le Vine and Campbell, 1972, p. 9). In other words, the test of the validity or "entitativity" of the construct of ethnocentrism lay in the convergence of different features of ingroup orientations and outgroup orientations; whether the assessment of ingroup-outgroup relations taken from different perspectives could be "mapped onto" one another to generate a single representation of intragroup acceptance–intergroup rejection.

The first section of this chapter provides a synopsis of the methodology of the CCSE project, as an illustration of Campbell's notion of methodological triangulation, along with a brief summary of the general results of the study and their impact on our conceptualization of the nature of ethnocentrism and ingroup bias. The second section represents a more speculative exercise on the prospects for extending the findings of CCSE and related work on ingroup bias to the domain of interpersonal trust.

The Cross-Cultural Study of Ethnocentrism

The design of the CCSE project represented an attempt at triangulation in two senses: (1) that of examining a single construct from multiple perspectives through cross-cultural replication and comparison involving non-Western societies; and (2) that of collecting data utilizing two very different research strategies—one involving extensive, quantitative methods, the other representing intensive, qualitative methodologies.

A thorough empirical investigation of ethnocentrism requires in-depth knowledge of the patterns of intragroup behavior within a particular society and the history of relations with and attitudes toward all salient outgroups. Yet such intensive study of a single social group precludes extensive sampling of a large number of groups or extensive sampling of observers within groups, as required for quantitative cross-validation. In order to satisfy both of these research requirements, the data collection efforts of the ethnocentrism project were divided between two separate but complementary studies: one using survey techniques to sample a large number of respondents within a particular region on a limited range of issues relevant to intergroup relations, and the other using ethnographic case-study methods to assess intensively a small number of groups from different regional settings on a wide range of relevant issues.

The survey study was conducted in East Africa in 1965, employing interviewers trained in techniques of public opinion polling. Standardized interviews were conducted with fifty respondents in each of thirty ethnic groups in Kenya, Uganda, and Tanzania. The survey was designed to provide data on reciprocal perceptions among groups, ten for each country. The content of the interview schedule was focused on assessing the respondent's personal contact with and perceptions of members of his or her own ethnic group and those of the other nine groups in the same country that were included in the survey sample. In order to stay within the constraints of a relatively brief, structured interview, the content dealt with only a limited range of intergroup orientations: (1) the respondent's familiarity and contact with each of the outgroups, (2) willingness to engage in social interactions with members of each outgroup, (3) liking and perceived similarity between the ingroup and each outgroup, and (4) the respondent's stereotypes of ingroup and outgroup characteristics. For each of these dimensions, the investigators could obtain a separate index of ingroup-outgroup differentiation by combining data from the fifty ingroup respondents.

By contrast, the ethnographic studies of the CCSE adopted anthropological data collection methodologies to obtain information from a small number of expert informants in one society at each of twenty field sites, ranging from Northern Canada to the South Pacific and West Africa. The application of ethnographic methods to a systematic, comparative study required some adaptation of traditional data collection techniques. As a compromise between the open-ended probe technique of the unstructured ethnographic interview and the need for comparability of information obtained from different cultures, the project relied on the use of a standardized field manual to structure the interviews undertaken at each field site (Campbell and LeVine, 1970). The manual contained an extensive interview schedule, which was to be followed in detail for each informant but which allowed for open-ended responding and intensive probing at the discretion of the ethnographer. All interviews were conducted with elderly local residents who were able to provide *retrospective* information on patterns of intragroup and intergroup behavior *prior to* extensive European contact. Interviews were conducted by trained anthropologists, whose own field stay in the area was extended to collect the data required for the ethnocentrism project. Within each society complete interviews were obtained from two to five informants, with the definition of the ingroup unit and the identification of outgroups determined within the particular cultural context. Thus, the ethnographic study differed from the survey study not only in intensiveness and historiocity of data but also in heterogeneity of social groupings and intergroup settings.

Results of analyses of the quantitative data from the survey study have been reported extensively elsewhere (Brewer, 1968; Brewer and Campbell, 1976; Brewer, Campbell, and LeVine, 1971). Detailed qualitative materials from the ethnographic case studies are available in archival form (Human Relations Area Files, 1972) and have also been subjected to content analyses for purposes of statistical summary parallel to the analyses of the survey data (Brewer, 1974, 1979a). What emerged from these complementary approaches was a consistent but complex representation of ingroup-outgroup perceptions and relationships that challenged the original formulation of the nature of ethnocentrism and ingroup identification.

Although affective attachment to an ingroup, as distinct from specifiable outgroups, was found universally, the pattern and size of correlations among different dimensions of intergroup differentiation were not at all of the order required by a unidimensional model. Instead, respondents' differentiation between their own group and particular outgroups appeared to be opportunistically variable, subject to changes in the nature of the response dimension and the salience of different bases of intergroup categorization. For example, in our analyses of the ethnograhpic data, we found that the extent of evaluative bias in stereotypes of the ingroup compared to a particular outgroup was not systematically related to an index of overt conflict between those groups, which in turn was not highly correlated with a measure of social distance (such as social exchange or intermarriage) between the groups. Similarly, analyses of the East African survey data suggested that some indices of intergroup attitudes were highly correspondent to current political party allegiances, while other measures were relatively insensitive to shifting political alignments and more indicative of historical ties and contacts between groups related to geographical factors.

In contrast to Sumner's imagery that "loyalty to the group, sacrifice for it, hatred and contempt for outsiders, brotherhood within, warlikeness without—all grow together, common products of the same situation" (1906, p. 12), our data indicate that ingroup favoritism is relatively independent of outgroup attitudes, which depend in turn on whether or not the particular outgroup is perceived as distinct from or an extension of the ingroup in a specific setting. Group members' "cognitive maps" of intergroup distinctions correspond imperfectly, at best, to any objective mappings of actual discontinuities of appearance, behavior, or interaction patterns among groups in a given region. Which differences are emphasized under what circumstances appears to be flexible and context dependent; this flexibility permits individuals to mobilize different group identities for different purposes.

The ethnic groups sampled in the CCSE studies could generally be said to have a high degree of entitativity as social aggregates, in the sense discussed by Campbell (1958). With respect to multiple criteria for defining social units—behavioral similarity, geographical proximity, frequency of co-occurrence, and common fate—members of the target groups could be distinguished from nonmembers. As a consequence, for each case in our study some "basic-level" ingroup identification could be specified, reflecting multiple redundant bases of categorization and represented by recognized labels and personal allegiance. Yet the boundaries of the ingroup with respect to a given set of outgroups could be shifted by altering the judgment context or the response dimension on which differentiations were to be made. Just what factors determine whether a particular categorization will be socially significant has yet to be clarified, but it is evident that the bases of group identity lie in shared conceptions of what distinctions are recognized and relevant in given contexts.

These conclusions from the CCSE research found convergent validation in both anthropological and experimental social psychological literature. In evaluating the nature of boundaries between social groups, anthropologists have found what Gulliver (1969, pp. 20-21) refers to as a "general principle of fission and fusion in a segmentary system": "Cultural differences change both in their actuality and in the degree of emphasis and importance given to them. Geographical boundaries are indistinct, and they change as interests and activities develop, and as populations shift, expand, and contract. . . . Differences can be ignored or consolidated, minimized or emphasized, as overriding interests demand. Cultural differences are in part the ideology and the symbolism of group distinction, and in part the concrete reality of it. Both can and do change; both have varying referents according to context."

Similar conclusions have been drawn from results of experimental laboratory research on ingroup bias in minimal intergroup situations (see, for example, Brewer, 1979b; Turner, 1978). Interpretation of these studies in terms of social identity theory (Tajfel, 1974; Tajfel and Turner, 1979) holds that an individual's identity is highly differentiated, based in part on membership in significant social categories. When a particular category distinction is highly relevant or salient, the individual responds with respect to that aspect of social identity, acting toward others in terms of their group membership rather than their personal identity. At this group level, the motivation for positive self-identity takes the form of pressure to differentiate the ingroup from the outgroup to the extent that favorable comparisons are available on dimensions relevant to the ingroup-outgroup distinction. As a result, the degree of bias expressed in favor of the ingroup relative to a particular outgroup will vary, depending on the relevance of that group as a basis of comparison and the evaluative significance of ingroup-outgroup differences.

Above and beyond this flexibility of social categorization, however, our own data from both field and experimental research do suggest one universal concomitant of the distinction between ingroup and outgroups. Informants from ethnic groups in the CCSE studies and subjects in minimal social groupings in the laboratory uniformly evaluate members of their own group as high relative to those from other groups on characteristics such as trustworthiness, honesty, and loyalty. This consistency in the *content* of intergroup perceptions has led to the speculation that, in addition to serving functions related to positive personal identity, ingroup bias also serves to solve a pervasive social dilemma—that of the problem of interpersonal trust.

Ingroup Identity and Trust

The Dilemma. For purposes of clarification, interpersonal trust may be defined in the context of a simplified two-person situation, where

I = an individual's outcomes independent of the other person
B = value of a benefit received by one person from the other
C = the cost to one person of giving benefit to the other

and where $B > C$. (For notational simplification, assume that values of B and C are the same for both parties.)

Assume further that each individual has a choice between trusting/trustworthy behavior (T) that benefits the other at some cost to self, versus untrusting/untrustworthy behavior $(\overline{T})$ that withholds the benefit. Putting these terms together in the form of an outcome matrix for one person in the dyad (self) yields the following:

Outcomes to Self

		Self-Behavior	
		T_s	$\overline{T}_s$
Other Behavior	T_o	$I + B - C$	$I + B$
	$\overline{T}_o$	$I - C$	I

As this matrix stands, in its deterministic form, it has the structure of a one-trial Prisoner's Dilemma Game, where, if choices are made independently, pure self-interest always dictates a $\overline{T}$ choice. No matter which behavior other chooses, self's outcomes are better in the $\overline{T}_s$ column, even though mutual $\overline{T}$ choices result in less than optimal outcomes for both parties. However, if the possibility of *contingent* social behavior is added to the situation, perceived contingent probabilities must be taken into account in determining optimal choices for each individual.

For the outcome matrix given above, the relevant contingent probabilities are:

$$\text{Self Behavior}$$

		T_s	$\overline{T}_s$
Other Behavior	T_o	$p(T_o|T_s)$	$p(T_o|\overline{T}_s)$
	$\overline{T}_o$	$p(\overline{T}_o|T_s)$	$p(\overline{T}_o|\overline{T}_s)$

where

$$p(T_o|T_s) + p(\overline{T}_o|T_s) = 1.00 \text{ and}$$
$$p(\overline{T}_o|\overline{T}_s) + p(T_o|\overline{T}_s) = 1.00.$$

If the behavior choices of self and other are nonindependent, then there is some degree of *reciprocity* (either sequential or simultaneous) in the system; reciprocity will be defined here as cases where $p(T_o|T_s) > p(T_o|\overline{T}_s)$, and the degree of reciprocity as:

$$p(T_o|T_s) - p(T_o|\overline{T}_s) = p(\overline{T}_o|\overline{T}_s) - p(\overline{T}_o|T_s)$$
$$= 1.00 - p(T_o|\overline{T}_s) - p(\overline{T}_o|T_s)$$

If reciprocity were perfect, then the probability of symmetric choices ($T_o|T_s$ and $\overline{T}_o|\overline{T}_s$) would be 1.00, and the probability of asymmetric choices ($T_o|\overline{T}_s$ and $\overline{T}_o|T_s$) would be .00. In such a case, the self-interest of both parties would dictate a T choice for mutual benefit. However, if reciprocity is imperfect, there is uncertainty in the system—that is, $p(T_o|T_s) < 1.00$—and risks are introduced that tend to favor distrust ($\overline{T}$) over trust (T).

Where reciprocity is imperfect, there are two types of decision errors an individual can make—misplaced trust (choosing T when the other does not intend to reciprocate) and misplaced distrust (choosing $\overline{T}$ when the other would have reciprocated T). In terms of expected values of outcomes, the potential costs associated with these two kinds of "wrong" choices are characterized by a fundamental asymmetry. The cost of misplaced trust is equal to the value of C with certainty, regardless of contingent probabilities. For misplaced distrust, on the other hand, the potential cost is one of benefits forgone, less costs avoided, which is *not* equal to $B - C$ with certainty but is probabilistic, depending on the value of $p(T_o|\overline{T}_s)$ (that is, the possibility of receiving the benefit without paying the cost). Thus, a trusting choice involves an uncertain benefit and a certain cost, whereas a distrusting choice involves an uncertain loss of benefit with a certain avoidance of cost.

Putting it more formally, the risk of incurring cost with a trusting choice is equal to:

$$C[p(\overline{T}_o|T_s)] - (B - C)\,p(T_o|T_s) + B[p(T_o|\overline{T}_s)]$$
$$= C - B[p(T_o|T_s) - p(T_o|\overline{T}_s)]$$

while the risk associated with distrust is

$$(B - C)p(T_o|T_s) - C[p(\overline{T}_o|T_s)] - B[p(T_o|\overline{T}_s)]$$
$$= B[p(T_o|T_s) - p(T_o|\overline{T}_s)] - C.$$

The latter will exceed the former only when B is very large relative to C and/or perceived reciprocity is very high. In other circumstances misplaced trust will be more costly than misplaced distrust. Yet if everyone chooses $\overline{T}$, no benefits are accrued and everyone loses.

By way of illustration that this type of decision dilemma is not limited to artificial game settings, the decision structure described above can be applied to a number of different situations of social interdependence that have been prominent in the social science literature.

One illustration comes from the sociobiological analysis of altruism (Trivers, 1971), which suggests that an individual can benefit from helping another genetically unrelated individual at some cost to the self (or some potential loss of reproductive success) if and only if there is a significant increase in the probability that the other will provide help in return. This concept of reciprocal altruism, which finds its sociological analogue in the general "norm of reciprocity" (Gouldner, 1960), is prototypic of the notion of contingent trust. In matrix form the decision to help another can be represented as follows:

Outcome Analysis of Helping

		Self	
		Help other	*Not help*
	Help	$I - C$ $p(H_o|H_s)$	I $p(H_o|\overline{H}_s)$
Other			
	Not help	$I - C - D$ $p(\overline{H}_o|H_s)$	$I - D$ $p(\overline{H}_o|\overline{H}_s)$

where I is some initial status, D is a disadvantage accrued in the absence of help from another, and C is the cost or risk incurred in giving that help. In this framework it is clear that, whether a tendency toward reciprocal helping is genetically or socially determined, its selective advantage depends on correct identification of those who will reciprocate. Otherwise, one who gives help is at a disadvantage relative to one who does not.

A similar analysis can be made of the decision to disclose negative personal information about oneself in the interests of interpersonal intimacy. Self-disclosure can be represented as a trust dilemma as follows:

Outcome Analysis of Self-Disclosure

		Self	
		Disclose	*Not disclose*
Other	*Disclose*	$S + I - R$ $p(D_o \vert D_s)$	$S + I$ $p(D_o \vert \overline{D}_s)$
	Not disclose	$S - R$ $p(\overline{D}_o \vert D_s)$	S $p(\overline{D}_o \vert \overline{D}_s)$

where S is the status of the individual without intimacy, I is the benefit of receiving intimate disclosures from another, and R is the risk to self-esteem of giving disclosures (risks such as disapproval or embarrassment). As with reciprocal altruism, the advantages of intimacy cannot be attained without confidence in a high probability of reciprocity.

Potential Solutions. The above examples illustrate the kinds of social situations in which each member of a dyad may *prefer* outcomes associated with mutual trust but be unwilling to engage in trusting behavior because the attendant risks are not low enough. One solution to such dilemmas involves long-term social exchange between two individuals who incur initially very small costs and then incrementally larger risks until a sufficient base is established for high-cost trust. Although this process has been generalized from the interpersonal level and applied to international relations in Osgood's GRID proposal (Osgood, 1962), in general the negotiations involved would be too limiting and time-consuming to be applicable to a multiperson system where all possible dyadic interactions are involved.

It is in this context that we can view ingroup identity as one solution, on a large scale, to the dilemma of trust. Common membership in a salient social category can serve as a rule for defining the boundaries of low-risk interpersonal trust that bypasses the need for personal knowledge and the costs of negotiating reciprocity with individual others. As a consequence of shifting from the personal to the social group level of identity, the individual can adopt a sort of "depersonalized trust" based on category membership alone. Within categories the probability of reciprocity is presumed, a priori, to be high, while between categories it is presumed to be low or subject to individual negotiation.

At least two mechanisms might operate to make trust in terms of social category membership viable. One derives from the cognitive consequences of unit formation; when self and others are included within a common boundary, or social unit, psychological distance is reduced; and this may have the effect of increasing orientations toward mutual or joint outcomes rather than individual gain. Common membership in a social category is one basis for arousal of what Hornstein (1972) calls "promotive tension," whereby one individual's goal orientation becomes coordinated to another's goal attainment. In addition, group membership provides a mechanism for increasing the perceived probability of sanctions against failures to reciprocate trust; if defection is regarded as a violation of *group* norms—as disloyalty to the collective rather than victimization of an individual—it may be detected and punished by any member of the group, not just the individual victim (who may be in no position to retaliate personally). Indeed, Campbell (1975) and others (for example, Trivers, 1971; Wilson, 1978) have postulated the existence of an inherent human tendency toward "moralistic aggression" against those who violate group-level social obligations even when one is not affected directly.

Interestingly, the fact that reciprocal trust involves normative prescriptions that apply to intragroup—as opposed to intergroup—behavior provides a "grain of truth" for the content of intergroup stereotypes. Since differential trust gives rise to preferential treatment of ingroup over outgroup members, it generates a set of universal complementary stereotypes (such as "we are loyal; they are clannish"; "we are honest and peaceful among ourselves; they are hostile and treacherous toward outsiders"), for which each group can legitimately place itself on the positive side of the scale, thus maximizing the conditions for perceived contrast and ingroup bias.

Flexibility of Ingroup Identity. On the one hand, one can imagine conditions under which a particular social categorization could become cross-situationally pervasive, as under pressures of intense intergroup conflict. Brewer and Campbell (1976) have discussed some of the historical and sociological factors that might lead to a kind of "boundary convergence," where group identity based on religious, political, and economic distinctions all correspond in a given system. In such an intergroup context, the unidimensionality of ethnocentrism postulated by Sumner might be expected to hold. On the other hand, however, are social systems characterized by cross-cutting categories where the basis for ingroup-mediated trust will be flexible and situation specific. Because trust based on common category membership is depersonalized, it is also subject to shifts in group identity—the same individual may be presumed trustworthy on one occasion and regarded as a distrusted outgrouper on another.

A complete understanding of the relationship between ingroup identity and interpersonal trust would require a theory of the functional significance and adaptive value of shifting group loyalties. If the mechanisms for ensuring ingroup reciprocity are to operate effectively, the size and boundaries of the relevant social category must be very clear and consensually validated. If the category is too large or amorphous, it is not likely to be psychologically "convincing," the belief in reciprocity will be undermined, and the decision to trust will again become problematic. Yet it is only very rarely that social categories are based on sharp discontinuities or that the various potential category memberships of a significant number of individuals are highly intercorrelated. The conditions promoting such clear, convergent group identity are highly restrictive, the most extreme requirements involving constant close physical proximity, face-to-face contact, common kinship, and minimal individual mobility. But if reciprocal trust were "locked in" to such restrictive boundaries, its role in the development of higher-order social systems would be minimal indeed.

To be effective in this latter sense, the unit of ingroup identity must be sensitive to the varying functional requirements of different situations.

The situational flexibility of social identity gives rise to an interesting conflict between different requirements for the effectiveness of ingroup-mediated trust. On the one hand, there is need for a certain amount of tolerance for ambiguity in defining the appropriate social category that is to be operative at a particular time. On the other hand, there is a need for unambivalent commitment to a given group identity if it is to serve as the basis for depersonalized trust. Recognition of this conflict fosters speculation on the possibility of a differentiation between cognitive and affective, or emotional, aspects of group identity. On the cognitive level, the individual recognizes the complexity of social categories, is opportunistically sensitive to situational demands, and responds to social cues as to the appropriate categorization of marginal individuals. On the emotional level, however, ingroup loyalty is engaged on an all-or-nothing basis, without ambivalence or consciousness of alternative social categorizations. Such a representation of the nature of social identity is portrayed rather dramatically by E. O. Wilson (1978, pp. 169-170):

> Moral aggression is most intensely expressed in the enforcement of reciprocation. The cheat, the turncoat, the apostate, and the traitor are objects of universal hatred. Honor and loyalty are reinforced by the stiffest codes. It seems probable that learning rules, based on innate, primary reinforcement, lead human beings to acquire these values and not others with reference to members of their own group. The rules are the symmetrical counterparts to the canalized development of territoriality and xenophobia, which are the equally emotional attitudes directed toward members of other groups. . . . Altruism is characterized by strong emotion and protean allegiance. Human beings are consistent in their codes of honor but endlessly fickle with reference to whom the codes apply. . . . The important distinction is . . . between the ingroup and the outgroup, but the precise location of the dividing line is shifted back and forth with ease.

In this representation, the affective underpinnings of in-group loyalty and trust are not particularly attractive or flattering. But it is important to recognize that the same processes of categorization and depersonalization that give rise to hostile inter-group behavior may also underlie more prosocial forms of behavior, on which our hopes for solving large-scale social dilemmas may depend.

References

Brewer, M. B. "Determinants of Social Distance Among East African Tribal Groups." *Journal of Personality and Social Psychology,* 1968, *10,* 279-289.

Brewer, M. B. "Cognitive Differentiation and Intergroup Bias: Cross-Cultural Studies." Paper presented at the 82nd annual convention of the American Psychological Association, Atlanta, 1974.

Brewer, M. B. "The Role of Ethnocentrism in Intergroup Conflict." In W. Austin and S. Worchel (Eds.), *The Social Psychology of Intergroup Relations.* Monterey, Calif.: Brooks/Cole, 1979a.

Brewer, M. B. "Ingroup Bias in the Minimal Intergroup Situation: A Cognitive-Motivational Analysis." *Psychological Bulletin,* 1979b, *86,* 307-324.

Brewer, M. B., and Campbell, D. T. *Ethnocentrism and Intergroup Attitudes: East African Evidence.* New York: Halsted Press, 1976.

Brewer, M. B., Campbell, D. T., and LeVine, R. A. "A Cross-Cultural Test of the Relationship Between Affect and Evaluation." In *Proceedings of the 79th Annual Convention of the American Psychological Association.* Washington, D.C.: American Psychological Association, 1971.

Campbell, D. T. "Common Fate, Similarity, and Other Indices of the Status of Aggregates of Persons as Social Entities." *Behavioral Science,* 1958, *3,* 14-25.

Campbell, D. T. "On the Conflicts between Biological and Social Evolution and Between Psychology and Moral Tradition." *American Psychologist,* 1975, *30,* 1103-1126.

Campbell, D. T., and LeVine, R. A. "Field Manual Anthropology."

In R. Naroll and R. Cohen (Eds.), *A Handbook of Method in Cultural Anthropology.* New York: Natural History Press/Doubleday, 1970.

Gouldner, A. "The Norm of Reciprocity: A Preliminary Statement." *American Sociological Review,* 1960, *47,* 73-80.

Gulliver, P. H. (Ed.). *Tradition and Transition in East Africa.* Berkeley: University of California Press, 1969.

Hornstein, H. A. "Promotive Tension: The Basis of Prosocial Behavior from a Lewinian Perspective." *Journal of Social Issues,* 1972, *28* (3), 191-218.

Human Relations Area Files. *Ethnocentrism Interview Series.* New Haven, Conn.: HRAFlex Books, 1972.

LeVine, R. A., and Campbell, D. T. *Ethnocentrism: Theories of Conflict, Ethnic Attitudes, and Group Behavior.* New York: Wiley, 1972.

Osgood, C. E. *An Alternative to War or Surrender.* Urbana: University of Illinois Press, 1962.

Sumner, W. G. *Folkways.* Boston: Ginn, 1906.

Tajfel, H. "Social Identity and Intergroup Behaviour." *Social Science Information,* 1974, *13,* 65-93.

Tajfel, H., and Turner, J. C. "An Integrative Theory of Intergroup Conflict." In W. Austin and S. Worchel (Eds.), *The Social Psychology of Intergroup Relations.* Monterey, Calif.: Brooks/ Cole, 1979.

Trivers, R. L. "The Evolution of Reciprocal Altruism." *Quarterly Review of Biology,* 1971, *46,* 35-57.

Turner, J. C. "Social Categorization and Social Discrimination in the Minimal Group Paradigm." In H. Tajfel (Ed.), *Differentiation Between Social Groups: Studies in the Social Psychology of Intergroup Behavior.* London: Academic Press, 1978.

Wilson, E. O. *On Human Nature.* New York: Bantam Books, 1978.

14 *Marshall H. Segall*

Cross-Cultural Research on Visual Perception

> Habit is second nature which destroys the first,
> but what is nature? Why is habit not natural? I am
> very much afraid that nature is itself only first habit
> as habit is second nature [Pascal, quoted in Skinner,
> 1969, p. 174].

Starting from a Brunswikian perspective which emphasizes
the functionality of the transactions between perceivers and envi-

Note: Thanks are due two graduate students in social psychology at
Syracuse University for sharing ideas with me as I prepared this chapter.
Caroline Keating discussed her dissertation data and her thoughts about her

ronmental events, Segall, Campbell, and Herskovits (1966) predicted, and found, cross-cultural differences in susceptibility to
several different geometric illusions. These findings were interpreted, in their original report and in subsequent citations of it,
as support for the empiricist side of the nativist-empiricist controversy. Although some subsequent findings (for example, by Berry,
1971; Jahoda, 1971; Pollack, 1970) point to an alternative—a particular nativistic interpretation, involving physiological differences
associated with eye pigmentation—the preponderance of studies
reported over the past fifteen years has added support to the argument that cultural differences in illusion susceptibility reflect the
acquisition of different habits of inference in response to environmental characteristics.

One objective of this chapter will be to review the studies
that have been reported since 1966 and to show that they add up
to impressive support for the learned-habits-of-inference theory
that was stressed by Segall, Campbell, and Herskovits (1966). But
the other objective will be to deemphasize the nature-versus-nurture
quality of the arguments that have long accompanied work on this
problem.

The latter objective may well be the more important one.
Donald Campbell's reflections on the themes of blind variation
and selective retention of "knowledge" make obvious the potential for defusing the tenacious, but probably naive, opposition between heredity and environment. Many psychologists, steeped in
a learning theory tradition, avoid evolutionary theorizing, on the
grounds that it implies biological determinism or is reductionistic.
This avoidance seems outmoded in light of Campbell's emphasis
on parallels between evolution and learning. Both evolution and
learning, which share adaptiveness, may contribute to significantly different behavioral tendencies across human groups who
occupy different ecological niches.

While either individual learning or changes in genetic dispositions of groups may be a more plausible mechanism than the
other to account for a particular behavioral difference across cul-

findings and allowed me to present a prepublication account of them. Thomas
Gamble analyzed several of Donald Campbell's writings on descriptive epistemology, including a transcript of his 1977 William James Lectures, and influenced significantly the interpretation of those writings that is presented
here.

tural groups, attempting to choose between them may not be a fruitful effort. It may be strategically more important to try to delineate specific links between ecological forces and resultant behaviors in particular societies. Intersociety differences may reflect group-wide biological modifications that occurred over centuries *or* changes in shared interpretations of reality that occur within individuals in the course of their own development, whereby they learn to live effectively in a particular environment—or both.

Cross-Cultural Research on Illusions: An Updating

Several threads of interest converged at Northwestern University in the late 1950s to produce the cross-cultural research that was first reported in a brief article in *Science* (Segall, Campbell, and Herskovits, 1963) and then more fully in a book by the same authors (1966). One was Herskovits's doctrine of cultural relativism, which implies unlimited flexibility of humans to acquire culturally specific ways to interpret reality. Another was anecdotal evidence —accumulated by many anthropologists, including Herskovits himself—of varying skills and habits relating to space. Notable among these were observations concerning difficulties for unacculturated persons living in non-Western societies in drawing straight lines. Still another thread was Campbell's interest, derived from the teachings of Egon Brunswik (see, for example, Brunswik, 1956), in the likelihood that perceptions of reality, while phenomenally absolute, were actually relative to prior experiences with that reality. While Campbell was, at the outset, skeptical of Herskovits's expectation that such basic experiences as perceived line lengths would vary systematically within the human species, he was aware of other findings (Rivers, 1901, 1905) of differences (in both directions) between Europeans and Asians on two geometric illusions and treated the matter as deserving multicultural empirical research. A number of specific hypotheses were derived from Brunswik's transactional functionalism. These all pointed to differences across cultures in illusion susceptibility.

In summary, three of these hypotheses were:

1. *The carpentered world hypothesis.* There is a learned tendency among people reared in carpentered environments to interpret nonrectangular figures

(especially parallelograms) as rectangular, to perceive such figures in perspective, and to interpret them as two-dimensional representations of three-dimensional objects. Such a tendency would be expected to enhance the Müller-Lyer illusion and the Sander parallelogram illusion (see Figure 1). Since the tendency

Figure 1. The Müller-Lyer Illusion (top) and the Sander Parallelogram Illusion (bottom)

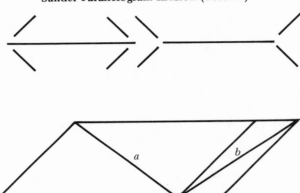

Note: In the Müller-Lyer figure shown here, the horizontal line on the left is 14 percent longer than the horizontal line on the right. In the Sander parallelogram, the right diagonal, *b*, is 16 percent longer than the left diagonal, *a.*

has more ecological validity for peoples in carpentered (Western or Westernized) environments, such groups should be more susceptible to this class of illusion.

2. *The foreshortening of vistas hypothesis.* The horizontal/vertical illusion (see Figure 2) is enhanced by a tendency to counteract the foreshortening of lines extended into space in front of a viewer. Thus, vertical lines, more "foreshortened" than horizontal lines, would be exaggerated. Since the tendency, which results in seeing vertical lines as representing distances, has more ecological validity for peoples living in open, spacious environments, such groups should be more susceptible to the horizontal/vertical illusion (that is, more likely to exaggerate the length of a vertical) than groups dwelling in urban environments. By the same logic, some rural dwellers—espe-

Figure 2. The Horizontal-Vertical Illusion

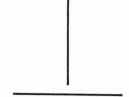

Note: In this figure, the horizontal line is 11 percent longer than the vertical line.

cially those in constricted environments, such as rain forests or canyons—should be less susceptible to this illusion.

3. *The acquired sophistication in representational conventions hypothesis.* Learning to interpret drawings and photographs should enhance all geometric illusions presented on two-dimensional surfaces. Thus, persons anywhere who are exposed to two-dimensional representations of three-dimensional objects (as happens in all Western-influenced schools and to varying degrees in locales where pictorial representation is employed) should be more susceptible to illusions than persons who have not had such experience.

Over a six-year period, standardized stimulus materials were administered to fourteen non-European samples (nearly all in Africa) and to three "European" samples, two in Evanston and one in Johannesburg, with responses to several different illusions collected from a total of 1,848 persons, including children and adults. These data provided clear support for the empiricist hypothesis that space perception involves the acquisition of habits of inference that are ecologically valid.

On both the Müller-Lyer and the Sander illusions, the three "European" samples were more susceptible than any of the non-European samples. On the two horizontal/vertical illusions, many, but not all, of the non-European samples were more susceptible than the "Europeans." Thus, as predicted, cross-cultural differences in illusion susceptibility occurred in two directions, an outcome that had been anticipated by Rivers (1901, 1905) and that made it impossible to attribute the differences in illusion susceptibility to some single characteristic of persons, such as general intelligence or sophistication. At the same time, the overall pattern of the cross-cultural differences accorded well with the set of specific hypotheses enumerated above, all of which attributed perceptual tendencies to ecologically valid inference habits.

However, the findings from this study could not eliminate all plausible alternative hypotheses. Nor were the findings in accord with empiricist theorizing in every respect. For example, for all illusions employed and in nearly every society sampled, illusion susceptibility declined with age; yet the empiricist line of argument entails an expectation of growth in susceptibility, at least for the Müller-Lyer and the Sander illusions, at least in carpentered environments. While it could be argued that the samples employed in this study did not include persons young enough for the expected growth to show itself, the failure to document an increase in susceptibility with age contributed to a search by other students of perception for an alternative to the empiricist theory.

The most successful effort of this kind was Robert Pollack's. He found (Pollack, 1963) that contour detectability declined with age. (The ability to detect contours is an acuity-like skill in reporting a boundary between two portions of a visual stimulus where each portion is, for example, of different hue.) Recalling that a decline with age in Müller-Lyer illusion susceptibility had been found by Binet (1895) and by Piaget and von Albertini (1950), among others, as well as by Segall, Campbell, and Herskovits, Pollack found it reasonable to suggest that illusion susceptibility was functionally related to contour detection ability, particularly in the light of findings from a study done in Chicago (Pollack and Silvar, 1967) of a correlation between the two. In that study, subjects with higher contour detection thresholds were less susceptible to the Müller-Lyer illusion. Race, in the sense of skin color, was implicated when Silvar and Pollack (1967) found that skin color was correlated both with retinal pigmentation and contour detection threshold. Since the non-European samples employed in the Segall, Campbell, and Herskovits study were all non-Caucasian (and nearly all African), Pollack could plausibly argue that what the earlier students of the problem had accepted as cultural differences might well have been racial differences. And, while Pollack's argument was based exclusively on findings with the Müller-Lyer illusion (and, an unusual variant of it, presented under special illumination), his argument could, potentially, be extended to illusions generally.

Pollack's work prompted searches for racial differences in illusion susceptibility. Since race and culture were confounded in the Segall, Campbell, and Herskovits study, its findings fit both an empiricist and racial explanation. An early attempt to unconfound race and culture was made by John Berry (1971), who administered the Müller-Lyer illusion to five samples that could be ordered along two dimensions—carpenteredness and skin pigmentation—and found that pigmentation, with carpenteredness controlled, correlated more highly with susceptibility than did carpenteredness, with pigmentation controlled. Although Berry's analytic technique is subject to the criticism of unreliability (since a change of rank for any two samples along either dimension would have affected his correlation coefficients), his study did not eliminate race as an alternative explanation for cross-cultural differences in susceptibility to the Müller-Lyer.

Another attempt was Jahoda's (1971) study. It took into account the fact that density of retinal pigmentation is negatively correlated with contour detectability specifically when bluish stimuli are used. This fact also underlay Pollack and Silvar's (1967) decision to present the Müller-Lyer illusion under blue lighting. Jahoda reasoned that if retinal pigmentation related functionally to illusion susceptibility as it does to contour detectability, then, for darkly pigmented subjects, Müller-Lyer susceptibility should be less when the stimulus appears in blue than when it appears in red. Employing both Scottish and Malawian university students, and both a red and a blue version of the Müller-Lyer illusion, Jahoda found, as the only significant difference in his study, that the African subjects were less susceptible to the blue illusion than to the red one. Color made no difference to the Scottish subjects, nor were the Scots significantly different from the Malawians in susceptibility to either the red or the blue versions of the illusion. That the familiar intergroup difference was not found was a puzzle, and it may have prompted Jahoda to extend the study. When he did (Jahoda, 1975), substituting Ghanians for Malawians and modifying both color values and exposure times, and employing perspective views of solid objects as stimuli, *all* subjects (African and Scottish) performed a spatial task less well when it involved blue stimuli. This "failure to replicate," as Jahoda termed it, or failure to demonstrate a link between stimulus color, retinal pigmentation, and spatial perception, as I would label it, casts doubt on the ability of a retinal pigmentation-contour detection mechanism to account for cross-cultural differences in illusion susceptibility. Admittedly, however, Jahoda's pair of studies are rather indirect approaches.

More directly relevant to the controversy sparked by Pollack's work are studies that seek racial differences in illusion susceptibility among persons living in a single environment, or studies that, in Stewart's (1973) terms, vary race while holding environment constant. In Evanston, Armstrong, Rubin, Stewart, and Kuntner (1970) administered the Müller-Lyer and the Sander illusions, as well as the Ames distorted room. Half of their sample was composed of black schoolchildren; the other, white. No differences across the racial groups in illusion susceptibility were found. Stewart (1973) extended this direct approach by holding race constant and allowing environmental carpenteredness to vary by employing several Zambian samples, ranging from a group of unschooled children from the very rural Zambezi Valley to a group of middle-class children going to school in the cosmopolitan capital city, Lusaka. She also worked in Evanston, with white and black schoolchildren, thereby once again allowing race to vary with environment constant. The illusions employed were the Müller-Lyer, the Sander, and the Ames distorted room. On all three, the Zambezi Valley sample was less susceptible than the Lusaka sample. On none were there racial differences. Blacks and whites in Evanston showed virtually identical levels of susceptibility. So Stewart's study, along with the research of Armstrong and his colleagues, provided strong support for the empiricist position and a very direct challenge to the contour detectability alternative.

So, too, did a dissertation by Donald Weaver (1974), who reasoned that, *if* illusion susceptibility were wholly a product of pigmentation-related characteristics of the eye, a difference in illusion susceptibility to the Müller-Lyer illusion between African and American samples should occur only when the illusion involves a figure produced by lightness contrast and not when the figure is produced by hue contrast. And, if pigmentation were the "cause" of differing illusion susceptibility, age declines should occur in samples of varying carpenteredness, but more so when the stimulus was produced by lightness contrast. Despite some technical difficulties with the hue-contrast stimuli, Weaver's findings, in the main, contradicted the pigmentation hypothesis. For example, with the Müller-Lyer illusion, susceptibility was greater among carpentered world subjects, regardless of type of stimulus. Illusion susceptibility declined steadily with age, but mainly within the carpentered sample.

Since Pollack's argument was based on research with the Müller-Lyer illusion, most studies which attempted to assess his argument employed either that illusion or others in the same class —namely, the Sander parallelogram and the Ames distorted room— to all of which the carpentered-world hypothesis is relevant. But empiricist theorizing applies to other illusions as well, with the foreshortening of vistas hypothesis offering an empiricist explanation for differences in susceptibility to the horizontal/vertical illusion. Two recent studies supported that hypothesis.

In one, Bolton, Michelson, Wilde, and Bolton (1975) employed two samples of rural Peruvians, equally noncarpentered and both living about 15 degrees south of the Equator, but one dwelling on a 13,000-foot-high plateau and the other living in a forested hillside at 4,500 feet of altitude. The former sample enjoyed a much broader and more open vista. Bolton and colleagues administered two versions of the horizontal/vertical illusion, along with the Müller-Lyer and the Sander parallelogram. They reasoned that a pigmentation hypothesis would predict less susceptibility to all four illusions for the high-altitude sample (dwelling closer to the sun), while empiricist theorizing would lead to a prediction of intersample differences only for the horizontal/vertical illusion and in the opposite direction—with the high-altitude sample, because of its open vista, being more susceptible. The only significant differences obtained were for the horizontal/vertical illusions, and in the direction predicted by the foreshortening of vistas hypothesis.

In the other study, with the amusing but misleading subtitle "Is the Carpentered World Hypothesis Seaworthy?" (Pollnac,

1977), a small sample of fishermen living near the Pacific Ocean in Costa Rica were found to be differentially susceptible to the horizontal/vertical illusion to degrees dependent on years of fishing experience and level of responsibility for navigation.

Another illusion for which the empiricist line of theorizing has gained recent support is the Ponzo illusion. That support relates to two sets of facts, one having to do with diverse versions of the illusion, the other with some complex age trends.

Concerning variations in the illusion itself, the key point is that the more perspective cues involved in the Ponzo stimulus, the more potent its illusory character. The Ponzo illusion—composed of two inward-pointing lines, both inclined from a vertical position, and two horizontal lines, one high, the other low on the figure, with the two horizontal lines to be compared—is an illusion to which experience with railroad tracks, straight roads, and similar environmental phenomena is very plausibly thought to contribute (see Figure 3). More so than for any other illusion that has been

Figure 3. The Ponzo Illusion

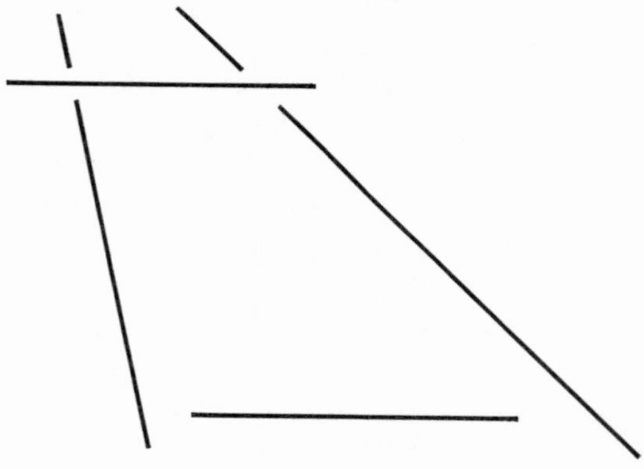

Note: This version was used by Segall, Campbell, and Herskovits (1963, 1966). In this figure, the bottom horizontal is 5 percent longer than the top horizontal.

employed in cross-cultural research, it is intuitively obvious how such experience should contribute to its strength. In addition to arguments presented by Segall, Campbell, and Herskovits, theoretical efforts by Wohlwill (1962), Gregory (1966), and Gillam (1980) —the first of which emphasizes the impact of perspective cues and the latter two "misplaced constancy"—apply to the Ponzo illusion more obviously than to any other. It easily follows that if more context-rich versions of the Ponzo induce stronger illusions than more abstract versions, such a finding accords with empiricist thinking, whether of the variety contained in Wohlwill, Gregory, Gillam or Segall, Campbell, and Herskovits.

Among studies which showed greater potency for more context-rich versions of the Ponzo illusion are one by Liebowitz, Brislin, Perlmutter, and Hennessey (1969) and another by Brislin (1974). In these two studies, all the variations were two-dimensional, and the respondents were from carpentered environments. In another study, performed in Hawaii by Brislin and Keating (1976), a three-dimensional version, composed of wooden planks displayed some 40 feet away from adult respondents, replicated the usual cross-cultural findings of differences that accord with environmental experience. Specifically, Brislin and Keating found that respondents from the Philippines and from the mainland United States (both rich in roads, tracks, and the like) were more susceptible to the three-dimensional version of the Ponzo illusion than were respondents from a variety of small Pacific islands. This finding demonstrates that ecologically valid inferences about perspective are elicited by three-dimensional objects in the form of the Ponzo, which is a phenomenon that heretofore had been assumed by empiricist theorists but not actually demonstrated.

The picture regarding age trends for the Ponzo illusion had, until 1977, been rather confused; some studies showed a decrease with age, some a decrease followed by an increase, some no trend at all and thus virtually no evidence for increased susceptibility with age (which is what the empiricist line of theorizing, of course, requires). Then a foray into Morocco by Daniel Wagner produced an important clarification. The key to it is that age trends are either positive or negative depending on the version of the illusion that is employed. The positive age trend (an increase in susceptibility with increasing age) that empiricist theorists had long sought seems to exist for the more context-rich version.

Wagner employed 384 boys in four different age groups, spanning 7 to 19 years of age. Half of them were schooled; half of them were not. Half lived in the urban, carpentered environment of Marrakesh, and the other half in a rural setting in the Atlas

Mountains. Wagner administered several versions of the Ponzo illusion, one of them maximally abstract and two others (one resembling a field and the other depicting a railroad track) quite rich in contextual perspective cues. For the abstract version, susceptibility decreased slightly but consistently, with increasing age for all subjects, whether schooled or unschooled, urban or rural. But for the two context-rich versions, susceptibility increased with increasing age. And it increased more for urban dwellers than for rural dwellers, more for schooled than for unschooled boys, and most for boys who were both urban and school-going. Also, the oldest school-going boys were more susceptible to the two complex illusions than to the abstract one; while for all groups combined, the abstract version was most potent. Thus, it appears that susceptibility to the more complex versions of the Ponzo illusion grows as a product of increasing perceptual activity of the kinds made probable by living in an urban environment and by going to school.

In more detail, Wagner's findings may be interpreted as follows: Because the context-rich figures were less potent than the abstract version for the Moroccan sample (which was generally from a less carpentered environment and which was partly composed of boys from a very uncarpentered environment), it is in accord with empiricist theorizing that the complex figures accured potency, especially for the urban, schooled boys, who clearly had the most opportunity to acquire illusion-enhancing inference habits. By the same token, the abstract version apparently reached its ceiling by age 7, resulting in negative age trends for this version of the illusion. If this interpretation is correct, Wagner's (1977) study goes a long way toward clarifying what had been rather confusing age trends.

The several studies that have been reviewed here, all employing geometric illusions as stimuli, and carried out subsequent to the Segall, Campbell, and Herskovits (1963, 1966) study, confirm its basic finding—namely, that systematic differences in illusion susceptibility prevail across cultural groups—and support its empiricist line of interpretation—namely, that illusion susceptibility is a reflection of acquired habits of inference, habits that are ecologically valid ones. The single most plausible alternative hypothesis, Pollack's (1963, 1970) pigmentation and contour detection theory, has not fared as well. Limited to the Müller-Lyer illusion, Pollack's hypothesis is, at best, no better able to account for cultural differences than is the empiricist hypothesis. With regard to other illusions—particularly those like the horizontal/vertical illusion, which belong, according to Segall, Campbell, and Herskovits, to a different class—the Pollack hypothesis is largely irrelevant. With the accrual of such additional studies as those reviewed here, the empiricist hypothesis seems at least as tenable in 1981 as it was in 1963.

Still, as was suggested at the outset of this chapter, it may well be more appropriate to focus on the existence of systematic differences across cultures in performance of these perceptual tasks—differences that are predictable on the basis of the kinds of environmental features on which the empiricist theory focuses— than it would be to interpret the adaptiveness revealed by such relationships as produced solely by individual learning. While Pollack's particular physiological theory may not hold up well, clearly the research to date has not eliminated all possible physiological mechanisms.

Cultural Differences in Facial Expression

A report of a somewhat related line of research is in order. The research resembles that of Segall, Campbell, and Herskovits in that it is cross-cultural, it employs standardized pictorial stimuli administered by cooperating field workers, and it has its theoretical roots in a biological-cultural controversy that resembles the nativist-empiricist controversy. The project, centering around a dissertation recently completed by Caroline Keating, began as a joint effort involving Alan Mazur and myself at Syracuse University (see Keating, Mazur, and Segall, 1977). The substantive concern of this research is the communication of status via facial gestures.

The central question in Keating's dissertation is whether eyebrow positions on human faces communicate dominance and submissiveness in a universal way. The research grew out of findings, accumulated mostly by ethologists, that differing eyebrow positions correspond to assertive and submissive behavior in a variety of primates. Lowered brows are usually displayed by threatening individuals, while receptive ones typically show raised brows (see Andrew, 1963; van Hoof, 1967), with only occasional exceptions (see Redican, 1975). In a preliminary study conducted with Syracuse University undergraduates (Keating, Mazur, and Segall, 1977), these human observers selected every one of twelve photographed human models as dominant significantly more often when the models displayed lowered eyebrows than when they displayed raised eyebrows, and this finding was also obtained with cartoon faces on which eyebrow position was varied systematically. Since these findings were obtained in a single culture, it remained to be seen whether the same kind of differential status information would be conveyed by brow position to persons in diverse cultures, in which case, Keating argued, a reasonable inference would be that the eyebrow-down gesture signifying dominance has phylogenetic origins.

Employing the same stimulus materials that had been used in the Keating, Mazur, and Segall (1977) preliminary study, Keating distributed her materials among cooperating field workers, who, collectively, obtained data from Germany, Colombia, the Canary Islands, Kenya, Zambia, Thailand, a sample of Chinese adults recently arrived in the United States, and two United States samples, one from Texas and one from Syracuse, New York. For many of the samples, respondents consistently designated models with downward-pointed eyebrows as dominant. This was clearly the case for the United States samples and for other samples that, in one respect or another, were culturally similar to the United States samples. In sharp contrast, however, were the behaviors of three other samples, two of them African and one Asian. The Kenyan and Zambian samples showed no significant disposition to assign apparent dominance on the basis of eyebrow position, and the Thai sample differed from all others by selecting models with *raised* eyebrows as dominant. Moreover, these cross-cultural differences occurred in the content of a cross-cultural sameness demonstrated by the fact that all samples in Keating's study selected smiling faces as happy when asked to make happiness judgments with the same stimulus materials—thus meeting the Campbellian dictum "Discrepancy can be noted and interpreted only against the background of an overwhelming proportion of nondiscrepant fit, agreement, or pattern repetition" (Campbell, 1964, p. 327). The happiness findings provide confidence that respondents in all cultures sampled understood and performed the general task of viewing pictures and selecting one from each pair that appeared either happier or more dominant. Of course, it is possible, even likely, that happiness is more easily communicated/detected than is dominance. But in most of the samples, dominance was linked with eyebrow position just as it was in the United States samples, so the failure of the Kenyan and Zambian samples to do the same has to be considered a substantive finding, albeit one whose explanation is not obvious.

It may be, for example, that in these two African samples the concept of dominance is complex and not easily communicated, whether by word or by gesture. If this were so, respondents in those two samples may not, in fact, have been judging dominance, despite the efforts taken in the instructions given them, efforts that included multiple examples of what was meant by the concept. On the other hand, if one assumes that respondents in these samples, like those in the other samples, were judging dominance, it would have to be concluded that, for the African respondents, eyebrow positions are not cues for dominance.

Data from the two African samples were collected by psychologists experienced in African research (Janet Kilbride in Zambia and the present writer in Kenya), and the respondents they employed were in the main educated and literate in English. While it is true that in the respondents' native languages the concept of dominance is difficult to communicate via a single word or simple phrase, there is in fact little reason to believe that the particular respondents employed did not have a concept of dominance (or power or the like) in mind when they made their choices. In view of this, the tendency of respondents in both of these samples (which are, not incidentally, culturally similar to each other) to make their choices on some basis other than eyebrow position challenges the view that eyebrow position universally communicates dominance-submissiveness. And, of course, the Thai sample's selection of models with raised eyebrows as dominant indicates that, while eyebrow position is relevant to status communication among Thais, lowered brows do not universally indicate dominance. The notion, then, that the eyebrow-down gesture signifying dominance has phylogenetic origins seems to meet a serious challenge in Keating's findings.

On the other hand, it seems imprudent to conclude that

there is not a continuity across diverse species of primates and humans in the communication of differential status via eyebrow gestures, despite the cross-cultural exceptions presented by the Kenyan, Zambian, and Thai samples to the tendency to detect dominance in lowered brows. After all, lowered brows were associated with dominance by a good number of samples from other cultures. And, more to the point, it seems reasonable that cross-cultural universality is not required as evidence for a genetic underpinning to any particular class of behavior. Besides, as was asserted earlier in this chapter, resolving the nature-nurture controversy ought not be the primary objective of cross-cultural research.

Implications of Evolutionary Epistemology for Cross-Cultural Psychology

Campbell's insights regarding evolution and behavior should direct us toward a search for the environmental factors that make adaptive and functional whatever behavior prevails in the world. To avoid circularity, of course, such a search should be informed by testable functionality hypotheses. Whether the mechanism that produces the adaptiveness is evolution *or* learning (or some combination of both) is not the question that should motivate cross-cultural research. Rather, such research should be primarily a quest for those specific environmental circumstances, both natural and cultural, that are plausibly hypothesized to predispose the humans who occupy the environments to behave the way they most consistently do.

Campbell's perspective on the natural selection model of evolution provides a coherent view of the social and behavioral sciences and links them with the biological sciences. It induces us to search for and to detect comparable processes in phenomena as diverse as molecular biology, operant conditioning, and the fit of cultural systems to their environments.

This model—contemporary biological evolution, which, unlike Social Darwinism, does not contemplate an overall evolutionary progress—emphasizes random variation and selective retention, tending toward adaptiveness in a circuitous manner and only in the long run. Genetic variability comes largely from random mutations, and it is only after the fact that environmental selection operates. The variations are nonprescient. In 1975 Campbell explained his prescription that this model be applied to psychology:

> My own fascination with evolutionary theory is centered around the general model for adaptive processes illustrated in natural selection. I date this fascination to my reading Ashby's *Design for a Brain*, 1952, in which the formal analogy between natural selection and trial-and-error learning is made clear. . . . This point is also clearly stated in my notes from Egon Brunswik's lectures of 1939 attributed to Karl Bühler. I have come to the conclusion that this model —which I summarize as blind variation and selective retention—is the only and all-purpose explanation for the achievement of fit between systems and for the achievement and maintenance of counterentropic form and order [Campbell, 1975, p. 1105].

One may question whether trial-and-error learning is the best appellation for the psychological analogue of the biological process, or whether Skinnerian operant behavior analysis provides the better analogy. However, the central point, that the learning of an organism tends toward adaptiveness, is the principle that should influence the planning of cross-cultural research in psychology.

The notion of "fit"—that is, adaptiveness deriving from selection among variations—applies to systems as diverse as species, individuals, and such sociocultural products as cultural institutions. No qualitative leaps are required to move from one system to another. The "embodied knowledge" of the environment that is inherent in the blossoming and bearing of northern fruit trees is characteristic also of humans in their diverse and changing environmental niches. Indeed, when features of the environment are so transitory that genetic decision rules have no time to develop, then behavioral adaptations must occur. Moreover, these learned behavioral modifications must be functional in the same way that genetic changes would have been functional had they had time to occur.

Thus, Campbell's extension of the natural selection description of systemic fit to human behavior places psychology on a coherent continuum with biology while not ignoring distinctly human modes of adaptation; notably, complex learning, both direct and vicarious. Additional meaning is thus given to the concept of adaptation, and learned achievements are thus understood in the same way evolutionary achievements are understood.

In this respect, his perspective is in tune with a dominant ecological emphasis in contemporary cross-cultural psychology (see, for example, Berry, 1976), whereby behaviors and cultural recipes for behaviors are cumulatively selected for their adaptive character and are transmitted intergenerationally by social means, including socialization and enculturation, rather than by genetic adaptations. These social processes are, however, also blind and nonprescient. "Just as the human and octopus eyes have a functional wisdom that none of the participating cells or genes have ever had self-conscious awareness of, so in social evolution we can contemplate a process in which adaptive belief systems could be accumulated which none of the innovators, transmitters, or participants properly understood, a tradition wiser than any of the persons transmitting it" (Campbell, 1975, p. 1107).

The main implication is that searching for and focusing on the adaptiveness of behavior should lead us to attach priority to studying the selective system, to ask what environmental conditions prevail that could have made more appropriate whatever behaviors our observations reveal. It is less important to inquire whether the behavior's adaptiveness is the long-term product of random, genetic mutation followed by natural selection or individually stumbled on and intergenerationally transmitted through one or another form of learning.

From this it follows that a cross-cultural difference in behavior—some departure from universality, as in the Thai sample's apparent perception of raised-eyebrow faces as dominant or the lesser susceptibility to the Sander parallelogram illusion among samples dwelling in less carpentered worlds—ought not to be taken, in itself, as evidence against a biologically based, genetically influenced behavioral tendency. Such findings should serve as impetus for a search for the behavior's antecedent conditions, to be found, one should expect, in the physical or cultural environments of the "discrepant" group. The goal of the search should be a plausible account of the fit of the behavior to those environmental conditions.

In the case of the differences in illusion susceptibility, such a plausible account was available from the beginning of the enterprise; and, as I tried to show earlier in this chapter, where relevant studies were reviewed, that account has survived well the competition from one plausible alternative. In the case of Keating's provocative findings concerning eyebrow gestures and dominance, the adaptiveness argument for the discrepant cases has yet to be developed. But the direction in which she and others interested in nonverbal communication of status ought to look is clear. In cultures where persons depart from other animals and from other human groups in the meaning they attach to particular facial expressions, we should expect to find intergenerationally transmitted rules that make sense in the overall social interaction norms of those cultures. And we should try to specify these in advance.

Genetic hypotheses and learning hypotheses as competing alternatives must both be assessed in the light of their competence in pointing to systemic fit. Their relative competence in this regard, and hence their relative plausibility, is likely to relate to time. Thus, it must first be granted that human groups could acquire perceptual modes that equip them to function well in their environments, with that acquisition a product of biological evolution extending over many generations. But, if possibly relevant environmental changes could have (or, better, are *known* to have) occurred over a few generations, rather than gradually over millennia, then learning is very likely involved. Such is clearly the case for the proliferation of carpenteredness, first in Western and later in Westernizing cultures. Therefore, the correlation between illusion susceptibility and environmental differences is plausibly explained in terms of individually acquired inference habits among all persons dwelling in such cultures.

Very likely, of course, are mixed models, since in most instances both biological evolution and individual learning ultimately move in the same direction. When we are confronted by findings like Keating's, for example—cross-cultural consistency and a few deviations therefrom—we should not consider the deviant cases as evidence against genetically based systemic fit. Rather, they should be taken as anomalies that trigger a search for the learned variations that could have been prompted by whatever socioenvironmental forces prevail in these particular instances. These variations would probably be viewed best as interacting with species-wide genetic factors. Nevertheless, whatever human behavior is found that departs from the patterns characterizing our primate relatives and the bulk of humankind comprises an empirical puzzle that demands theoretical ingenuity, and a cross-cultural psychology that ignores biology does so at its peril.

Conclusion

The meta-methodological message that resides in Campbell's stress on systemic fit is that we should abandon the view that behavioral differences across cultures in themselves necessarily mean that individual habit acquisition alone has produced them and that such differences cannot, in part at least, reflect genetic influences. Learned habits of perceptual inference may satisfactorily be demonstrated, as I believe they have been in research on geometric illusions, but many cross-cultural differences may have to be attributed to genetic differences, especially when a satisfactory account of a genetic process exists and when an equally satisfactory account of a learned process cannot be given. We must examine the evidence that relates to both. To do so, we must develop a comparative psychology in the fullest sense, one that is both cross-species *and* cross-cultural.

This call is not for reductionism but for an expansion of perspective. It is consistent with other of Campbell's teachings, including his discussions of perspectival relativism and his appeals for methodological triangulation. Biology and psychology are, in any case, not competing perspectives but complementary ones. While hypotheses that reflect the disciplinary predispositions of each—evolution in the case of contemporary biology and learning in the case of contemporary psychology—should compete on the grounds of plausibility as suggested above, scientists in both disciplines need the perspective of the other in order to detect how their hypotheses are mutually compatible accounts of an incremental tendency toward adaptiveness of all living organisms. Only when that adaptiveness is accounted for, whether via evolution, learning, or both, can behavior be said to be understood.

It is perhaps a touch of poetic irony that the empirical domain in which Campbell's systemic-fit-search ideology has most clearly been applied is that of the geometric illusion. The concept of illusion has, as is well known, played a central role in epistemology through the ages. It is important in Campbell's own evolutionary epistemology, wherein he has cogently argued that perspectival relativism, rather than justifying ontological nihilism, presupposes a knowable reality, but one that is imperfectly known from any particular perspective. What the empirical research on geometric illusions has shown, as I have tried to argue in this chapter, is that an instance of imperfect knowledge, as in "knowing" one line to be longer than another, has an explanation that is rooted in the overall, long-term adaptiveness of that knowledge, even if wrong in particular instances. That cross-cultural research in visual perception yielded such an explanation is its real achievement and not that it resolved a nativist-empiricist controversy. Campbell's major unifying contribution to the study of behavior will reap its richest dividends when additional behavior domains are similarly understood.

References

Andrew, R. J. "Evolution of Facial Expression." *Science,* 1963, *142,* 1034-1041.

Armstrong, R. E., Rubin, E. V., Stewart, M., and Kuntner, L. "Susceptibility to the Müller-Lyer, Sander Parallelogram, and Ames Distorted Room Illusions as a Function of Age, Sex, and Retinal Pigmentation Among Urban Midwestern Children." Evanston, Ill.: Department of Psychology, Northwestern University, 1970. (Mimeograph.)

Berry, J. W. "Müller-Lyer Susceptibility: Culture, Ecology, or Race?" *International Journal of Psychology,* 1971, *6,* 193-197.

Berry, J. W. *Human Ecology and Cognitive Style: Comparative Studies in Cultural and Psychological Adaptation.* Beverly Hills, Calif.: Sage, 1976.

Binet, A. "Les Mésure des Illusions Visuelles chez les Enfants" ["The Measure of Visual Illusions with Children"]. (R. H. Pollack and F. K. Zetland, Trans.) In *Perceptual and Motor Skills,* 1965, *20,* 917-930. (Originally published 1895.)

Bolton, R., Michelson, C., Wilde, J., and Bolton, C. "The Heights of Illusion: On the Relationship Between Altitude and Perception." *Ethos,* 1975, *3,* 403-424.

Brislin, R. "The Ponzo Illusion: Additional Cues, Age, Orientation and Culture." *Journal of Cross-Cultural Psychology,* 1974, *5,* 139-161.

Brislin, R., and Keating, C. "Cultural Differences in the Perception of a Three-Dimensional Ponzo Illusion." *Journal of Cross-Cultural Psychology,* 1976, *7,* 397-411.

Brunswik, E. *Perception and the Representative Design of Psychological Experiments.* Berkeley: University of California Press, 1956.

Campbell, D. T. "Distinguishing Differences of Perception from Failures of Communication in Cross-Cultural Studies." In F. S. C. Northrop and H. H. Livingston (Eds.), *Cross-Cultural Understandings: Epistemology in Anthropology.* New York: Harper & Row, 1964.

Campbell, D. T. "On the Conflicts Between Biological and Social Evolution and Between Psychology and Moral Tradition." *American Psychologist,* 1975, *30,* 1103-1126.

Campbell, D. T. "Descriptive Epistemology: Psychological, Sociological, and Evolutionary." William James Lectures, Harvard University, 1977.

Gillam, B. "Geometrical Illusions." *Scientific American,* January 1980, pp. 102-111.

Gregory, R. L. *Eye and Brain: The Psychology of Seeing.* New York: McGraw-Hill, 1966.

Jahoda, G. "Retinal Pigmentation, Illusion Susceptibility and Space Perception." *International Journal of Psychology,* 1971, *6,* 199-208.

Jahoda, G. "Retinal Pigmentation and Space Perception: A Failure to Replicate." *Journal of Social Psychology,* 1975, *97,* 133-134.

Keating, C. F. "A Cross-Species, Cross-Cultural Approach to Human Facial Gestures of Dominance and Submission." Unpublished doctoral dissertation, Syracuse University, 1979.

Keating, C. F., Mazur, A., and Segall, M. H. "Facial Gestures Which Influence the Perception of Status." *Sociometry,* 1977, *40,* 374-378.

Liebowitz, H. W., Brislin, R., Perlmutter, L., and Hennessey, R. "Ponzo Perspective Illusion as a Manifestation of Space Perception." *Science,* 1969, *166,* 1174-1176.

Piaget, J., and von Albertini, B. "L'illusion de Müller-Lyer." *Archives de Psychologie* (Genève), 1950, *33,* 1-48.

Pollack, R. J. "Contour Detectability Thresholds as a Function of Chronological Age." *Perceptual and Motor Skills,* 1963, *17,* 411-417.

Pollack, R. H. "Müller-Lyer Illusion: Effect of Age, Lightness Contrast and Hue." *Science,* 1970, *170,* 93-94.

Pollack, R. H., and Silvar, S. D. "Magnitude of the Müller-Lyer Illusion in Children as a Function of Pigmentation of the Fundus Oculi." *Psychonomic Science,* 1967, *8,* 83-84.

Pollnac, R. B. "Illusion Susceptibility and Adaptation to the Marine Environment: Is the Carpentered World Hypothesis Seaworthy?" *Journal of Cross-Cultural Psychology,* 1977, *8,* 425-533.

Redican, W. K. "Facial Expressions in Nonhuman Primates." In L. A. Rosenblum (Ed.), *Primate Behavior: Developments in Field and Laboratory Research.* Vol. 4. New York: Academic Press, 1975.

Rivers, W. H. R. "Vision." In A. C. Haddon (Ed.), *Reports of the Cambridge Anthropological Expedition to the Torres Straits.* Vol. 2, Pt. 1. Cambridge, England: Cambridge University Press, 1901.

Rivers, W. H. R. "Observation on the Senses of the Todas." *British Journal of Psychology,* 1905, *1,* 321-396.

Segall, M. H., Campbell, D. T., and Herskovits, M. J. "Cultural Differences in the Perception of Geometric Illusions." *Science,* 1963, *193,* 769-771.

Segall, M. H., Campbell, D. T., and Herskovits, M. J. *The Influence of Culture on Visual Perception.* Indianapolis: Bobbs-Merrill, 1966.

Silvar, S. D., and Pollack, R. H. "Racial Differences in Pigmentation of the Fundus Oculi." *Psychonomic Science,* 1967, *7,* 159-160.

Skinner, B. F. *Contingencies of Reinforcement.* New York: Appleton-Century-Crofts, 1969.

Stewart, V. M. "Tests of the 'Carpentered World' Hypothesis by Race and Environment in America and Zambia." *International Journal of Psychology,* 1973, *8,* 83-94.

van Hoof, J. A. R. A. M. "The Facial Displays of the Catarrhine Monkeys and Apes." In D. Morris (Ed.), *Primate Ethology.* London, Weidenfeld and Nicolson, 1967.

Wagner, D. A. "Ontogeny of the Ponzo Illusion: Effects of Age,

Schooling, and Environment." *International Journal of Psychology*, 1977, *12*, 161-176.

Weaver, D. B. "An Intra-cultural Test of Empiricistic Versus Physiological Explanations for Cross-Cultural Differences in Geometric Illusion Susceptibility Using Two Illusions in Ghana." Unpublished doctoral dissertation, Northwestern University, 1974.

Wohlwill, J. F. "The Perspective Illusion: Perceived Size and Distance in Fields Varying in Suggested Depth, in Children and Adults." *Journal of Experimental Psychology*, 1962, *64*, 300-310.

15

Barry E. Collins

Hyperactivity:
Myth and Entity

Conceptual (theory-testing) research in social psychology during the 1960s and 1970s typically began with a theoretical idea; the researcher then crafted a laboratory context, an independent variable, and a dependent variable—each ideally suited to test the conceptual idea. Some of us, however, have reversed that process. We begin with a context, an independent variable, and a dependent variable (such as helping or obesity) and then sift through available conceptual analyses to find a theoretical idea that illuminates the applied problem and vice versa. In my own effort to integrate basic

Note: Preparation of this paper was facilitated by NIDA Grant DA 01070 and NIMH Grant MH 29475.

385

and applied social psychology, I initially tried several different applied problems. Research on police, urban transit, and psychotherapy did not, for me, produce an opportunity for advance on the conceptual issues in psychology. Then, in the fall of 1975, Barbara Henker and Carol Whalen gave me a copy of their paper on hyperactivity (Whalen and Henker, 1976), which was then under editorial review by the *Psychological Bulletin*. I was fascinated by the issues raised in that paper. As we now have spelled out elsewhere (Collins, Whalen, and Henker, 1980), hyperactivity research is an ideal arena in which to explore both applied problems and also basic science issues.

This chapter is divided into three sections. The first section highlights a few of the lacunea uncovered by the Whalen and Henker (1976) analysis of hyperactivity. The second section presents an empirical answer to some of these questions. The third section illustrates and elaborates the six themes in the thinking of Donald Campbell, identified in Chapter One of this volume.

The process to be described here resembles a prototypic "who dunnit." The research problem presented in the first section is a mystery—a puzzling set of irreconcilable clues (Who killed the Major and how was it done, since all the windows and doors were locked from the inside?). The data/solutions in the second section are the denouement, or last chapter of the detective story. The final section corresponds to the middle chapters of the story. Here the would-be knower (detective or researcher) uses a variety of search strategies and slowly accumulates bits of information which, in concert, fill in the gap between puzzle and answer.

Chapter One: The Mystery of Hyperactivity

Although hyperactivity is a relatively recent medical diagnostic category, there has been a rapid increase in its use. Clinical descriptions of the hyperactive boy—perhaps 80 percent of the children diagnosed as hyperactive are male—contain words like *disruptive, inattentive,* and *constantly fidgeting.* Far and away the most common operational definition of hyperactivity is based on responses to a ten-item questionnaire completed by teachers who are observing classroom behavior. The Conners Abbreviated Symptom Questionnaire (Conners, 1973; Goyette, Conners, and Ulrich,

1978) instructs teachers to use a four-point scale ("not at all," "just a little," "pretty much," and "very much") to rate a child's behavior for each of ten descriptive statements:

1. Restless and overactive.
2. Excitable, impulsive.
3. Disturbs other children.
4. Fails to finish things he starts—short attention span.
5. Constantly fidgeting.
6. Inattentive, easily distracted.
7. Demands must be met immediately—easily frustrated.
8. Cries easily and often.
9. Mood changes quickly and drastically.
10. Temper outbursts, explosive and unpredictable behavior.

These ratings represent *global* impressions formed by the child's *teacher* observing *school* behavior. The correlations among teacher, parent, and pediatric ratings on this same questionnaire are quite low. Clinical wisdom specifies that it is the teacher ratings of school behavior that are valid; the tasks and/or contexts sampled in the home and pediatric office are, presumably, less relevant to the diagnosis of hyperactivity.

   The most common treatment-intervention for hyperactivity is psychostimulant medication—methylphenidate (Ritalin) or dex-troamphetamine (Dexadrine). Although the interobserver reliability among teacher, pediatrician, and parent Conners ratings are puzzlingly low, a number of carefully conducted double- and triple-blind studies—including our own (Henker, Whalen, and Collins, 1979)—demonstrate substantial improvement in a child's ratings from placebo to medication conditions.

   Additional data continue to supply other seemingly ill-fitting clues in the mystery. Almost all clinical impressions of these hyperactive boys describe a high level of motor activity ("He's off the wall"); but objective measures of activity—such as time-lapse photography and motion-sensitive instruments—often do *not* detect a decrease in activity level with psychostimulant medication. These objective measures directly contradict the experience of those involved with hyperactive boys. Furthermore, there appears to be something about the grammar school classroom that makes the unmedicated hyperactive boy particularly problematic; medication is often not administered on weekends, evenings, or school vacations. Many observers expect to find an organic etiology for hyperactivity, because a psychopharmacological intervention (medication) appears to provide such dramatic amelioration. However, the search for organic brain dysfunctions that separate hyperactive children from others has not been particularly successful. (Dubey, 1976, provides a critical review of research on organic factors in hyperactivity.)

There are some myths *about* hyperactivity (as there are surely myths about any entity, scientific and otherwise). Henker and Whalen (1980, pp. 322-324) identify several—for instance, that "hyperactivity vanishes with adolescence" or that "hyperactive children respond paradoxically to medication." Others write about the myth *of* hyperactivity—for example, Schrag and Divoky (1975) in *The Myth of the Hyperactive Child*. In contrast, still others take the drug-placebo differences as evidence that the concept of hyperactivity has an objective referent. Henker and Whalen (1980, p. 322), for instance, list the notion that hyperactivity is entirely in the eyes of the beholder as their Myth 2: "The converse of the child deficit view is the presumption that hyperactivity itself is a myth, and that the diagnostic process is used primarily to satisfy the personal, social, economic, or political needs of the people doing the labeling (see Conrad, 1975; Schrag and Divoky, 1975)."

If hyperactivity does exist objectively, what is it? And where is it? In the child? In the curriculum? In an unresponsive school system? In parents? In an oppressive society? In a conspiracy by the drug firms? And what is it about the behavior of the unmedicated boy that makes him such a problem to his teacher and peers in the classroom? What is it about the teacher and peers that render them so vulnerable to whatever it is that the hyperactive boys do? And what is it about the classroom environment that magnifies and multiplies the problems generated by (and faced by) the unmedicated hyperactive boy?

Chapter 14: Who Dunnit?

This section skims over, for the moment, the process of investigation and detective work in order to present some illustrative answers—a little bit like reading the last page of the novel. Both research design and data are presented with more detail and with different perspectives elsewhere (Henker and Whalen, 1980; Henker, Whalen, and Collins, 1979; Whalen, Collins, Henker, Alkus, Adams, and Stapp, 1978; Whalen and Henker, 1976, 1980a, 1980b; Whalen, Henker, Collins, Finck, and Dotemoto, 1979; Whalen, Henker, Collins, McAuliffe, and Vaux, 1979; Whalen, Henker, and Dotemoto, 1980).

A Working Conception of Hyperactivity. The Campbellian philosopher of science avoids talking about definitions in order to avoid the implication that any single measurement procedure can exhaustively define a concept. Not even a set of multiple measurement operations can define a concept exhaustively, since such a set is made up of imperfect pointers, method-bound indicators, and incomplete manifestations. Therefore, when we choose one or two operational realizations for use in research design, we realize that we have chosen only one or two of many possible perspectives and that we have an imperfect grip on the phenomena being examined.

It was suggested previously that the best single operational indicator of hyperactivity is teacher ratings of classroom behavior on the Conners scale. The best evidence for the validity of the Conners scale is found in the ability of these ratings to distinguish drug and placebo conditions. (The search for independent measures in activity level in the brain structure or physiology of the boys, however, has been largely unsuccessful.) Thus, the best guess I could make in 1975 was that the entity of hyperactivity is, in part, "whatever it is that is improved by psychostimulant medication and is also correlated with teacher ratings on the Conners scale." Quantifying a concept in terms of a response to medication is not as unreasonable as it may seem on first reading. Had penicillin, for instance, been discovered before bacteria, medical scientists would have had to define Type A pneumonia (bacterial) as any feverish cough which gets better with the administration of an antibiotic medicine; and Type B pneumonia (viral) as any feverish cough which does not improve with administration of penicillin. A less fanciful example is the use of nitroglycerin to diagnose heart pain (angina). Even in the best-equipped coronary care units, current medical practice distinguishes angina from other chest pains, in part, by defining angina as "any chest pain that gets better with nitroglycerin."

The decision to create such a working definition invests the drug-produced improvement in teacher ratings with a temporary "most likely to be true" status. This tactic of scientific investigation is close kin to what Campbell calls the doubt/trust ratio. At any one time in the progress of the scientific method, the scientist chooses to be trusting of some pieces of information and skeptical of others. In contrast to the present study, investigators at other points in the study of hyperactivity should anchor their research designs elsewhere and focus their skepticism on the medication-produced improvement in teacher ratings. (My collaborators and I, for instance, have searched for individual differences among teachers that would lead one teacher to assign higher ratings than another teacher assigns to the same child.) For the purposes of this investigation, however, we will explore the implications of thinking of hyperactivity as that which improves with medication and also correlates with teacher ratings of classroom behavior on the ten-item Conners scale.

Overview of the Summer School Research Design. The data to be used for illustration here are drawn from those collected in a research summer school in 1977, where Henker, Whalen, and I studied a total of 61 boys—22 hyperactive and 39 comparison. Hyperactive children were recruited through local pediatricians. An unselected, heterogeneous group of normal comparison children were recruited from the local community.

The boys spent two mornings a week in a classroom setting, two mornings a week participating in various structured assessment modules, and one day a week on a field trip. Parents were supplied with envelopes containing medication for the critical weeks of the program at the dosage levels previously established by the referring physician. During the third week of the project, a randomly assigned half of the hyperactive boys received a placebo, and half received medication. During the fourth week, these conditions were reversed. Ratings from both double- and triple-blind staff (Henker, Whalen, and Collins, 1979) demonstrate significant differences between medication and placebo conditions. The staff had no information about the experimental design, the research focus on hyperactivity, or our intention to study medication effects. They knew that they were working in an experimental summer school, but they did not know that the focus was on hyperactivity and psychostimulant medication.

Medication Effects on Classroom Behaviors: What Is It That Hyperactive Boys Do? The medication effects uncovered in the classroom generally indicate that hyperactive boys behave more conventionally when medicated than when on a placebo. The most typical data pattern is for the medicated hyperactive boys to be almost indistinguishable from the comparison boys on our classroom behavior ratings. We did find evidence of undesirable side effects of the medication outside of the classroom. For instance, hyperactive boys on medication—in contrast to those on placebo—were observed to direct more negative affect toward themselves while they were engaged in our Space Flight experiment (Whalen, Henker, Collins, Finck, and Dotemoto, 1979). Furthermore, outside the classroom we found differences between hyperactive and comparison boys which were unremediated by medication. For instance, hyperactive boys—whether medicated or placebo—were more likely to disagree inappropriately with their partner in the Space Flight experiment, even though only the partner knew the correct solution.

Our major measures in the classroom derived from our Classroom Observation System and from teacher ratings. The observers of classroom behavior focused on a child for a ten-second interval before filling out the Classroom Observation System observation card. The observation procedure was specifically designed to eliminate all contexts not contained in the ten-second observation interval. (For detailed presentations of the data, a more complete description of the observation system, and tests of statistical significance, see Whalen, Collins, Henker, Alkus, Adams, and Stapp, 1978; and Whalen, Henker, Collins, Finck, and Dotemoto, 1979. Collins, Whalen, and Henker, 1980; and Henker and Whalen, 1980, provide an intermediate level of detail.)

All seven of the observation categories defined below reflect the same empirical pattern: (1) Placebo hyperactive boys are different from both medicated hyperactive boys and comparison boys. (2) A boy's score on the observation category correlates with his score on the Conners scale. (3) Medicated hyperactive boys and comparison boys are similar. Thus, all seven of the following observation categories meet our working definition of hyperactivity (points 1 and 2 above).

1. *Off task.* Off-task behavior is behavior other than that predicted from or expected on the basis of two things: the assign-

ment specified in the curriculum (for example, "Spell a word out loud") or explicit instructions from the teacher (for example, "During this period, work with a partner, but no talking" or "John, close the door" or "Now it is time to clean up"). Off-task percentages are quite comparable for comparison and medicated hyperactive boys (5 percent and 6 percent) but are elevated for unmedicated placebo hyperactive boys (13 percent). Off-task ratings correlate with teacher ratings on the Conners scale (about .65). A similar pattern occurs for the remaining six observer categories.

2. *Stand out.* As the observer training manual explains, normative behaviors are predictable from the implicit social rules that specify appropriate, typical, or socially desirable behaviors in a given setting; such behaviors form the background for stand-out behaviors. Behaviors that "stand out" probably violate the observer's expectation of what should happen or is likely to happen in a particular social setting. The requirements of the curriculum are not used to define these informal social rules; otherwise, "stand out" and "off task" would be the same judgment.

3. *Sudden.* "Sudden" refers to an abrupt change from previous behavior (specifically, the previous five seconds of the child's behavior). It reflects a change in the class of behavior—that is, a change in direction of flow, quality of behavior, or kind of behavior—and does not necessarily reflect a change in intensity.

4. *Intense.* High intensity is scored when a child's behavior is effortful, intense, vehement, rapid, or loud. Energy is judged directly from the act itself; it is thus independent of task requirements, local or momentary social norms, and adjacent behaviors.

5. *Consequences (disruption).* Disruption is rated whenever the child's behavior has observable consequences that interrupt other people's behavior. For example, if the child's behavior induces intervention from the teacher, disruption is scored.

6. *Squirm-wiggle.* This category refers to movements made while the child is in a relatively stationary position: wiggling, squirming, swaying, spinning around, rocking while seated, rocking from one foot to another while standing, and stretching. (In contrast, translocation—walking, running, skipping, moving self

from one place to another while seated on a chair, crawling and rolling around on the floor—does *not* show a medication effect. Therefore, the widely held view that hyperactive behavior involves a walking around or running about is not upheld. As Imber and Collins (1981) demonstrate in a laboratory study, behaviors which do not match the context and behaviors associated with loud noise create the *illusion* that the actor has "moved around.")

7. *Noisy.* Three kinds of noise were included in the Classroom Observation System: *verbalization* (actual speech), *vocalization* (nonverbal noises, such as laughing, whistling, humming, intentional sneezing), and *noise* (audible sounds other than verbalization or vocalization; for example, slapping face with own hand, clapping, banging chair against adjacent desk, pounding, or slamming desk).

Chapters 2-13: Illuminating Major Themes in Campbellian Thought

The following section represents a partial reconstruction of the thought processes used to get from the mysteries encountered in Whalen and Henker (1976) to my current conceptualizations of hyperactivity. It is a sort of intellectual autobiography, so I will make increasing use of the first person. I am inclined to think that the following reconstruction tracks the process of discovery reasonably well. I used a number of colloquium presentations to garner free consulting on our research design from my colleagues at other universities. The hypotheses and design had to be stated explicitly for the research proposals, for the colloquia, for site visits, and for various project staff and graduate students.

Pattern Matching and the Contextual Basis of Knowledge. Our classroom observers were asked to judge whether a child's behavior was predictable by, expected on the basis of, or "fit" the task context; the Classroom Observation System required them to attend first to one context and then another. Observers were required to judge whether the behavior observed in a ten-second interval was, or was not, in sharp contrast with several contexts: (1) the curriculum and explicit teacher instructions, (2) the implicit social rules which specify the typical or socially desirable behavior in that setting, (3) the boy's previous stream of behavior, and (4) the normal or typical level of intensity or effort. Thus, the meaning of a "stand-out" behavior, for instance, lies as much in the context of the act as in the act itself—as much in the ground as in the figure. Identical behaviors (a broad smile and chuckles) might "stand out" at church but not at Sunday school, at a street corner accident but not at a demolition derby.

Campbell (1966, p. 82) states: "Imagine the task of identifying 'the same' dot of ink in two newspaper prints of the same photograph. The task is impossible if the photographs are examined by exposing only one dot at a time." Analogously, the meaning of a particular child behavior could not be rated by one of our observers unless the assigned tasks, social rules, or prior behaviors were known. It is the *pattern* of the behavior-context *relationship* that contains the meaning defined by the first three observational categories.

At another level, some sort of theory is needed to integrate the medication effects and the correlations between the behaviors and the teacher ratings. It is not obvious that each of these particular seven observation categories would be affected by medication. Nor is it obvious that this particular set of behaviors should correlate with teacher ratings of hyperactivity on the particular ten items of the Conners scale. One might summarize the results empirically and say that hyperactivity is being rated as off task, standing out, sudden, intense, disruptive, squirming, and noisy; but such a list is unsatisfying. The list provides no metaphor that turns the separate empirical findings into a coherent whole or Gestalt; the list of seven classroom behaviors is hard to remember—much less assimilate and integrate. More important, the list provides no conceptual matrix that would allow one to form expectations about additional places and additional forms in which the entity of hyperactivity might be manifested (additions such as new observation categories, organic concomitants, or adult behaviors). And, argues Campbell, such a state is typical where one is provided only with mean differences, figures, correlations, or empirical relationships. As Campbell (1966, p. 97) puts it: "The formal theory becomes one 'pattern,' and against this pattern the various bodies of data are matched, in some overall or total way. . . . The data are not required to have an analytical coherence among themselves, and indeed cannot really be assembled as a total except upon the skeleton of theory."

Similarly, in the words of Webb, Campbell, Schwartz, and Sechrest (1966, pp. 28-29):

> The goal of complete description in science is particularly misleading when it is assumed that raw data provide complete description. . . . Theories are more complete descriptions than obtained data, since they describe processes and entities in their unobserved as well as in their observed states. . . . The raw data, observations, field notes, tape recordings, and sound movies of a social event are but transient superficial outcroppings of events and objects much more continuously and completely (even if abstractly) described in the social scientist's theory. Tycho Brahe and Kepler's observations provided Kepler with only small fragments of the orbit of Mars, for a biased and narrow sampling of times of day, days, and years. From these he constructed a complete description through theory. The fragments provided outcroppings sufficiently stubborn to force Kepler to reject his preferred theory.

As an effort to move toward such a theory pattern, the seven observation categories discussed in the previous section can be expanded with four more words from everyday language. The behaviors of hyperactive children are, among other things, *inappropriate, salient, conspicuous,* and *intrusive*. All seven of the behavior attributes—off task, stand out, sudden, intense, disruptive, squirm-wiggle, and noisy—observed with placebo (unmedicated) hyperactive boys are inappropriate, salient, conspicuous, and intrusive. These four words are not yet a formal theory; yet they do begin to construct a concept far more encompassing than would be the case if the concept of hyperactivity was defined entirely by seven mean differences and seven correlations.

The concept denoted as hyperactivity can be further expanded to encompass meanings beyond those endowed by these four words. The remaining sentences in this paragraph are composed of words far less precise than the words in a computer language or the elements of mathematical formulas. Even with this limitation, the rest of this paragraph is intended to impregnate the concept of hyperactivity with loose connections to other concepts and other operational realizations: Persons who are behaving appropriately modulate their behavior in accord with some external situational cue or discriminative stimulus. Appropriate, nonsalient, and inconspicuous behaviors are matched to the current contexts. This appropriate modulation requires at least three separate tasks. The first is a search-and-monitor activity: the organism must monitor the current environment for the relevant discriminative stimuli. Second, the organism must know the rules which define appropriate behavior in that particular circumstance. Third, the organism must comply with the relevant rules. An appropriate driver adjusts automobile speed according to discriminative stimuli posted on the maximum speed limit signs and to discriminative stimuli contained in road conditions and in the movement of other cars. An appropriate student modulates question asking according to the teacher's nonverbal cues indicating that he is very interruptable versus noninterruptable and cues in the current task assignment which specify whether questions are appropriate or not.

Inappropriate, salient, conspicuous, and intrusive behaviors are of particularly great consequence and significance in the grade school classroom. A major prerequisite for mastery and rehearsal of reading, writing, and arithmetic skills is an environment in which it is possible to ignore distractions and thus to concentrate on the assigned work: good conduct is good citizenship. The orienting response induced by salient behavior is often relatively primitive, automatic, and autonomic (Coombs, 1938; Edwards, 1975; Graham, 1973; Harding and Rundle, 1969; Lovibond, 1969; McCubbin and Katkin, 1971; Porter, 1938). Perhaps your freedom ends not with the end of my nose but with the beginning of my orienting reflex. Salient behavior is difficult if not impossible to ignore. The presence of a cohort of age peers may also magnify the ramifications of hyperactive behaviors in a grade school classroom. Any parent will agree that nothing brings out the essence of a 9 year old better than a stationwagon full of them on the way to a birthday party. Yet a room full of 9 year olds is routine for a teacher who must live with the contagion of sundry salient behaviors.

Thus, both classroom discipline and academic instruction often stretch the teacher's capacity for concentration, and the tasks assigned the students often challenge or exceed the capacity of the students, at that developmental level, for focused attention. In summary, the classroom ecology is vulnerable to distraction. Inappropriate behaviors, perhaps particularly in the classroom, are thus intrusive and noxious to others. Furthermore, the orienting response to loud, sudden, and moving objects may be so automatic and primitive that teachers and/or peers may be unable to inhibit their negative response to the salient, hyperactive behaviors.

The concept of hyperactivity can be yet further embellished with an anecdote. I agreed to help out a 9-year-old friend by taking a group of his buddies to a birthday party at a "grown-up restaurant, one with a tablecloth." As soon as we entered the restaurant, I found myself spouting an endless stream of restrictions: "Don't take the ice out of the glass with a spoon." "Don't blow your straw wrapper across the table, you're hitting people at the next table." "Only one kid at a time in the bathroom." "Put your shirt back on." "Leave the sugar packets alone." "Keep the plates flat on the table." "No ketchup on the table cloth." I thought my specific list of *Verboten* would never end. I was stuck with a list of forbidden behaviors in the same way I would be stuck with a list of hyperactive behaviors in the preceding section of this chapter; I did not have an integrating principle or metaphor.

Soon I got what I regarded as a reasonable complaint: "Hey! I'm the birthday boy! How come you're making all those rules?" I was wondering about the same thing myself; the necessity for new rules was increasing more rapidly than I could articulate them. Then I had an idea: "Hey, have you looked at the prices on the menu? This place is more expensive than McDonald's, isn't it? Why do you think they charge more money here? Well, I'll tell you. All these people at the other tables are paying extra money so they can pretend that there are no kids in this restaurant. You guys can do whatever you want to do except for one thing. You have to be *invisible*. The people at the next table have to be able to pretend you're not even here."

It worked. The boys were able to go back and forth between the table and the bathroom in pairs and trios. They shot all the straws they wanted, but the targets were confined to our table and the protestations of the victim were confined to silent but sinister looks. I suppose I could have told them: "You have to modulate your behavior in accord with the external discriminative stimulus. You have be be appropriate, nonsalient, and inconspicuous." I suppose I might have said that they had to stay on task, inhibit their squirming, make few sounds, display low energy, not stand out, not act suddenly, not disrupt, and . . . But then my 9-year-old friend probably would have said, "Modulate yourself! This is a kid's birthday party, not a *Festschrift* chapter!"

Let us depart for a moment from this discussion of the role of context in the definition of hyperactivity and turn to the importance of context in diagnosis and/or impression formation more generally. Data from studies of forced compliance and impression formation (Bem, 1965, 1972; Collins, 1973; Collins and Hoyt, 1972; Jones and Davis, 1965) clearly indicate that not all actor behaviors are used when an observer forms an impression of an actor. Some behaviors are discounted because those behaviors seem clearly caused by a pressure external to the actor—a pressure out there in the situation or context (see Kelley, 1967, 1971, for more on internal-external locations of causation). Socially desirable behaviors, for instance, are largely ignored when an observer is asked to make ratings which reflect a stable, dispositional characteristic of the actor (Jones and Harris, 1967). Presumably this is because socially desirable behaviors are explained by norms which provide a context sufficient to explain the behavior without recourse to causal factors within the actor. Similarly, behaviors about which the actor apparently had no choice, behaviors for which there is a clear external justification, and behaviors which induce few if any consequences are ignored both by an observer and the actor (Cooper and Worchel, 1970; Collins and Hoyt, 1972; Calder, Ross, and Insko, 1973). It is the context-behavior relationship which observers use when they form global personality impressions; they use only behaviors not explained by the context.

Returning to the case of hyperactivity, the behavior of hyperactive children does not follow from the various contexts; there are no apparent causes for the behavior external to the child in the context. Thus, teachers and other observers will make dispositional, internal attributions to explain the child's failure to behave unintrusively; observers will believe that the behavior of the hyperactive child is caused by an entity within the child (his personality, minimal brain damage, moral disorder, or failure to take his medicine). As we will soon see, the behaviors of hyperactive and nonhyperactive boys are sometimes *equally* responsive to external, environmental interventions. Nonetheless, observers will often mistakenly locate the cause of the noxious, intrusive behavior within the personality of the child, because possible circumstantial excuses are judged implausible. Whatever it is that adults or peers experience as a consequence of hyperactive behavior, the cause of those experiences will be located within the child rather than located outside the child in the environmental context.

The preceding paragraphs were intended to sketch the beginnings of a theory pattern which brings some coherence to (1) the seven statistically significant placebo-medication mean differences; (2) the seven correlations with teacher ratings on the Conners scale; (3) the words *inappropriate, salient, intrusive,* and *conspicuous*; (4) comments about the modulation of behavior in accord with external cues; (5) comments on the levels of concentration required in a grade school classroom; (6) speculation about the orientation to loud, abrupt, and moving objects; and (7) an anecdote about a birthday party. The theory pattern so imperfectly indicated by intersection among all these points is intended to suggest something about what it is that hyperactive boys do, what makes grade school teachers and peers so vulnerable to hyperactive boys, and why the school environment magnifies all these problems. And, at yet an even more abstract context, the theory pattern illustrates one major theme in Campbell's thinking: pattern matching and the contextual bases of knowledge.

Hypothetical, Critical Realism. The data reviewed in Whalen and Henker (1976) were enough for me; since that first contact, I have been personally convinced that hyperactivity is really out there. No "myth of the hyperactive child" (Schrag and Divoky, 1975) for me. Since that first contact, I have faith that it has a thingness or entitativity which exists independently of the teachers who label it in the classroom, of the pediatricians who prescribe Ritalin to cure it, and of those who theorize about and quantify it. If challenged, or if I think about it explicitly, I have no aspiration to prove conclusively or analytically that hyperactivity (or anything) is real. My realism is the Campbellian hypothetical and critical realism, which can be analyzed into three components: (1) the *realism*—the claim that some form of realism (a belief that an ontological world of objects and entities exists independently of the knower) is the best philosophy and faith for the working practitioner of science; (2) the *hypothetical* qualifier—an acknowledgment that the philosophy of realism, like any other conclusion from inductive argument, cannot be justified analytically or logically; and (3) the *critical* qualifier—the claim that all knowledge is a joint product of the posited reality and the method-process by which the knowledge is obtained.

Of these three components in a Campbellian philosophy of science, the *critical* qualifier is clearly the most pregnant with implications for the day-to-day practices of a working scientist. In fact, in my own intellectual history, I first embraced the value of multiple conceptions and multiple operational definitions of the social attitude concept (Campbell, 1963). Next I adopted the very concrete rules requiring multiple methods for the measurement of personality traits (Campbell and Fiske, 1959). Only later did I come to my Campbellian realism because it seemed to be the more general and encompassing conceptual foundation of these specific tactics for producing integrated research in psychology.

Because it is unqualified itself, the *critical* qualifier to the realism presents the strongest claim of the three components. The *critical* qualifier states that knowledge (reliable, replicable information) represents a composite integration of reliable information about entities out there in reality and reliable information about the method-process used to gather that knowledge. The implications of a critical realism for strategies of measurement and discovery (namely, the normative quest for multiple perspectives and triangulation) are discussed in the next section.

It remains, here, to discuss the realism itself. Other authors, in this volume and elsewhere, have discussed the philosophical and history-of-science contexts for a qualified realism. These next few paragraphs, however, explore *psychological*, functional implications of realism as a belief system.

The psychological implications following on the embracement of a realistic belief system are uncovered by such questions as "How do professional behaviors change as a consequence of endorsing the beliefs associated with hypothetical realism?" "How does the personality of a realist express itself in problem selection, hypothesis generation, research design, data analysis, interpretation, and other professional behaviors of the working scientist?" "What are the emotional consequences of different cognitive world views?" These questions imply an evaluation of the philosophy-of-science curriculum in the same way one would evaluate the forced feeding (readings assigned and comprehension tested) of any other element in a professional curriculum. An introduction to critical realism could be evaluated conjointly with an introduction to statistics or symbolic logic; one would analyze the professional activities of professionals who had (and had not) been randomly assigned to instruction in these tools with which to make sense of the universe.

What kinds of answers might one get to these questions? Surely the convictions that "There is something objective out there to discover" and "Science cumulates so that my small contribution will live after me" help motivate the individual scientific practitioner. Surely the thought "Today I will take another small step toward truth" helps an empirical scientist fight off our endemic professional depression and get out of bed in the morning. The belief that the phenomena one studies will live on once one's attention is turned elsewhere is more compatible with enthusiasm and perseverance in the empirical endeavor than with doubt and paralysis. It is easier to gather data if day-to-day activities are construed as part of an important larger enterprise, which marches, more often than not, toward an objective truth.

These implications of psychological support for the working scientist are not inadvertent. Campbell is self-conscious in his efforts to craft a philosophy of science which contributes to vigor, enthusiasm, and methodological sophistication among science's practitioners. It is probably not an accident that Campbell's critical realism was invented by a working scientist and that many of its more severe critics are also, and perhaps even primarily, critical of the entire empirical, scientific enterprise and its claim to a special epistemological status.

Multiple Perspectives and Triangulation. My Campbellian epistemology of hypothetical realism leads me to regard hyperactivity as an entity in an objective ontological world that exists independently of my thinking about it, my efforts to measure it, and the literature on the subject. The critical component of my qualified realism can be restated as a belief that any one measurement simultaneously contains information about the objective entity of hyperactivity and also contains information deposited by the method of measurement. In fact, I am inclined to think that method variance is often greater than trait (entity) variance; any one operational realization tells us more about the method of measurement and perspective of view than it tells us about the entity under study. One factor analysis of hyperactivity measures, for instance, produced method or source of information factors rather than entity factors (Langhorne, Loney, Paternite, and Bechtoldt, 1976).

This critical realism has direct and immediate methodological implications: The quest for increasingly accurate information about reality requires multiple sources of knowledge which share a focus on the entity of hyperactivity but which are heterogeneous with respect to method and perspective bias. Research designs should seek heterogeneity of irrelevancies and diversification of bias. When multiple operational realizations are combined to form a composite image of hyperactivity, the *heterogeneous* method biases will tend to cancel out, just as the random error variance associated with a single test item tends to cancel out when a total test score is computed. To use a metaphor of Campbell's, the various measures triangulate in on the entity.

Only the random, uncorrelated error variance cancels out when multiple items are combined to form a composite score. Anything in common to two measures, most particularly including method bias, increases the observed correlation between the two measures. This humility about the mixture of bias and truth contained in any one point of view is at the center of my hypothetical, critical realism—so much so that I think any view of Campbell's epistemology which does not include a heavy emphasis on multiple perspectives and triangulation misses the point, or at least *my* point, completely. From my perspective, the most important manifestation of Campbell's qualified realism is the claim that any measurement procedure is an imperfect pointer, constitutes a method-bound indicator, captures only a partial manifestation of the real-world entity, and contains much that is unique to the perspective used.

Our research program was quite explicitly and self-consciously designed to be convergent on the entity of hyperactivity and maximally divergent in methods of assessment; we explicitly strove for a maximum heterogeneity of irrelevancies (diversification of bias) in our assessment procedures. We observed the boys in a quasi-naturalistic classroom over a variety of academic tasks, and we also observed them during classroom juice time. We quantified the reactions of involved peers and teachers and less involved observers. We observed the boys individually as they worked on several computer tasks. We observed them working in pairs and in groups of four; we observed them while they were working on a discussion-to-consensus problem; and we observed them pulling ropes on one of Madsen's cooperation-competition tasks. Three separate behavior observation schemes were developed. Our Classroom Observation System was applied to classroom behaviors and, where appropriate, to the videotapes generated in our experimental tasks. We developed a global rating scheme, in which observers made a summary judgment for each sixty seconds of videotape; and there was a system to content-analyze the communication in some experimental modules. We had observers who formed global impressions while escorting the boys from one part of campus to another, and observers who observed the boys as hall monitors when the boys were being transported from one experiment to another.

It is hard for me to imagine a philosophy of science other than hypothetical, critical realism that could justify such a far-ranging fishing expedition. It is hard for me to understand what other faith would have set me free to place bets on measures other than the tried-and-true Conners scale—far and away the most successful single operational definition of hyperactivity. It never occurred to me, for instance, to create a multivariate design crafted to account for as much variance as possible in the Conners ratings. And that, it seems to me, is what one would do if one believed, with the logical positivists, that hyperactivity *is,* and only is, "whatever it is that the Conners scale measures." Rather, the design was tailored to uncover moderate correlations among diverse and heterogeneous independent and dependent variables. We did not begin our research program by fishing at random in hopes of hooking some manifestation of hyperactivity; nor did we cast all our fishing hooks immediately adjacent to the Conners scale.

This is not to say our study is free of method and perspective biases. We explicitly chose to focus on that facet of the hyperactivity entity which is manifested in behavior and to focus on the joint contribution of social context and medication on those behaviors; and this is because I like to study social context as an independent variable and human behavior as a dependent variable. In fact, a major reason for choosing psychostimulant medication as a research problem was because the obvious leads from an organic perspective had been explored without notable success (Dubey, 1976) and because I thought it time to explore hyperactivity with my own favorite (biased, method-bound, and provincial) tools of experimental social and personality psychology. (How nice it was to be able to randomly assign personalities to boys with the psychostimulant medication-placebo design!) But, timely or not, this social psychological perspective is, as are all perspectives, a biased and incomplete source of knowledge.

Entitativity. The previous section focused on measurement processes; the following paragraphs will focus on the flip side of measurement processes—the entities or objects of measurement. Any discussion of measurement processes and objects of measurement will, of course, overlap. Partitioning observations into components traced to the method of measurement and the objective entity requires a sophisticated theory of both entity and measurement process.

Campbell's first paper on entitativity (1958) deals explicitly with aggregates of persons as social entities, but his philosophy of science expands the concept to include all the objects that are hypothesized to exist in a real world. His analysis of entitativity begins, as is typical with Campbell, with the assertion that the data generated by various measurement procedures are to be explained, in part, by positing an external ontogeny composed of real entities; and these can be known only by biased methods of gathering information: "Science is the most distal form of knowing. . . . The processes and entities posited by science (for example, radio, stars, . . . atoms) are all very distal objects very mediately known via processes involving highly persumptive pattern matchings at many stages. . . . The formal theory becomes one 'pattern,' and against this pattern the various bodies of data are matched" (Campbell, 1966, pp. 96-97). "Distal knowledge means . . . knowledge of external objects [entities] and events, [it means] knowledge of predictable ... [entities] reidentifiable as the same, [and it means] hypothetical knowledge optimally invariant over points of observation and observers" (1966, p. 82). Entitativity is defined as "the degree of being entitative. The degree of having the nature of an entity, of having real existence" (1958, p. 17).

The data presented earlier depict a high degree of social science entitativity for hyperactivity. The correlations between (1) judgments made by several different observers on the basis of isolated ten-second observation intervals and (2) global ratings by the teachers typically range from the .30s to the .50s. These correlations are substantial when compared to other such heteromethod correlations found elsewhere in social science (see, for example, Campbell and Fiske, 1959). Furthermore, these correlations appear in spite of the fact that the words used to label and define the seven categories of behavioral observation are often very different from the words in the teacher rating scale.

Our images of social science entities undergo major transformations as the entity is viewed at different levels of analysis, at different points in time, and from different perspectives. The frog-tadpole-frog entity manifests itself in quite diverse forms depending on the time perspective from which it is viewed; the mountain entity manifests itself with different profiles when viewed from east and north perspectives. Hyperactivity manifests one facade to teachers as they reflect over the entire day (the Conners ratings) and a different facade to observers as they pluck a ten-second interval of behavior from its context (the behavior observations).

Consider, for example, the low correlations between self and peer responses to the same adjective. The present argument suggests that self-ratings will not necessarily generalize to peer ratings in a straightforward, face-valid way. Campbell (1967) cites an example from Bruner (1956) to make this point. Both Hidatsa Indians and local ranchers of European extraction reported accurately on group differences, but a transformation was necessary to adjust for a positive, ingroup, evaluative bias. The Indians described themselves as generous, unselfish, and sharing good fortune immediately with relatives and friends. In contrast, the ranchers described the same qualities by the adjectives *spendthrifts* and *improvident*.

The entity of hyperactivity is, at once, more than and less

than teacher responses to the ten-item Conners scale and observed judgments of classroom behavior. The entity of hyperactivity is, in part, obscured from sight to behavior observers in the classroom. Even if we maintain a behavioral perspective (and ignore, for instance, manifestations in blood chemistry and dream content), there must be facets of hyperactivity which become manifest only when the child is eating at home, delivering newspapers, traveling long distances in a car, having teeth filled at the dentist's. These comments indicate ways in which our measures capture far less than the full entity of hyperactivity. On the other side of the coin, these same measures reflect a great deal more than the hyperactivity entity; the measurement outcomes are full of information generated by method and perspective bias; this bias is unrelated to the entity of hyperactivity. The observers are, for instance, asked to observe a child for ten seconds and then make a wide variety of judgments simultaneously. These judgments—or meter readings, as Campbell analogizes—will fluctuate extensively as subtle and minor changes are made in the measurement process—and this is so even though the hyperactivity entity is held constant. Thus, for instance, meter readings would probably fluctuate if the various judgment tasks were assigned to separate observers, if the length of the observation interval were increased or decreased, if the training manual were rewritten with different examples, if the observations were made from videotapes, if the observations were made by members of a different cultural or linguistic community.

In summary, entities will vary in degree of entitativity. Entities can never be described totally at any one level of analysis or from any one perspective. They are manifested (observable) over many levels of analysis, from many points of view, and with widely divergent measurement procedures. A Campbellian epistemologist rejects the logical positivists' notion that an entity is defined exhaustively by any one measurement procedure. Finally, the image or conceptualization of the entity will often undergo a substantial transformation as the entity is viewed from different perspectives and at different levels of analysis. It is necessary that we know enough about the entity so that we can specify what will remain invariant and what will be transformed as the entity is viewed from different points of view, at varying levels of analysis, and with different measurement techniques.

The Natural Selection Model of Knowing, Knowledge Accumulation, and Moral Tradition. The fact that no comment is made here on the natural selection model—whereas entire chapters are devoted to the same topic elsewhere in this volume—illustrates the provincial bias of this chapter. Aspects of the natural selection conceptual entity, which loom large for philosophers and historians of science, fall out of sight when the model is viewed from the perspective of a working scientist trudging along with the day-to-day chores of empirical research on hyperactivity.

Composite Dispositions. As already indicated in the section on triangulation, Campbell has argued that good knowledge comes from a wide variety of channels of information and processes of knowing. He has argued that social attitudes (1963) and conformity (1961) represent knowledge acquired through several channels of information—some individual and some social or vicarious. Knowledge is often achieved when information obtained through individual modes is combined with information obtained through social-vicarious modes, so that a composite disposition is formed.

Analogously, the behaviors associated with hyperactivity can best be understood as composite behaviors reflecting (1) elements within the child (for example, personal characteristics such as age, sex, and hyperactivity diagnosis), (2) the current pharmacological status of the child (psychostimulant medication versus placebo), (3) contextual factors in the classroom ecology separate from the child, and (4) reactions of others in the situation to the child's responses.

In an effort to uncover some of the ecological determinants of hyperactive behaviors, my colleagues and I systematically varied environmental and curricular variables within each school day. Noise level was one of the variables manipulated. During the two periods when ambient noise levels were high (compared to the two periods when ambient noise levels were low), *all* boys (medicated, placebo, comparison) were more likely to be off task, more likely to make sudden movements, more likely to squirm, more likely to display intensity or high energy, and more likely to generate noise. In some cases the ecological independent variable was more powerful than the psychopharmacological independent variable. In another classroom experiment, task difficulty was manipulated. Task difficulty produced a significant main effect on many of the same behaviors affected by psychostimulant medication and ambient noise. These data require us to think of hyperactive behavior as a composite. Behaviors modified by medication can also be manipulated by environmental variables in the classroom; there are multiple modes through which the behaviors associated with hyperactivity can be elevated and/or depressed.

Conclusion

Our efforts to shuttle back and forth between basic conceptual issues and an applied problem have been successful. Hyperactive boys behave in ways that do not fit the contexts provided by task instructions, social norms of appropriate behavior, their own prior stream of behavior, and the effort-intensity appropriate to the task at hand. Because of these behavior-context discrepancies—and because unmedicated boys make more noise and provoke social consequences—the behavior of unmedicated hyperactive boys is inappropriate, salient, conspicuous, and intrusive. The widely held view that hyperactive children walk around or run about is based, in part, on an *illusion* of movement. Behaviors that are context discrepant or associated with loud noise produce the illusion that the actor has "moved around" (Imber and Collins, 1981).

Such behaviors create a particular problem in the grade school classroom because teachers and students must struggle constantly to maintain focused attention on the assignment at hand; pupils and teachers are, in other words, vulnerable to distraction. This fundamental importance of the behavior-context relationship illustrates one major theme in Campbell's thinking. The heteromethod procedures of the Whalen, Henker, and Collins Hyperactivity Summer School Project (Collins, Whalen, and Henker, 1980; Whalen and Henker, 1980a, 1980b) illustrate other elements of Campbell's philosophy of science.

References

Bem, D. "An Experimental Analysis of Self-Persuasion." *Journal of Experimental Social Psychology,* 1965, *1,* 199-218.

Bem, D. "Self-Perception Theory." In L. Berkowitz (Ed.), *Advances in Experimental Social Psychology.* Vol. 6. New York: Academic Press, 1972.

Bruner, E. M. "Primary Group Experience and the Process of Acculturation." *American Anthropologist,* 1956, *58,* 605-623.

Calder, B. J., Ross, M., and Insko, C. A. "Attitude Change and Attitude Attribution: Effects of Incentive, Choice, and Consequences." *Journal of Personality and Social Psychology,* 1973, *25,* 84-99.

Campbell, D. T. "Common Fate, Similarity, and Other Indices of the Status of Aggregates of Persons as Social Entities." *Behavioral Science,* 1958, *3,* 14-25.

Campbell, D. T. "Conformity in Psychology's Theories of Acquired Behavioral Dispositions." In I. A. Berg and B. M. Bass (Eds.), *Conformity and Deviation.* New York: Harper & Row, 1961.

Campbell, D. T. "Social Attitudes and Other Acquired Behavioral Dispositions." In S. Koch (Ed.), *Psychology: A Study of a Science.* Vol. 6. *Investigation of Man as Socius.* New York: McGraw-Hill, 1963.

Campbell, D. T. "Pattern Matching as an Essential in Distal Knowing." In K. R. Hammond (Ed.), *The Psychology of Egon Brunswik.* New York: Holt, Rinehart and Winston, 1966.

Campbell, D. T. "Stereotypes and the Perception of Group Differences." *American Psychologist,* 1967, *22,* 817-829.

Campbell, D. T. "On the Conflicts Between Biological and Social Evolution and Between Psychology and Moral Tradition." *American Psychologist,* 1975, *30,* 1103-1126.

Campbell, D. T., and Fiske, D. W. "Convergent and Discriminant Validation by the Multitrait-Multimethod Matrix." *Psychological Bulletin,* 1959, *56,* 81-105.

Collins, B. E. "Public and Private Conformity: Competing Explanations by Improvisation, Cognitive Dissonance, and Attribution Theories." In B. E. Collins (Ed.), *Public and Private Conformity: Competing Explanations by Improvisation, Cognitive Dissonance, and Attribution Theories.* Andover, Mass.: Warner Modular Publications, 1973.

Collins, B. E., and Hoyt, M. F. "Personal Responsibility-for-Consequences: An Integration and Extension of the 'Forced Compliance' Literature." *Journal of Experimental Social Psychology,* 1972, *8,* 558-593.

Collins, B. E., Whalen, C. K., and Henker, B. "Ecological and Pharmacological Influences of Behaviors in the Classroom: The Hyperkinetic Syndrome." In J. Antrobus, S. Salzinger, and J.

Glick (Eds.), *The Eco-System of the "Sick Kid."* Academic Press: New York, 1980.

Conners, C. K. "Rating Scales for Youth in Drug Studies with Children." *Psychopharmacology Bulletin,* 1973, (Special issue: *Pharmacotherapy of Children*), 24-84.

Conrad, P. "The Discovery of Hyperkinesis: Notes on the Medicalization of Deviant Behavior." *Social Problems,* 1975, *23,* 12-21.

Coombs, C. H. "Adaptation of the Galvanic Response to Auditory Stimuli." *Journal of Experimental Psychology,* 1938, *22,* 244-268.

Cooper, J., and Worchel, S. "Role of Undesired Consequences in Arousing Cognitive Dissonance." *Journal of Personality and Social Psychology,* 1970, *16,* 199-206.

Dubey, D. R. "Organic Factors in Hyperkinesis: A Critical Evaluation." *American Journal of Orthopsychiatry,* 1976, *46,* 353-366.

Edwards, D. C. "Within Mode Quality and Intensity Changes of Habituated Stimuli." *Biological Psychology,* 1975, *3,* 295-299.

Goyette, C. H., Conners, C. K., and Ulrich, R. F. "Normative Data on Revised Conners Parent and Teacher Rating Scales." *Journal of Abnormal Child Psychology,* 1978, *6,* 221-236.

Graham, F. K. "Habituation and Dishabituation of Responses Innervated by the Autonomic Nervous System." In H. V. S. Peeke and M. J. Herz (Eds.), *Habituation.* Vol. 1. New York: Academic Press, 1973.

Harding, G. B., and Rundle, G. R. "Long-Term Retention of Modality and Nonmodality-Specific Habituation of the GSR." *Journal of Experimental Psychology,* 1969, *82,* 390-392.

Henker, B., and Whalen, C. K. "The Changing Faces of Hyperactivity: Retrospect and Prospect." In C. K. Whalen and B. Henker (Eds.), *Hyperactive Children: The Social Ecology of Identification and Treatment.* New York: Academic Press, 1980.

Henker, B., Whalen, C. K., and Collins, B. E. "Double-Blind and Triple-Blind Assessments of Medication and Placebo Responses in Hyperactive Children." *Journal of Abnormal Child Psychology,* 1979, *7,* 1-13.

Imber, L., and Collins, B. E. "Effects of Inappropriate and Conspicuous Behaviors on Perceptions of Hyperactivity." Paper presented at meeting of Western Psychologists, Los Angeles, 1981.

Jones, E. E., and Davis, K. "From Acts to Dispositions: The Attribution Process in Person Perceptions." In L. Berkowitz (Ed.), *Advances in Experimental Social Psychology.* Vol. 2. New York: Academic Press, 1965.

Jones, E. E., and Harris, V. A. "The Attribution of Attitudes." *Journal of Experimental Psychology,* 1967, *3,* 1-24.

Kelley, H. H. "Attribution Theory in Social Psychology." In D. Levine (Ed.), *Nebraska Symposium on Motivation,* 1967, *15,* 192-238.

Kelley, H. H. *Attribution in Social Interaction.* Morristown, N.J.: General Learning Press, 1971.

Langhorne, J. E., Jr., Loney, J., Paternite, C. E., and Bechtoldt, H. P. "Childhood Hyperkinesis: A Return to the Source." *Journal of Abnormal Psychology,* 1976, *85,* 201-209.

Lovibond, S. H. "Habituation on the Orienting Response to Multiple Stimulus Sequences." *Psychophysiology,* 1969, *5,* 435-439.

McCubbin, R. J., and Katkin, E. S. "Magnitude of the Orienting Response as a Function of Extent and Quality of Stimulus Change." *Journal of Experimental Psychology,* 1971, *88,* 182-188.

Porter, J. M., Jr. "Adaptation of the Galvanic Skin Response." *Journal of Experimental Psychology,* 1938, *23,* 553-557.

Schrag, P., and Divoky, D. *The Myth of the Hyperactive Child and Other Means of Child Control.* New York: Pantheon Press, 1975.

Webb, E. J., Campbell, D. T., Schwartz, R. D., and Sechrest, L. B. *Unobtrusive Measures: Nonreactive Research in the Social Sciences.* Chicago: Rand McNally, 1966.

Whalen, C. K., Collins, B. E., Henker, B., Alkus, S. R., Adams, D., and Stapp, J. "Behavior Observations of Hyperactive Children and Methylphenidate (Ritalin) Effects in Systematically Structured Classroom Environments: Now You Can See Them, Now You Don't." *Journal of Pediatric Psychology,* 1978, *3,* 177-187.

Whalen, C. K., and Henker, B. "Psychostimulants and Children: A Review and Analysis." *Psychological Bulletin,* 1976, *83,* 1113-1130.

Whalen, C. K., and Henker, B. (Eds.). *Hyperactive Children: The Social Ecology of Identification and Treatment.* New York: Academic Press, 1980a.

Whalen, C. K., and Henker, B. "The Social Ecology of Psychostimulant Treatment: A Model of Conceptual and Empirical Analysis." In C. K. Whalen and B. Henker (Eds.), *Hyperactive Children: The Social Ecology of Identification and Treatment.* New York: Academic Press, 1980b.

Whalen, C. K., Henker, B., Collins, B. E., Finck, D., and Dotemoto, S. "A Social Ecology of Hyperactive Boys: Methylphenidate (Ritalin) Effects and Medication by Situation Interactions in Systematically Structured Classroom Environments." *Journal of Applied Behavior Analysis,* 1979, *12,* 65-81.

Whalen, C. K., Henker, B., Collins, B. E., McAuliffe, S., and Vaux, A. "Peer Interaction in a Structured Communication Task: Comparisons of Normal and Hyperactive Boys and of Methylphenidate (Ritalin) and Placebo Effects." *Child Development,* 1979, *50,* 388-401.

Whalen, C. K., Henker, B., and Dotemoto, S. "Methylphenidate and Hyperactivity: Effects on Teacher Behaviors." *Science,* 1980, *208,* 1280-1282.

16

Norman Miller

Changing Views About the Effects of School Desegregation: *Brown* Then and Now

Included in the documents submitted by the plaintiffs in the *Brown* case was an appendix (Allport and others, 1954) entitled "The Effects of Segregation and the Consequences of Segregation: A Social Science Statement," drafted, according to counsel, "by some of the foremost authorities in sociology, anthropology, psychology, and psychiatry who have worked in the area of American race relations." Although undoubtedly written to obtain broad endorsement among the social science community, although it may have represented to some degree the views of social scientists whose bias was "liberal" (Friedman, 1968; Kluger, 1976), although it failed to emphasize data that qualified and limited its conclusions to the degree warranted by existing evidence (Miller,

413

1979; Hennigan, Flay, and Cook, 1980), and although it sometimes extrapolated beyond existing data, it nevertheless approxi-
mates rather closely the truth as known at the time. In any case,
the counsel for the appellants certainly seemed to think it did.
"As a summary of the best available scientific evidence relative to
the effects of racial segregation on the individual we file it here-
with." So too did its endorsers: "We are nonetheless in agreement
that this statement is substantially correct and justified by the
evidence" (p. 18).

Expectations About School Desegregation in *Brown*

I will begin by briefly summarizing its major points and, to
some degree, providing hints about the logic presented in it. I will
then show that to a large extent its interpretation and conclusions
have subsequently been shown to be ill taken. In presenting the
accumulation of information that unfolded in the ensuing years, I
will also discuss the emergence of the lateral transmission of values
hypothesis, which in the mid 1960s was the major model used to
explain how desegregation might produce academic benefit to minority children. More specifically, I will discuss theorizing and
data about the effects of school desegregation on children's self-
esteem, achievement motivation, academic performance, social acceptance, and intergroup attitudes, and the causal relations among
these variables. This analysis raises fundamental questions about
causal inference, the relation between social climate and scientific
fact, and the relation between social science and social policy.
These issues are relevant in the context of this volume because
they have been focal points among Donald Campbell's diverse interests.

The social science statement primarily attended to the dele-
terious effects of a segregated and prejudiced society (and of segregated schooling) on the minority child's self-concept or feeling of
self-worth. It also expressed considerable concern about the role
of segregation in the development of hostile interracial attitudes
and negative stereotypes among both blacks and whites. The effects
of segregation on academic achievement and ambition are discussed much more sparingly, but the statement does make specific
reference to the idea that "segregated schools impair the ability of

the [minority] child to profit from the educational opportunities provided him" (p. 9); the "low level of aspiration and defeatism so commonly observed in segregated groups" and the debilitating effects of segregation on "one's expectations with regard to achievement" (p. 9); and the reduced learning opportunities created by homogeneous grouping (p. 13).

In addition, it recognized the gap between the academic achievement of black and white students (p. 12) and argued against the notions that desegregated education will have negative consequences for the academic mastery of white pupils (by lowering educational standards) or of black children (by increasing the level of academic competition). Instead, it took the position that desegregated schooling will reduce the academic achievement gap, basing its argument, by analogy, on the beneficial effect that a more pervasive, broadly defined desegregation (implicit in comparisons of northern versus southern states) had had on black IQ.

As Stephan (1978) points out, these ideas, taken in conjunction with other testimony given by social scientists in the several component legal cases of the *Brown* decision, can be cast into a single causal model, in which white prejudice toward blacks, low black self-esteem, impaired black motivation and achievement, and black prejudice toward whites are linked in a circular feedback system. Although Kenneth Clark had argued that official sanction of racial segregation damaged the personality of white as well as black children, the Supreme Court primarily responded to arguments emphasizing the effects of segregation on black children (Bower, 1980). Among the various harms of segregation discussed in the *Brown* decision and in the social science statement, impaired self-esteem of minority children receives the most prominent treatment and was viewed then as causally antecedent to its other harms.

Later Research Findings

Self-Esteem. The empirical consensus of contemporary research comparing black to white self-esteem stands in very sharp contradiction to the conclusions reached in the social science statement. Although some exceptions do appear, the majority of contemporary studies report either no differences, mixed results, or higher self-esteem among blacks. Recent reviewers concur in this assessment (Epps, 1975, 1979; Porter and Washington, 1979; St. John, 1975; Stephan, 1978; Wylie, 1979). Furthermore, they find that school desegregation, if it has any effect, lowers rather than raises the self-esteem of black children. St. John (1975, p. 54), who reviews some forty studies, concludes that "the effect of school desegregation on the general or academic self-concept of minority group members tends to be negative or mixed more often than positive." Summarizing twenty studies, Stephan (1978) states that they provide "strong evidence" against the notion that school desegregation increases self-esteem." Epps, in two reviews (1975, 1979), takes a more middling view: "There is little evidence that desegregation itself enhances self-esteem" (1979, p. 312). Wylie (1979), who has followed the research on self-concept for more than three decades and has amassed two major reviews of it in book form, believes that the burden of proof rests on those who assert that children in segregated schools have lower self-esteem. Porter and Washington (1979) conclude that "school integration correlates negatively with self-esteem." Finally, Gray-Little and Applebaum (1979) point out that most investigators, including themselves, who found higher self-esteem for white students or no racial differences conducted their research in an integrated setting, whereas those who found more positive self-concepts among black children typically conducted their research in segregated schools or in desegregated schools where the black students were a majority.

What can account for this disparity between the conclusions reached in the social science brief and subsequent research? Taylor and Walsh (1979) studied black and white adults to test four possible explanations. One, which assumes that the institutionalized racism found in American society *must* have harmful effects on blacks (see, for example, Adams, 1979; Cook, 1979; Miller, 1979), contends that white dominance and oppression manifest themselves in either of two extreme ways: It pushes some blacks toward excessive compliance and lowered self-esteem, and it instigates excessive militancy and exaggerated self-esteem in others. Consequently, the mean level of self-esteem among blacks and whites will not differ, because this comparison hides its opposing effects on blacks. This explanation suggests, instead, a comparison of variability of self-esteem scores; if the explanation is correct, greater variance should appear among black than among white comparison groups, reflecting the two opposing adaptations of blacks—compliance or militancy.

A second explanation points to cultural changes and the emergence of black power in the 1960s, thereby accounting for the difference in studies performed recently in comparison to those performed some decades ago. This explanation implies that the typical developmental trends of decreasing self-esteem would be stronger among blacks than among whites.

A third focuses on the social class level of blacks who are studied. As argued by McCarthy and Yancey (1971), lower-class and working-class blacks maintain their self-esteem by "blaming the system"; they point to white-dominated social institutions as the source of their own problems. Middle-class blacks, on the other hand, being less insulated from mainstream American culture, tend to internalize its values and, consequently, less readily engage in system blame. Thus, according to this explanation, the self-esteem of working-class blacks exceeds that of their white counterparts, whereas among middle-class people the racial effect will be reversed. A more elaborated view of this explanation (Heiss and Owens, 1972) argues for a more specialized buffering function of "system blame"; high self-esteem among lower- and working-class blacks appears only for the spheres of self-esteem concerned with family life or social activity, where white dominance is absent or where social comparison to whites is largely irrelevant. Other dimensions of self-esteem, such as those concerned with school or work, suffer because in these latter areas whites continue to exert important control over black outcomes. This variation of the McCarthy and Yancy explanation requires multidimensional measurement of self-esteem, thereby enabling comparison of blacks to whites on different aspects of their self concepts.

Taylor and Walsh test these views with data from black and white workers at three occupational status levels: park workers (low), mail carriers (medium), and teachers (high). Their self-esteem scale contained an occupational factor, a family factor, and a sociability factor, as well as a single global measure. Global self-esteem varied directly with occupational status. With status controlled, there was no racial difference; the nonsignificant direction of ef-

fect favored black respondents. This outcome confirms the consensus of other recent findings on children. More important, however, the tests of the various explanations of lack of racial difference in self-esteem show no support for any of them. Taylor and
Walsh conclude that "black individuals live through threatening
situations without experiencing radical damage to a more stable
self-picture" and that "the fact that the significant others of most
black Americans are fellow blacks rather than prejudiced whites is
certainly one critical resource serving to sustain black self-esteem."
Their conclusion fits nicely with other data cited above, showing
that black self-esteem is higher in segregated than in desegregated
schools. It is in the former setting that blacks are less likely to be
exposed to the conscious and/or unconscious prejudice, bias, and
rejection of white teachers and white students.

What else might explain the disparity between recent and
earlier conclusions? Much of the early research contributing to the
view that impaired self-esteem is the inevitable consequence of a
segregated society rested on measures which were projective and
indirect (see, for example, Clark and Clark, 1939; Kardiner and
Ovesey, 1951), whereas, virtually without exception, the research
cited above uses direct measures that allow respondents to readily
present whatever self-image they wish. Therefore, some scholars
(Cook, 1979; Miller, 1979) have suggested that, despite the spate
of recent studies showing black self-esteem to equal if not exceed
that found among whites, the debilitating effects of living in a
prejudiced society would manifest themselves if indirect or projective methods were used.

In much of the earlier literature relying on indirect or projective assessment of black self-esteem, researchers used some version of the Doll Preference Technique developed by Clark and
Clark (1939). Children are presented with black or white dolls, or
pictures, or animals, and asked questions such as "Which do you
prefer?" "Which one is nice?" "Which one is most like you?" Although their outcomes are sometimes inconsistent, these studies
most often show a preference on the part of both black and white
children for the white stimuli. Typically, though, this preference
for white stimuli was more true of white than black youngsters.
Stephan and Rosenfield (1979) point out several implicit assumptions in the interpretation of such studies: that the black and

white stimuli respectively represent black and white people; that choosing the white stimuli implies a rejection of the black stimuli and consequently of black people; and, finally, that rejection of black people implies a rejection of the self. Little evidence exists with regard to any of these assumptions. In addition to the empirical inconsistencies that it sometimes yields, the Doll Preference procedure has also been subject to numerous criticisms (see Brand, Ruiz, and Padilla, 1974); for instance, the rejection of black dolls may largely reflect a distaste for strange and unfamiliar toys. Finally, Banks (1976) argues that prior summaries of this literature reflect empirical misinterpretation. Choices by black children do not significantly depart from the chance expectation of 50 percent black and 50 percent white choices. If the Doll Preference literature does indeed reveal anything about racial attitudes, it shows that white children are ethnocentric, whereas black children are unprejudiced.

Whether other indirect or projective measures that are methodologically sound would reveal impaired or lowered self-concepts in the black children of today remains unknown. Clearly, however, some researchers doubt that they would: "The black child tends to be surrounded by other blacks. Thus, those persons who matter most to him—parents, teachers, and peers—tend to be black and to evaluate him as highly as white parents, teachers, and peers evaluate the white child. In addition, although his race, family structure, or socioeconomic status may be devalued in the larger society, in his immediate context most mothers share these characteristics. Comparing himself to other economically disprivileged blacks, the black child does not feel less worthy as a person on account of race or economic background. In fact, encapsulated in a segregated environment as are most urban black children, they may be less aware of societal prejudice than is assumed" (Simmons, 1978, p. 56).

Academic Motivation. Contemporary commentators on school desegregation frequently suggest that less attention be directed to its effects on academic achievement as measured by standardized tests. Instead, they urge researchers to study educational or occupational attainment (for example, how many years of schooling a student completes or where an individual's adult work position lies on the occupational prestige dimension). A more cynical observer may contend that such calls for a shift in emphasis merely reflect the fact that studies of the effects of desegregation on educational achievement yield such mixed results. Regardless, these issues merit attention in their own right. Underlying them are the motivational factors that impel attainment. The social science statement directly asserts that the aspirations of minority children are lower than those of whites and implies that desegregated schooling will remedy this disparity. To what extent, then, do black and white children differ in their aspirations or academic motivation, and what effect does desegregation have on them?

Reviews of research comparing the educational aspirations of minority and white students again seem to contradict the social science statement. Segregated black children typically report higher aspirations than do white students (Epps, 1975; Proshansky and Newton, 1968; Weinberg, 1975). If anything, blacks seem to value education even more strongly than do whites (Wilson, 1970). Among fifteen social goals, blacks rank increased educational opportunities second only to better occupational opportunities, whereas for whites the ranking of education was midway among the fifteen social goals. More recently, DeBord, Griffin, and Clark (1977), in a study of over 3,000 students in twenty-three racially mixed Mississippi schools, confirm the picture of black aspirations derived from studies conducted in segregated settings. The educational aspirations of black students in these southern desegregated schools exceeded those of their white peers.

Consideration of the effect of desegregated schooling, however, shows a less consistent overall picture. Curtis (1968), Fisher (1971), Knapp and Hammer (1971), and Reniston (1973), as well as DeBord, Griffin, and Clark (1977), found the aspirations of black students to be higher when they are enrolled in predominantly white schools. However, St. John (1966), White and Knight (1973), and Wilson (1959) found the opposite effect. Still other investigators found little difference between the aspirations of black students attending predominantly white schools in comparison to those in predominantly black schools (Curtis, 1968; Falk, 1978; Hall and Wiant, 1973). In sum, the aspirations of blacks equal or exceed those of whites, and the effect of school desegregation on them seems mixed.

Academic Achievement. Academic performance or achievement is the outcome measure most frequently examined in studies of school desegregation, undoubtedly because school districts routinely measure it. In assessing the effects of desegregation on it, one needs a proper definition of "educational improvement" or "benefit." In response to the fact that the courts view school desegregation as a vehicle for increasing the social mobility of minority members, I have argued that the dependent measure of proper interest is the extent to which desegregation reduces the academic achievement gap (Miller, 1975a, 1980). Other researchers, more vocally identified with a strong prodesegregation stand (Crain and Mahard, 1978; Pettigrew, Smith, Useem, and Norman, 1973; Weinberg, 1975), interpret any minority improvement as demonstrating benefit, even if such improvement is accompanied by more substantial gains among white children than among minority children. In somewhat more isolated instances, it is possible that a decrease in the achievement gap is taken as a positive effect even though it might have reflected declines for both black and white children. Similarly, in some instances no change has been counted as a positive outcome. Reviewers clearly differ in their definition of benefit. Furthermore, individual reviews often fail to apply a single definition consistently (see, for example, Weinberg, 1975). With these caveats in mind, I will quickly summarize the reviews in the mid 1970s.

St. John (1975) and Bradley and Bradley (1977), in relatively comprehensive reviews, find overall no achievement gains as a consequence of desegregation. In a review prepared for the National Institute of Education, Clement, Eisenhart, and Wood (1976) conclude that "race mixing alone has little consistent effect on black-white outcomes" (p. 47). Stephan (1978) examines fifteen published studies and nineteen unpublished studies and concludes that desegregation was rarely associated with decrease in achievement and in better than 25 percent of the cases was associated with increased achievement among blacks. Weinberg (1975) and Crain and Mahard (1978), who review larger arrays of studies than do either Stephan, St. John, or Bradley and Bradley, both interpret more than half as showing improved scholastic achievement for black students.

To assess this literature more formally, Krol (1978) applied quantitative aggregating procedures to it. In essence, these procedures provide a means of pooling the statistical effects of an array of independent studies. The "effect size" for Krol's total population of studies (129) reveals that the mean academic achievement of minority children was .16 standard deviations higher as a consequence of desegregation. How significant or important is a positive average effect size of this magnitude? Procedures have not been developed for applying a test of significance to the average effect size. In its policy implications, however, it suggests that desegregation, as typically implemented, at best has a very modest impact. Although Krol's comparison of the effects of desegregation on academic performance in the elementary grades (.22) and the effects found at the junior high and high school grades (.01) does not reach statistical significance, it is noteworthy that for the latter grades the mean effect size is close to zero. One wishes that he had also made a comparison of the early to later elementary grades. As expected, less benefit tends to occur in short (.11) versus longer periods of desegregation (.21), and in methodologically strong (.10) versus weaker studies (.21); but neither of these differences attains significance. Finally, it is important to note that Krol uses a lenient definition of benefit. His procedures do not exclusively reflect minority student gains relative to those of white students; instead, they reflect in part the absolute gains of minority students, independent of what has happened to the white students in those same school districts or classrooms. Lastly, most of the studies comprising this literature remain unpublished and are probably best viewed in the context of Wilson's laws, developed in specific response to them: (1) "All policy interventions in social problems produce the intended effect if the research is carried out by those implementing the policy or their friends." (2) "No policy intervention in social problems produces the intended effect if the research is carried out by independent third parties, especially those skeptical of the policy" (Wilson, 1973, p. 133).

In sum, the effects of school desegregation on minority academic performance suggest that, as a social reform, its benefits have been, at best, meager.

Interracial Attitudes. Another issue as important to social scientists and the courts as the previous three, and prominent in the social science statement, is the effect of desegregation on interracial attitudes. Many viewed the increase in interracial contact that school desegregation presumably creates as a major ameliorative step.

The major theoretical perspective relating racial attitudes to school desegregation has been called contact theory (see Allport, 1954; Cook, 1969). It argues that increased contact with others provides accurate information about them, which in turn decreases prejudice and tension and improves intergroup relations. It also specifies, however, that benefits from the mixing of white and minority children occur *only under certain conditions*: (1) equal status contact, (2) mutual interdependence (cooperation), (3) a social climate with norms favoring intergroup contact, (4) attributes among group members which contradict prevailing stereotypes, and (5) contact which promotes personal or intimate association (Cook, 1969). Another possible stipulation is the presence of goals that all share and that require joint effort; namely, superordinate goals (Sherif and Sherif, 1953).

Amir (1969, 1976), Carithers (1970), Cohen (1975), Cook (1969), McConahay (1978), St. John (1975), Schofield (1978), and Stephan (1978) provide recent reviews of relevant literature. Those by Amir (1969, 1976) are the most extensive; they broadly examine intergroup contact, focusing on its effects in the school environment as a subset of the literature. He suggests two time periods. The first, which centers around the early 1950s, emphasized change in the attitudes of the majority group, expected and sought a reduction in prejudice, studied contacts that occur infrequently in our society and tend to produce only casual interactions, and generally reported positive effects. The second, encompassing the late 1960s and the 1970s, focused more on changes in minority persons, studied situations more typical of everyday life, and found much less favorable results.

Although not all reviewers agree on any one point, a number of major conclusions can be culled from their syntheses. People prefer to interact with others like themselves. In order to change attitudes, meaningful contact, joint participation, interdependence, and superordinate goals are important; cooperation facilitates, and competition hinders. Contact increases the intensity of prevailing attitudes more often than it changes their valence; those with "intense" attitudes typically show no change. When contact does produce a change in attitudes, it rarely generalizes to other areas and situations. Contact is more likely to reduce prejudice if it also contradicts prevailing stereotypes, shows that members of "other" groups possess valued traits, leads to positive perceptions of the other group, is pleasant, and/or has institutional support. Contact that is unpleasant, lowers the prestige of one group, and/or involves unequal status tends to strengthen prejudice. Even under optimal conditions, changes may be minor. Work settings (and, seemingly, school settings) often produce only superficial contact. Even when social class is matched, whites are viewed more favorably than blacks. Since contact has not typically taken place in favorable conditions, it has not typically reduced prejudice. The "best" results seem to occur when children are desegregated early. Junior high especially seems to be a difficult time. Mandatory desegregation most frequently produces "unfavorable" attitude change.

As Campbell (1967) observed in his discussion of stereotypes, although a particular behavior is interpreted positively when exhibited by an ingroup member, an outgroup member observed to perform the same act receives derogation. Whites see an ambiguous shove as more violent when performed by a black as opposed to another white, regardless of the victim's race (Duncan, 1976). And Lewin (1948) pointed out long ago that minorities often internalize negative views held by the majority. Blacks also view ambiguously aggressive acts as more threatening, even if performed by blacks (Patchen, Hoffman, and Davidson, 1976; Schofield and Sagar, 1977). Black children were more frequently chosen by both blacks and whites as exemplifying negative attributes (Bartel, Bartel, and Grill, 1973). Black students also perceived other blacks as physically tougher than whites, and whites as less helpful and pleasant to them than to other whites, as insincere, superior, and expecting special privileges. White students viewed black students as more likely to possess negative traits, as physically tougher, as unfriendly toward whites, as less academically oriented, and as less socially desirable (Patchen, Hoffman, and Davidson, 1976). Such stereotypes obviously preclude any possibility of equal status contact (Cohen,1980; Cohen, Lockheed, and Lohman, 1976).

Theoretical controversy focuses on the importance of the average social class difference that exists between minority and white students, on the correlated disparities in their academic performance, and on whether or not these differences inevitably prevent "equal status contact" in the desegregated classroom. Whereas Pettigrew (1967), McConahay (1978), and others take the view that these racial-ethnic differences in social class, ability, and achievement do not play a preemptory role, others (Armor, 1972, Cohen, 1975, 1980; Miller, 1979) to varying degrees take an opposing view. In his research on contrast effects and the exaggeration of differences between categories, Campbell (Campbell, 1956; Clarke and Campbell, 1955) seems to fall comfortably in this latter camp.

Given that it is proper for schools and teachers to focus on academic performance and to coordinate their reward system with it, I believe that the differences in academic performance that exist on the average between minority and white students do constitute a very serious obstacle to equal status contact. So do the social class differences. Although these same differences also existed in other desegregated settings—such as the armed services, the merchant marines, housing projects, or other examples cited in the social science statement as instances in which desegregation improved interracial attitudes—they were virtually irrelevant to the common goals that blacks and whites faced in each of these situations; it is manifestly obvious that black and white soldiers facing a common enemy can contribute equally to their own survival and their defeat of opposing armies. In the classroom, however, the existing average scholastic achievement differences are very often critical to the task at hand and typically do constitute a potent discrepancy in the resources that students bring to it. Rist (1978) vividly describes their consequences for a group of lower-class black children embedded in an achievement-oriented white school as a result of a desegregation plan. Their feelings of bitter rejection, isolation, and intellectual incompetence can hardly make us proud of or content with the social reform that produced such a result. Cohen (1980) has studied and documented these social disasters more thoroughly than any other researcher and pleads with force and sensitivity for new procedures to deal with status and power differences and the human relations climate in the desegregated school. It takes much skill and ingenuity for the ordinary teacher to restructure classroom learning activities in such a way that these differences do not form the basis for *unequal* rather than equal status contact. The standard approach to the academic curriculum exacerbates negative stereotypes and unequal status.

Behavior. Although the social science statement seemed to

imply a close relation between racial attitudes and behavior, and perhaps a causal one as well, some recent research questions this assumption. Measures of interracial attitudes at times show minimal relation to each other or to behavior (Katz and Zalk, 1978; Patchen, Davidson, Hoffman, and Brown, 1977; Stephan and Rosenfield, 1978; Clore, Bray, Itkin, and Murphy, 1978).

Post-*Brown* studies of behavioral effects, although sometimes positive, as in camp settings (Campbell and Yarrow, 1958; Yarrow, Campbell, and Yarrow, 1958), nevertheless suggest skepticism about assumptions of ubiquitous benefit. "The simple and attractive view that face-to-face contact per se leads to a favorable change in intergroup relations has been found inadequate" (Campbell and Yarrow, 1958, p. 37). Virtually all show a strong preference for others from one's own group (see Silverman and Shaw, 1973; Shaw, 1973; Stephan and Rosenfield, 1978). Beyond this one finding, the evidence is mixed. For example, amount of contact seems to increase (Justman, 1968) or to decrease (Shaw, 1973) as numbers of minority students increase; small-town blacks seem not to share the intergroup contact difficulties of their urban peers (Schmuck and Luszki, 1969); contact seems more likely in heterogeneous classes where racial balance matches school racial balance (Koslin, Koslin, Pargament, and Waxman, 1972; Schofield and Sagar, 1977); white peers more readily accept minority children who are higher achieving and from a higher social class (Lewis and St. John, 1975). Further, school desegregation has mildly increased cross-ethnic choices (Silverman and Shaw, 1973) and, in other instances, ethnic encapsulation (Gerard, Jackson and Conolley, 1975; Shaw, 1973).

In summary, if recent research adds any further clarity, it is that school desegregation will not automatically produce the desired attitudinal or behavioral changes, even in schools that seem to come very close to meeting the conditions stipulated by Allport (1954) as necessary for positive and beneficial contact (Schofield and Sagar, 1977).

Causal Relations Among the Variables. As indicated, the social science statement viewed racial differences in self-esteem as pivotal and as causally antecedent to differences in academic performance. The static cross-sectional data of the Coleman report, showing correlated self-esteem and academic performance differences, were often interpreted as supporting this view (see Pettigrew, 1967). Purkey (1970, p. 27) concluded that the literature indicates "a strong reciprocal relationship and gives us reason to assume that enhancing the self-concept is a vital influence in approving academic performance." The problem with most of the earlier studies that Purkey cites, however, is that, at the time they were conducted, no acceptable methodology existed for assessing the causal relation between the two variables.

The cross-lag panel correlation technique, developed by Campbell and his associates (Campbell and Stanley, 1963; Cook and Campbell, 1976; Crano, Kenny, and Campbell, 1972; Kenny, 1973, 1975; Rozelle and Campbell, 1969), relies on the temporal sequencing of the two respective measures to assess the "causal preponderance" of effect. Several studies that use this technique (Calsyn, 1973; Calsyn and Kenny, 1977; McGarvey and Miller, 1979) provide little evidence that self-esteem will affect subsequent academic achievement. Instead, they indicate that, if there is any direction of causal effect, it flows from academic achievement to subsequent adjustment as indexed by personality measures.

Using path analysis, Bachman and O'Malley (1977) confirm this general outcome. Level of self-esteem in high school exerted no causal influence on either educational or occupational attainment. Instead, both self-esteem and attainment were consequences of academic ability and academic achievement. In both cross-sectional (Maruyama, 1977) and longitudinal (McGarvey, 1977) analyses using Joreskog's (1973) structural equation procedures for causal modeling, which are superior to ordinary path analysis and supersede the cross-lag panel technique (Cook and Campbell, 1976), little or no evidence was found that self-esteem is causally antecedent to educational achievement or ability measures. This was true among black and Mexican-American as well as white children. Using these same procedures, though not examining racial differences, Maruyama, Rubin, and Kingsbury (in press) assess the relation between self-esteem and achievement among 715 children born between 1961 and 1965. Social class measures were obtained at the time of birth; intellectual ability measures at age 7; achievement measures at ages 9, 12, and 15; and self-esteem measures at age 12. Self-esteem exerted little causal influence on subsequent achievement or ability; nor did achievement affect later measures of self-esteem. Instead, it seemed to be primarily a consequence of prior ability.

In summary, whereas researchers initially saw self-esteem as an antecedent of academic achievement, many now accept the opposite causal conclusion, concluding instead that it reflects past achievement (Scheirer and Kraut, 1979) or ability (Maruyama, Rubin, and Kingsbury, in press).

One theoretical view of the relation between interracial attitudes and academic performance is the lateral transmission of values hypothesis (Coleman, Campbell, Hobson, McPartland, Mood, Weinfeld, and York, 1966). As does contact theory, this hypothesis suggests that in the desegregated school setting improved intergroup relations will subsequently improve the academic achievement of minority children as a consequence of their internalization of the values of the majority. The logic behind this process is deeply rooted in experimental social psychology. The notions of (1) specific values facilitating achievement (McClelland, 1961), (2) social influence processes resulting in the norms of the majority being passed on to the minority (Asch, 1952), and, within limits, (3) performance levels being responsive to standards (Atkinson, 1964) have all been well documented, at least within the laboratory. Evidence also suggests that achievement-related values are in fact more prevalent in white than in minority children (Mussen, 1953). Given the laboratory evidence supporting the notions comprising the lateral transmission of values hypothesis, social scientists readily interpreted their findings as supporting it (Coleman and others, 1966; Crain and Weisman, 1972; Katz, 1964; Pettigrew, 1967).

The "value" dimensions in the lateral transmission of values hypothesis presumably bring about achievement change in two different ways. First, they can provide *informational social influence* (Deutsch and Gerard, 1955; Kelley, 1952) that will affect achievement; the presence of high-achieving students who possess adaptive values enables other children to learn to behave in ways that are consistent with these values, since the high-achieving children will model specific behaviors that are directly instrumental to better academic performance or achievement. This process might occur without measurable changes in basic value orientations or personality characteristics. An alternative version of this hypothesis sees changes in basic values and/or personality as critical for subsequent improvement in achievement. This latter sequence, which can be drawn more clearly from the social psychology literature cited earlier, can be called *normative social influence* (Deutsch and Gerard, 1955; Kelley, 1952). It appears to be the one implied by Coleman and his associates (1966); Crain and Weisman (1972); Katz (1964); and Pettigrew (1967).

Two additional variables might enhance the value transmission stage that is critical for the second version of the model. First, unless minority children are accepted by their white peers, value transmission seems unlikely to occur. Even informational social influence, however, would be minimized if children have no friendly contact with others; without friendships children would have less opportunity to interact with, observe, and exchange information with others. Consequently, friendships should facilitate both informational and normative social influence. And, as minority students adopt white children's values and attitudes, white children should like them better.

Second, teachers' values may produce social influence effects that parallel those produced by peers. However, insofar as there is little reason to believe that teachers in white schools hold basic educational values that differ from those held by teachers in minority schools, this latter source of influence seems less important than peer acceptance.

The U.S. Commission on Civil Rights (1967) provides indirect support for the importance of peer acceptance. It found the achievement of black children to be related to (1) lack of racial tension, (2) having a close friend who is white, and (3) the percentage of white students in the classroom. Coleman and his associates (1966) present similar findings. Therefore, peer acceptance seems to occupy a pivotal position in the transmission of values hypothesis.

Lewis and St. John (1975) provided the first direct empirical test of this causal model, using longitudinal data to focus on the relation between popularity and acceptance of black students in the desegregated classroom. They proposed a path model, with "socioeconomic status" (SES) and "percentage of white students" as background variables external to the model; either "past grade point average" (GPA) or "IQ" (Otis group-administered test) and "popularity with whites" as intervening variables; and "present GPA" or "reading achievement" as the criteria. The results of their analyses did seem to support the model; popularity was related to GPA (though not to reading achievement). They suggest that "the social process that best explains the beneficial effect of acceptance by white peers on black achievement is probably the lateral transmission of achievement-oriented norms and/or skills" (Lewis and St. John, 1975, p. 89).

Two major problems arise from their findings. First, reanalyses of their data using a superior methodology (Maruyama and Miller, 1979) reversed their conclusions, showing the "causal" effect of popularity on GPA as well as on reading achievement to be nonsignificant. Instead, popularity was "caused by" achievement. Second, Lewis and St. John (1975) did not test the model implied by Coleman and associates (1966), Katz (1964), and Pettigrew (1967), namely, one that views personality measures as important intervening variables.

A more extensive longitudinal study, however, designed to examine carefully the relation between cross-racial peer acceptance, personality, and achievement, measured black, white, and Mexican-American children in Riverside, California, as their schools underwent desegregation (Gerard and Miller, 1975). With its numerous measures, including those tapping peer acceptance, a wide array of personality dimensions, and achievement, the results must seriously disappoint anyone who advocates the lateral transmission of values model. Riverside voluntarily implemented a desegregation program backed by community support; further, the percentage of minority children in Riverside was 20 percent, a proportion presumably beneficial for school desegregation (see U.S. Commission on Civil Rights, 1967). Given these and other factors (see Hendrick, 1975), Riverside provided a relatively good environment both for producing successful desegregation and for examining the model.

Analyses revealed no overall positive effect of desegregation on the achievement test scores for minority children; their achievement deficit continued to increase as they progressed through school. Desegregation lowered their grades, whereas those of white children improved. This stemmed from two facts: (1) Since white children on the average perform at a "higher" scholastic level than do minority children, the desegregated classrooms contained a wider range in childrens' performance levels. (2) Teachers typically tend to employ a single set of norms when they grade their students, assigning approximately the same numbers of A's, B's, and so forth, from one year to the next; this grade normalization consequently resulted in poorer grades for the minority students and better grades for the white students in desegregated classes.

The achievement results, however, do not by themselves disconfirm the model; it was still possible that the few minority students who did in fact improve academically were those who had been accepted and did adopt the values of the white children. Consistent with the model, those minority children accepted by whites did in fact perform better than their peers who were not accepted (Gerard, Jackson, and Conolley, 1975). Nevertheless, the implications of this finding are ambiguous because it was paralleled by superior achievement among minority children who had previously been most popular in the segregated classrooms. This latter finding calls into question the causal sequence implied by the model; it suggests that *prior* ability and academic performance affect subsequent popularity in the desegregated classroom.

The teacher data also seem to provide some support for the lateral transmission of values model (Johnson, Gerard, and Miller, 1975). Classrooms in which teachers were less discriminatory toward minority pupils were more likely to foster cross-racial acceptance. Insofar as teachers' acceptance was correlated with achievement, the teacher can be viewed as playing an important role in lateral transmission of values.

In contrast to the peer acceptance and teacher measures, the personality measures provide absolutely no support for the model. The few effects consistent with it were so minimal in size that they preclude viewing personality measures as important predictors of achievement and as mediators of background effects (Miller, 1975a). Reanalyses of the Riverside data using "causal model" approaches (Maruyama, 1977; McGarvey, 1977) have reaffirmed my own earlier (Miller, 1975a, 1975b; Miller and Gerard, 1976) conclusion about personality variables. They have, however, produced results concerning the relation between peer popularity and achievement that on first glance seem inconsistent. Examining only cross-sectional data, Maruyama's (1977) results suggested that acceptance by whites did cause achievement for both black and Mexican-American as well as white children. In retrospect, this finding merely reflects the model's a priori assumption that acceptance causes achievement. Had the opposite causal sequence between acceptance and achievement been posited, it too would have been confirmed in an analysis of cross-sectional data. McGarvey (1977), on the other hand, testing a two-wave longitudinal model, found that for white children achievement was "caused by" popularity but that for minority children popularity and achievement were not "causally" related.

In a more recent study that applies structural equation modeling to three waves of data collected across a five-year span (Maruyama and Miller, 1980), we provide further evidence against the lateral transmission of values model. Confirming our (Maruyama and Miller, 1979) reanalysis of Lewis and St. John (1975), we find no evidence for the existence of peer influence processes that could support lateral transmission of values in a desegregated classroom. (For whites, acceptance by peers exerted a mildly *negative* influence on achievement.) For the minority sample, it is predesegregation achievement that affects postdesegregation peer acceptance; achievement in the desegregated classroom did *not* affect later peer acceptance. This suggests that a minority child's reputation among his own peers in a segregated school as one who is a good student may be critical when that child is desegregated. Perhaps this reputation later provides to white children a basis for exempting that child from their prevailing academic stereotypes of the minority child, and thereby allows friendship to develop.

The preceding discussion should not be construed as suggesting that peer acceptance cannot strongly influence school achievement. Rather, it suggests only that desegregation per se will not commonly produce cross-racial peer acceptance and thereby cause value changes and enhanced performance. It puts the final nail into the coffin of a model of benignly beneficial desegregation, which argues that merely putting together children whose racial-ethnic backgrounds differ will result in positive changes; it provides additional empirical validation for the shift of research emphasis away from defining the consequences of desegregation to examining instead how different procedures and techniques for implementing it can affect its outcomes. As Bronfenbrenner noted over a decade ago, many community leaders tacitly assume that "once the Negro child finds himself in an integrated classroom with a qualified teacher and adequate materials, learning will take place, and with it the deficiencies of the American Negro, and the judgments of inferiority which they in part encourage, will be erased" (1967, p. 910). He goes on to conclude: "Regrettably, this is not the case. Neither the scars of slavery which the child still bears nor the skills and self-confidence of his white companion rub off merely through contact in the same classroom" (p. 910).

Research relating achievement motivation and aspirations to antecedent background factors, to contemporary social psychological influences in the school setting, to subsequent academic performance, and to both educational and occupational attainment also tends to confirm the picture presented above. Most of it takes as its starting point the Wisconsin model developed by Sewell and his associates (Sewell, Haller, and Portes, 1969; Hauser, 1973; Sewell and Hauser, 1975). Developed from data on whites, this model has been shown to have considerable validity (Alexander, Eckland, and Griffin, 1975; Picou and Carter, 1976; Sewell and Hauser, 1975; Sewell, Haller, and Ohlendorf, 1970). Basically, the model shows that among white students encouragement of parents and the plans of peers shape ambitions more directly and with greater impact than do other sources of influence, such as one's scholastic aptitude or previous academic achievement. They exert considerable stronger impact than does the direct influence of social origin. The effect of teachers' influence on aspirations is considerably weaker than that of either parents or peers (Sewell and Hauser, 1975; Alexander, Eckland, and Griffin, 1975). Indeed, some have argued (Williams, 1976) that student's ambitions exert a greater influence on teacher's expectations or evaluations than vice versa. Finally, the causal impact of ambition on scholastic performance is relatively trivial (Williams, 1972, 1975; Alexander and Eckland, 1974; Kerckhoff and Campbell, 1977). In studies which apply the Wisconsin model to blacks, however, important disparities seem to arise repeatedly. Among black students the role of family background factors in predicting educational or occupational attainment is much less important than it is for whites. Moreover, differences in academic ability and academic performance also play a much more subdued role than they do for whites. Instead, motivational factors or aspirations exert greater impact among blacks (DeBord, Griffin, and Clark, 1977; Hout and Morgan, 1975; Kerckhoff and Campbell, 1977; Porter, 1974; Portes and Wilson, 1976).

One of the problems in attempting to assess the reasons why the Wisconsin model fits whites whereas it "simply doesn't work well for blacks" (Kerckhoff, 1976, p. 372) rests in the well-known fact that race and social class origins are strongly related. When procedures are used to "reduce" or "correct" the socioeconomic differences between blacks and whites, the model does fit both groups equally well (Howell and Frese, 1979). Other relatively recent data also support this view by showing that family background has a smaller effect on the educational decisions of whites from low socioeconomic origins than is the case for their peers from higher-status origins (Kerckhoff and Campbell, 1977).

Wilson (1979) compares the suitability of the Wisconsin model in predicting the educational attainment of blacks attending segregated as opposed to integrated schools. In general, the model is a better predictor for blacks attending integrated, as opposed to segregated, schools. To the extent that the model is generally more appropriate for those with middle- as opposed to lower-class origins (Howell and Frese, 1979; Kerckhoff and Campbell, 1977), these differences may reflect the higher social class standing among those blacks who typically attend such schools.

Spenner and Featherman (1978) comprehensively summarize much of this recent work. Many of their conclusions parallel those in the preceding paragraphs. In regard to my earlier discussion of the lateral transmission of values, however, one point needs to be emphasized. These studies consistently show that social psychological influences, such as the plans or values of peers, are much weaker among black than among white adolescents. If so, the promise of the social psychological influence processes implicit in the lateral transmission of values model—if not a causal consequence of good academic performance, as I argued in the preceding section—is at best seriously constrained for black children.

Do We Know More Today Than Yesterday?

Counter to the arguments of the 1954 social science statement, the data summarized above suggest that blacks do not have poorer self-concepts and lower aspirations than whites. Nor does school desegregation generally produce the direction of effect on these personality dimensions that those arguments led us to expect. The same conclusion applies to the effect of desegregation on interracial attitudes and behavior. Its effects on academic performance, though generally positive if any effect is found, certainly lack the robustness expected. Finally, evidence on the causal relations among these variables fails to confirm earlier theorizing and, more often than not, directly contradicts it.

Campbell has devoted considerable attention to the process by which science "progresses" in its knowledge, whereas other modes of study merely engage in substituting one "knowledge" for another (Campbell, 1974, 1979). Does the summary above, with its rather striking discrepancy from the conclusions of the social science statement, reflect this cumulative process? Is it the result of a triangulation toward an objective reality (Campbell and Fiske, 1959; Webb, Campbell, Schwartz, and Sechrest, 1966), as gleaned through different operationalizations of key concepts, tests of scientific propositions in different experimental arenas, and a distillation or winnowing of truth from research performed by those with alternate or conflicting theoretical views? Or, instead, as victims of change in the dominant political and value orientations, the social climate, or the popular theoretical perspective, do we simply continue to shift from one erroneous position to another as equally wrong as the preceding one?

Intrusion of these biasing factors into the school desegregation literature can readily be documented. In 1965, when we first designed our own research (Gerard and Miller, 1975), the prevailing theoretical view regarding the dynamics of beneficial school desegregation—the personality mediational version of the lateral transmission of values model—dominated our thinking. It dictated the design of our research by specifying the selection of dependent measures to be studied. Quite independently from Coleman and his associates, we too measured an array of personality dimensions directly related to achievement motivation (including, for instance, direct and projective measures of achievement motivation, internal control, tolerance for deferred gratification, school anxiety, self-esteem, level of aspiration, and goal-setting orientation). Similarly, we too judged it important to assess peer influence and therefore included measures of peer relations, susceptibility to peer pressure,

responsiveness to social reinforcement, and the like. Studying desegregation in a single moderate-sized school district, we could
assess these aspects of personality and social behavior much more
intensively than Coleman and his associates could in their nation-
wide survey design. So fixed were we in our adherence to some
form of the lateral transmission of values hypothesis that even
after our data forced us to abandon the personality mediator version of it (Miller, 1975a; Miller, 1975b; Miller and Gerard, 1976),
we persisted in our commitment to its normative influence components. Thus, in Maruyama's (1977) dissertation, which applied
structural equation causal modeling procedures to a cross-sectional
component of our data, we found "evidence" that "confirmed"
the model by specifying that the causal paths flow from peer acceptance to academic achievement. Although our later work was
eventually corrective and rather definitively rejected this model, it
is noteworthy that this dissertation subsequently received a dissertation award from Division 9 of the American Psychological Association, the Society for the Psychological Study of Social Issues.
Did this judgment of merit reflect in part confirmation of a theoretical model that was manifestly coordinate with the dominant
social political value orientation of this organization?

   Earlier, Crain and Weisman (1972) exhibit a similar unwarranted adherence to the personality mediator model in their cross-
sectional comparison of the effects of segregated and desegregated
schooling on black educational achievement. Using retrospective
data from black adults in twelve northern cities, they found relevant personality differences only among those who attended either
segregated or desegregated *elementary* schools, whereas academic
performance differences emerged only among those who differed
in type of *high school* attended. No personality differences were
found as a consequence of type of high school, and type of elementary school did not affect academic performance. Yet, despite
these discontinuities between the effects of desegregation at the
lower and upper grade levels, they found "support" for the role of
personality in *mediating* the effects of desegregated schooling
upon academic achievement! How could the personality traits that
presumably caused superior academic performance have disappeared after causing it?

   That one's political orientation and other related individual biases similarly intrude into interpretations hardly causes surprise. Did the disappointing underutilization of information about the complexities of engineering positive interracial interaction (Hennigan, Flay, and Cook, 1980) reflect subtle political concerns of those who authored the social science statement? Does the fact that Stuart Cook was one of the three authors of the social science statement (Kluger, 1976, p. 555) underlie his rather favorable assessment of its accuracy today (Cook, 1979)? Is my own current rejection of its conclusions merely the self-serving consequence of my courtroom testimony for school districts? Robert Crain, who, as do many contemporary social scientists, holds a liberal, pro-desegregation orientation and has testified in legal proceedings for the American Civil Liberties Union and the National Association for the Advancement of Colored People, finds the studies of the effects of desegregation on academic achievement to be much more positive than do most other reviewers (Crain and Mahard, 1978). That review argues in addition for the superiority of early versus late desegregation. In a painfully detailed, picky, and often repetitive analysis, Kurtz (1979), who holds an opposing political orientation, forcefully challenges all the conclusions of Crain and Mahard. He points to some cases of probable misclassification as to whether a particular instance of desegregation was voluntary rather than mandatory and positive rather than negative in its outcome. In other instances, he shows that "positive" outcomes were insupportable on the basis of the types of data provided in the study; the apparent gain or improvement was probably an artifact of the type and timing of testing employed by the school district; and/or the study contains problems of sufficient magnitude to preclude any conclusion.

Other "findings" seem to parallel, in a somewhat suspicious manner, aspects of a more general change in social climate. In their review of the literature on black identity and self-esteem, Porter and Washington (1979) recognize this covariation and state: "The stress on low black self-esteem in the early 1960s was used as a lever in the struggle for civil rights. In the 1970s, black pride having increased, emphasis on low self-esteem among blacks has been less popular" (p. 70). Kirk and Goon (1975), in their review of the desegregation literature, reflect the growing tendency in the early 1970s to reject a cultural deficit model for explaining characteristics of minority group behavior. Although our own analyses contesting the validity of the lateral transmission of values model were unavailable to them, as were any other empirical disconfirmations, and although they cite no opposing data, they nevertheless reject the model as a viable explanation. Though they are partially correct in their assertation that it rests on findings from studies that suffered from methodological problems, their rejection of a cultural deficit model underlies their conclusion that the lateral transmission of values model is wrong and that racial differences in academic performance are due instead to the failure of teachers to expend the needed effort to teach minority children. The point here is not to argue the validity of this or that view but, instead, to emphasize the impact of prevailing social climate, values, and culture (Fleck, 1935; Kuhn, 1962).

Although numerous other instances and hints of bias from one source or another can easily be added to this list, they would only belabor the point. Despite such evidence of human fallibility, I am inclined to think that we do make progress toward understanding some objective and ascertainable external reality. Our view today certainly acknowledges a greater complexity than heretofore. (1) Recent studies raise renewed interest in methodological issues about the relation among attitudinal measures (Katz and Zalk, 1978), their meaning (Patchen, Davidson, Hoffman, and Brown, 1977; Stephan and Rosenfield, 1978; Stephan and Rosenfield, 1979; Amir, Sharan, Bizman, Rivner, and Ben-Ari, 1978), and their power (Slavin and Madden, 1979). (2) The need for a distinction between racial and personal self-esteem now more clearly warrants attention (Porter, 1974; Porter and Washington, 1979). (3) Researchers now recognize the possibility that the meaning of high scores on a particular variable may differ among races. In the past, for instance, psychologists often viewed feelings of internal control as indicating a healthy or normal response. Respondents in research settings apparently do so too. When attempting to create favorable impressions, whites respond in a more internal direction on Rotter's Internal-External Scale (Jellison and Green, in press). Recently the Gurins (Gurin, Gurin, Lao, and Beattie, 1969; Gurin, 1970; Gurin and Epps, 1975) suggest an opposite view, arguing that for blacks a well-developed belief that life is externally controlled is normative and healthy. (4) Although the social science statement does acknowledge some of the complexities involved in creating positive change in racial attitudes, the difficulties in achieving equal status contact in desegregated schools is now manifestly recognized by most reviewers. (5) Emphasis has now shifted from considering the quality of tangible resources as an input variable that affects pupil achievement to the study of the process of education. (6) Whereas our own work seriously challenges the validity of the lateral transmission of values model for desegregated schools, the model may be considerably more adequate for explaining the superior educational attainment of minority students who attend desegregated high schools. Nor have we explored the importance of distinguishing between normative and informational influence in desegregated settings. The question of whether sociometric and/or other measures of friendship constitute the best approach to assessing normative influence also remains unexplored. It is conceivable that the social climate of desegregated schools, in which minority students do happen to show academic gains, may normatively emphasize academic values and exert its impact directly, rather than transmitting it via cross-race friendships. (7) Finally, the recent recognition that school desegregation is a complex process rather than a unidimensional variable (Cook, 1979; Miller, 1980) has shifted the focus from observing intergroup relations to attempting to create and improve them, for it is clear that such relations do not naturally occur when schools are desegregated. Following the prescriptions of Cook (1962), recent studies attempt to be productive in several ways: by defining and comparing the conditions under which desegregation occurs; by isolating and focusing on the effects of specific variables; by testing specific aspects of contact theory, social influence theory, and expectation theory; by developing specific "treatments" for improving intergroup relations; and by examining methodological issues related to the measurement of attitudes.

Thus, despite my generally cynical outlook, I clearly "see" progress in our scientific understanding of issues related to school desegregation. I recognize the fact that I can never evaluate the question of progress in a direct, unbiased manner. Yet, even while intellectually aware of the pitfalls in accepting it, I find it difficult to reject my phenomenological experience as sheer mental trickery and delusion. I offer these thoughts not as final proof of anything;

perhaps they are best viewed as merely self-indulgent public expression of my commitment to scientific values.

Evaluation of School Desegregation Programs

The body of research discussed above raises another issue that has been strongly affected by Donald Campbell's thinking, namely, the relation between social science and social policy. Campbell (1971), advocating a role of "benign servant" for social scientists, suggests that they provide retrospective description of the consequences of social policy and its programs. Social science research methods, applied after policymakers have had their day, can enrich the information available about the effects of their decisions. Others (for example, Binstock, 1980) urge social scientists to take a much more proactive role, that of the "meddlesome protagonist," in which they attempt actively to determine the direction of policy by providing direct input to the policymaker. Whereas those who take Campbell's position spend their day collecting data about a specific program from the people it affects, those in the latter camp cull generalizations, consensus, or even offbeat ideas from the work of others and attempt to influence the policymaker's behavior. This oversimplified dichotomy obviously ignores the fact that in the real world people often have several different pairs of shoes. To the extent that a researcher actively serves as a "meddlesome protagonist," however, I increasingly question his or her ability to avoid the types of bias mentioned in the previous section.

The preceding dichotomy oversimplifies in other ways as well, in that there is more than one empirical approach to program evaluation. Ward Edwards is prominent for his advocacy of empirical procedures that in several respects stand in opposition to Campbell's. Edwards' approach (Edwards, 1980; Edwards, Guttentag, and Snapper, 1975), derived from applied decision theory, uses *Multi Attribute Utility Measurement* (MAUM). In it, evaluation means how well the program measures up to a set of value attributes. The value attributes are the things that people expect or want the program to achieve. These value attributes must be elicited from the stakeholders, those who have a personal interest in the potential effects of the program. Typically, such people have thought enough about the relevant issues to have meaningful values with respect to it. Stakeholders then assign importance weights to the value attributes. As a form of evaluation, MAUM consists of quantitatively describing experts' perceptions of the likelihood that a specific program or a specific implementation of a policy will achieve what it should, with "should" defined in terms of each experts' individual value structure. Edwards' procedures can be used to evaluate a program (or several alternative programs) prior to their implementation. Thus, it can be more proactive than Campbell's approach. It is most appropriate when there is not a concrete criterion for evaluating the effectiveness of a particular program, when outcome measures on ideal criteria are very difficult or expensive to obtain, or when their meaning may be unclear.

Despite its amenability to proactive advice, I see Edwards' form of evaluation as acting to maintain the status quo, whereas Campbell's prescriptions, despite their retrospective emphasis, are more likely to promote change. What can create this rather ironical outcome? Edwards' evaluation procedures rest on experts' application of what they know. To the degree that what they know consists of scientific information, that information will contribute to their evaluation. Yet their evaluation per se can have little impact on what is scientifically known tomorrow. Instead, in the absence of new scientific information, temporal changes in the political values of the experts and changes in cultural climate would provide major impetus to drift in evaluations performed at different points in time. Further, where the evaluations of the experts using MAUM evaluation strongly disagree, the policymaker must decide whose to attend to, in that there are no totally acceptable procedures for aggregating diverse evaluations. More likely than not, this choice will be oriented toward ingroup members, toward older or higher-status experts, or toward those among the stakeholders who represent "more respectable" domains. Consequently, it will augment the likelihood of conservative decision making. In contrast, Campbell's position that we should use experimental social science methodology (or quasi-experimental when necessary) to evaluate programs, and at the same time should keep ourselves removed from the policy decision phase, maximizes the likelihood of bias-free additions to and changes in that which is scientifically known. Had Edwards' procedures been substituted for Campbell's form of program evaluation, the inputs provided by many of the studies cited in the preceding sections would never have been available to alter our approach in designing effective school desegregation programs.

References

Adams, B. D. "Inferiorization and Self Esteem." *Social Psychology,* 1979, *41,* 47-53.

Alexander, K. L., and Eckland, B. K. "Sex Differences in the Educational Attainment Process." *American Sociological Review,* 1974, *39,* 668-682.

Alexander, K. L., Eckland, B. K., and Griffin, L. J. "The Wisconsin Model of Socio-economic Achievement: A Replication." *American Journal of Sociology,* 1975, *81,* 324-342.

Allport, G. W. *The Nature of Prejudice.* Reading, Mass.: Addison-Wesley, 1954.

Allport, G. W., and others. "The Effects of Segregation and the Consequences of Segregation: A Social Science Statement." Appendix to *Brown* v. *Board of Education of Topeka,* 347 U.S. 483 (1954).

Amir, Y. "Contact Hypothesis in Ethnic Relations." *Psychological Bulletin,* 1969, *71,* 319-342.

Amir, Y. "The Role of Intergroup Contact in Change of Prejudice and of Ethnic Relations." In P. Katz (Ed.), *Towards the Elimination of Racism.* Elmsford, N.Y.: Pergammon Press, 1976.

Amir, Y., Sharan, S., Bizman, A. M., Rivner, M., and Ben-Ari, R. "Attitude Change in Desegregated Israeli High Schools." *Journal of Educational Psychology,* 1978, *10,* 129-136.

Armor, D. J. "The Evidence on Busing." *Public Interest,* 1972, *28,* 90-126.

Asch, S. E. *Social Psychology.* Englewood Cliffs, N.J.: Prentice-Hall, 1952.

Atkinson, J. W. *An Introduction to Motivation.* New York: Van Nostrand, 1964.

Bachman, J. G., and O'Malley, P. M. "Self-Esteem in Young Men: A Longitudinal Analysis of Educational and Occupational Attainment." *Journal of Personality and Social Psychology,* 1977, *35,* 365-380.

Banks, W. C. "White Preference in Blacks: A Paradigm in Search of a Phenomenon." *Psychological Bulletin,* 1976, *83,* 1179-1186.

Bartel, H. W., Bartel, N. R., and Grill, J. J. "A Sociometric View of Some Integrated Open Classrooms." *Journal of Social Issues,* 1973, *29* (4), 159-176.

Binstock, R. H. "Public Policy and Aging: A Need to Shift Focus." Paper presented at the Symposium on Public Policy and the Social Sciences, Memphis State University, April 10-11, 1980.

Bower, S. "The Supreme Court and Racial Integration: An Historical Review 1849-1979." In F. Aquria (Ed.), *Race Equity in Education: The History of School Desegregation 1849-1979.* Bloomington: School of Education, Indiana University, 1980.

Bradley, L. A., and Bradley, G. W. "The Academic Achievement of Black Students in Desegregated Schools: A Critical Review." *Review of Educational Research,* 1977, pp. 399-449.

Brand, E. S., Ruiz, R. A., and Padilla, A. M. "Ethnic Identification and Preference." *Psychological Bulletin,* 1974, *81,* 860-890.

Bronfenbrenner, U. "The Psychological Costs of Quality and Equality in Education." *Child Development,* 1967, *38,* 909-925.

Calsyn, R. J. "The Causal Relationship Between Self-Esteem, Locus of Control, and Achievement: A Cross-Lagged Panel Analysis." Unpublished doctoral dissertation, Northwestern University, 1973.

Calsyn, R. J., and Kenny, D. A. "Self-Concept of Ability and Perceived Evaluation of Others: Cause or Effect of Academic Achievement?" *Journal of Educational Psychology,* 1977, *69,* 136-145.

Campbell, D. T. "Enhancement of Contrast as Composite Habit." *Journal of Abnormal and Social Psychology,* 1956, *53,* 350-355.

Campbell, D. T. "Stereotypes and the Perception of Group Differences." *American Psychologist,* 1967, *22,* 817-829.

Campbell, D. T. "Methods for the Experimenting Society." Paper presented at 79th annual meeting of the American Psychological Association, Washington, D.C., Sept. 5, 1971.

Campbell, D. T. "Evolutionary Epistomology." In P. A. Schilpp (Ed.), *The Philosophy of Karl Popper.* La Salle, Ill: Open Court, 1974.

Campbell, D. T. "A Tribal Model of the Social System Vehicle Carrying Scientific Knowledge." *Knowledge*, 1979, *2*, 181-201.

Campbell, D. T., and Fiske, D. W. "Convergent and Discriminant Validation by the Multitrait-Multimethod Matrix." *Psychological Bulletin*, 1959, *56*, 81-105.

Campbell, D. T., and Stanley, J. C. "Experimental and Quasi-Experimental Designs for Research on Teaching." In N. L. Gage (Ed.), *Handbook of Research on Teaching.* Chicago: Rand McNally, 1963.

Campbell, J. D., and Yarrow, M. R. "Personal and Situational Variables in Adaptation to Change." *Journal of Social Issues,* 1958, *14,* 29-46.

Carithers, M. "School Desegregation and Racial Cleavage, 1954-1970: A Review of the Literature." *Journal of Social Issues,* 1970, *26,* 24-37.

Clark, K., and Clark, N. "Development of Consciousness of Self and the Emergence of Racial Identification in Negro Pre-school Children." *Journal of Social Psychology*, 1939, *10,* 591-599.

Clarke, R. B., and Campbell, D. T. "A Demonstration of Bias in Estimates of Negro Ability." *Journal of Abnormal and Social Psychology*, 1955, *51,* 585-588.

Clement, D. C., Eisenhart, N., and Wood, J. W. "School Desegregation and Educational Inequality: Trends in the Literature, 1960-1975." In *The Desegregation Literature—A Critical Appraisal.* Washington, D.C.: U.S. Department of Health, Education, and Welfare, 1976.

Clore, G. L., Bray, R. M., Itkin, S. M., and Murphy, P. "Interracial Attitudes and Behavior at a Summer Camp." *Journal of Personality and Social Psychology,* 1978, *36,* 107-116.

Cohen, E. G. "The Effects of Desegregation on Race Relations." *Law and Contemporary Problems,* 1975, *39,* 271-299.

Cohen, E. G. "Design and Redesign of the Desegregated School: Problems of Status, Power, and Conflict." In W. G. Stephan and J. R. Feagin (Eds.), *Desegregation: Past, Present, and Future.* New York: Plenum, 1980.

Cohen, E. G., Lockheed, M. E., and Lohman, M. R. "The Center for Interracial Cooperation: A Field Experiment." *Sociology of Education,* 1976, *49,* 47-58.

Coleman, J. S., Campbell, E. Q., Hobson, C. J., McPartland, J., Mood, A. M., Weinfeld, F. D., and York, R. L. *Equality of Educational Opportunity.* Washington, D.C.: U.S. Government Printing Office, 1966.

Cook, S. W. "The Systematic Analysis of Socially Significant Events: A Strategy for Social Research." *Journal of Social Issues,* 1962, *18* (2), 66-84.

Cook, S. W. "Motives in a Conceptual Analysis of Attitude-Relevant Behavior." In W. J. Arnold and D. Levine (Eds.), *Nebraska Symposium on Motivation.* Lincoln: University of Nebraska Press, 1969.

Cook, S. W. "Social Science and School Desegregation: Did We Mislead the Supreme Court?" *Personality and Psychology Bulletin,* 1979, *5* (4), 420-437.

Cook, T. D., and Campbell, D. T. "The Design and Conduct of Quasi-Experiments and True Experiments in Field Settings." In M. D. Dunnette (Ed.), *Handbook of Industrial and Organizational Psychology.* Chicago: Rand McNally, 1976.

Crain, R. L., and Mahard, R. L. "Desegregation and Black Achievement: A Case-Survey of the Literature." *Law and Contemporary Problems,* 1978, *42* (3), 17-56.

Crain, R. L., and Weisman, C. S. *Discrimination, Personality, and Achievement: A Survey of Northern Blacks.* New York: Seminar Press, 1972.

Crano, W. D., Kenny, D. A., and Campbell, D. T. "Does Intelligence Cause Achievement? A Cross-Lagged Panel Analysis." *Journal of Educational Psychology,* 1972, *63,* 258-275.

Curtis, B. "The Effect of Segregation on the Vocational Aspirations of Negro Students." *Dissertation Abstracts,* 1968, *29,* 772.

DeBord, L. W., Griffin, L. J., and Clark, M. "Race and Sex Influences in the Schooling Processes of Rural and Small Town Youth." *Sociology of Education,* 1977, *50,* 85-102.

Deutsch, M., and Gerard, H. G. "A Study of Normative and Informational Social Influence upon Individual Judgment." *Journal of Abnormal and Social Psychology,* 1955, *51,* 629-636.

Duncan, B. L. "Differential Social Perceptions and Attribution of Intergroup Violence: Testing the Lower Limits of Stereotyping of Blacks." *Journal of Personality and Social Psychology,* 1976, *34,* 590-598.

Edwards, W. "Multiattribute Utility for Evaluation: Structures, Uses, and Problems." In M. W. Klein and K. S. Teilmann (Eds.), *Handbook of Criminal Justice Evaluation.* Beverly Hills, Calif.: Sage, 1980.

Edwards W., Guttentag, M., and Snapper, K. "A Decision-Theoretic Approach to Evaluation Research." In M. Guttentag and E. Struening (Eds.), *Handbook of Evaluation Research.* Beverly Hills, Calif.: Sage, 1975.

Epps, E. G. "Impact of School Desegregation on Aspirations, Self-Concepts and Other Aspects of Personality." *Law and Contemporary Problems,* 1975, *39,* 300-313.

Epps, E. G. "The Impact of School Desegregation on the Self-Evaluation and Achievement Orientation of Minority Children." *Law and Contemporary Problems,* 1979, *43,* 57-76.

Falk, W. W. "School Desegregation and the Educational Attainment Process: Some Results from Rural Texas Schools." *Sociology of Education,* 1978, *51,* 282-288.

Fisher, J. E. "An Exploration of the Effects of Desegregation on the Educational Plans of Negro and White Boys." *Dissertation Abstracts,* 1971, *31,* 5548.

Fleck, L. *Genesis and Development of a Scientific Fact.* (F. Bradley and T. J. Trenn, Trans.) Chicago: University of Chicago Press, 1979. (Originally published in German 1935.)

Friedman, L. (Ed.). *Argument.* New York: Chelsea House, 1968.

Gerard, H. B., Jackson, T. D., and Conolley, E. S. "Social Contact in the Desegregated Classroom." In H. B. Gerard and N. Miller, *School Desegregation.* New York: Plenum, 1975.

Gerard, H. B., and Miller, N. *School Desegregation.* New York: Plenum, 1975.

Gray-Little, B., and Applebaum, M. I. "Instrumentality Effects in the Assessment of Racial Differences and Self-Esteem." *Journal of Personality and Social Psychology,* 1979, *37,* 1221-1229.

Gurin, G. "An Expectancy Approach to Job-Training Programs." In V. Allen (Ed.), *Psychological Factors in Poverty.* Chicago: Markham, 1970.

Gurin, P., and Epps, E. *Black Consciousness, Identity, and Achievement.* New York: Wiley, 1975.

Gurin, P., Gurin, G., Lao, R., and Beattie, M. "Internal/External Control and the Motivational Dynamics of Negro Youth." *Journal of Social Issues,* 1969, *25,* 29-53.

Hall, J. A., and Wiant, H. V. "Does School Desegregation Change Occupational Goals of Negro Males?" *Journal of Vocational Behavior,* 1973, *3,* 175-179.

Hauser, R. M. "Socioeconomic Background and Differential Returns to Education." In L. C. Solomon and P. J. Taubam (Eds.), *Does College Matter: Some Evidence on the Impacts of Higher Education.* New York: Academic Press, 1973.

Heiss, J., and Owens, S. "Self-Evaluation of Blacks and Whites." *American Journal of Sociology,* 1972, *78,* 360-370.

Hendrick, I. G. "The Historical Setting." In H. B. Gerard and N. Miller (Eds.), *School Desegregation.* New York: Plenum, 1975.

Hennigan, K., Flay, B., and Cook, T. " 'Give Me the Facts': Some Suggestions for Using Social Science Knowledge in National Policy Making." In R. Kidd and M. Saks (Eds.), *Advances in Applied Social Psychology.* Hillsdale, N.J.: Erlbaum, 1980.

Hout, N., and Morgan, W. R. "Race and Sex Variation in the Causes of Expected Attainments of High School Seniors." *American Journal of Sociology,* 1975, *81,* 364-394.

Howell, F. N., and Frese, W. "Race, Sex, and Aspirations: Evidence for the 'Race Convergence' Hypothesis." *Sociology of Education,* 1979, *52,* 34-46.

Jellison, J. M., and Green, J. "A Self-Presentation Approach to the Fundamental Attribution Error." *Journal of Personality and Social Psychology,* in press.

Johnson, E., Gerard, H. B., and Miller, N. "Teacher Influences in the Desegregated Classroom." In H. B. Gerard and N. Miller (Eds.), *School Desegregation.* New York: Plenum, 1975.

Joreskog, K. G. "A General Method for Estimating a Linear Structural Equation System." In A. S. Goldberger and O. D. Duncan (Eds.), *Structural Equation Models in the Social Sciences.* New York: Seminar Press, 1973.

Justman, J. "Children's Reactions to Open Enrollment." *Urban Review,* 1968, *3,* 32-34.

Kardiner, A., and Ovesey, L. "The Mark of Oppression: Explorations in the Personality of the American Negro." New York: Meridan, 1951.

Katz, I. "Review of the Evidence Relating to Effects of Desegregation on the Intellectual Performance of Negroes." *American Psychologist,* 1964, *19,* 381-399.

Katz, P. A., and Zalk, S. R. "Modification of Children's Racial Attitudes." *Developmental Psychology,* 1978, *14,* 447-461.

Kelley, H. H. "The Two Functions of Reference Groups." In G. E. Swanson, T. M. Newcomb, and E. L. Hartley (Eds.), *Readings in Social Psychology.* New York: Holt, Rinehart and Winston, 1952.

Kenny, D. A. "Cross Lagged and Synchronous Common Factors in Panel Data." In A. S. Goldberger and O. D. Duncan (Eds.), *Structural Equation Models in the Social Sciences.* New York: Seminar Press, 1973.

Kenny, D. A. "Cross Lagged Panel Correlatons: A Test for Spuriousness." *Psychological Bulletin,* 1975, *82,* 887-903.

Kerchoff, A. C. "The Status Attainment Process: Socialization or Allocation?" *Social Forces,* 1976, *55,* 368-381.

Kerchoff, A. C., and Campbell, R. T. "Black-White Differences in the Educational Attainment Process." *Sociology of Education,* 1977, *50,* 15-27.

Kirk, D., and Goon, S. "Desegregation and the Cultural Deficit Model: An Examination of the Literature." *Review of Educational Research,* 1975, *45,* 599-611.

Kluger, R. *Simple Justice.* New York: Knopf, 1976.

Knapp, N., and Hammer, E. "Racial Composition of Southern Schools and Adolescent Educational and Occupational Aspirations and Expectations." Paper presented at annual meeting of the Association of Southern Agricultural Workers, Memphis, January 1971.

Koslin, S., Koslin, B., Pargament, R., and Waxman, H. "Classroom Racial Balance and Students' Interracial Attitudes." *Sociology of Education,* 1972, *45,* 386-407.

Krol, R. A. "A Meta-Analysis of Comparative Research on the Effect of Desegregation on Academic Achievement." Unpublished doctoral dissertation, Western Michigan University, 1978.

Kuhn, T. S. *The Structure of Scientific Revolutions.* Chicago: University of Chicago Press, 1962.

Kurtz, H. "Review of Crain and Mahard—1978." Unpublished paper, 1979.

Lewin, K. *Resolving Social Conflicts.* New York: Harper & Row, 1948.

Lewis, R. G., and St. John, N. H. "Race and the Social Structure of the Elementary Classroom." *Sociology of Education,* 1975, *48,* 346-368.

McCarthy, J. D., and Yancey, W. L. "Uncle Tom and Mister Charlie: Metaphysical Pathos in the Study of Racism and Personal Disorganization." *American Journal of Sociology,* 1971, *76,* 648-672.

McClelland, D. C. *The Achieving Society.* New York: Van Nostrand, 1961.

McConahay, J. "The Impact of School Desegregation upon Student Racial Attitudes and Behavior: A Critical Review of the Literature and a Prolegomenon to Future Research." *Law and Contemporary Problems,* 1978, *42* (3), 77-107.

McGarvey, W. E. "Longitudinal Factors in School Desegregation." Unpublished doctoral dissertation, University of Southern California, 1977.

McGarvey, W. E. and Miller, N. "Causal Relations Among Personality Variables and Academic Achievement: A Cross-Lagged Panel Analysis." Unpublished manuscript, University of Southern California, 1979.

Maruyama, G. "A Causal-Model Analysis of Variables Related to Primary School Achievement." Unpublished doctoral dissertation, University of Southern California, 1977.

Maruyama, G., and Miller, N. "Reexamination of Normative Influence Processes in Desegregated Classrooms." *American Educational Research Journal,* 1979, *16,* 273-284.

Maruyama, G., and Miller, N. "Does Popularity Cause Achievement? A Longitudinal Test of the Lateral Transmission of Values Hypothesis." Unpublished manuscript, University of Minnesota, 1980.

Maruyama, G., Rubin, R. A., and Kingsbury, G. G. "Self-Esteem and Educational Achievement: Independent Constructs with a Common Cause?" *Journal of Personality and Social Psychology,* in press.

Miller, N. "Summary and Conclusions." In H. B. Gerard and N. Miller, *School Desegregation.* New York: Plenum, 1975a.

Miller, N. "Trouble for the 'Personality Mediator Model' of Beneficial Desegregation Effects." Lecture delivered at the Maxwell School of Citizenship and Public Affairs, Syracuse University, 1975b.

Miller, N. "The 'Social Scientist's Brief' from the Perspective of 1979." Paper presented at 87th annual meeting of the American Psychological Association, New York City, September 1, 1979.

Miller, N. "Making School Desegregation Work." In W. Stephan and J. Feagin (Eds.), *School Desegregation: Past, Present, and Future.* New York: Plenum, 1980.

Miller, N., and Gerard, H. B. "A Longitudinal Study of School Desegregation." *Psychology Today,* 1976, *10* (1), 66-67, 69-70, 100.

Mussen, P. H. "Differences Between the TAT Responses of Negro and White Boys." *Journal of Consulting Psychology,* 1953, *17,* 373-376.

Patchen, M., Davidson, J. D., Hoffman, G., and Brown, W. R. "Determinants of Students' Interracial Behavior and Opinion Change." *Sociology of Education,* 1977, *50,* 55-75.

Patchen, M., Hoffman, G., and Davidson, J. D. "Interracial Perceptions Among High School Students." *Sociometry,* 1976, *39,* 341-354.

Pettigrew, T. F. "Social Evaluation Theory Convergence and Applications." In D. Levine (Ed.), *Nebraska Symposium on Motivation.* Lincoln: University of Nebraska Press, 1967.

Pettigrew, T. F., Smith, N., Useem, E. L., and Norman, C. "Busing: A Review of the Evidence." *Public Interest,* 1973, *30,* 88-118.

Picou, J. S., and Carter, T. N. "Significant Other Influence and Aspirations." *Sociology of Education,* 1976, *49,* 12-22.

Porter, J. D. R. "Race, Socialization, and Mobility in Educational and Early Occupational Attainment." *American Sociological Review,* 1974, *39,* 303-316.

Porter, J. D. R., and Washington, R. E. "Black Identity and Self-Esteem: A Review of Studies of Black Self-Concept, 1968-1978." *Annual Review of Sociology,* 1979, *5,* 53-74.

Portes, A., and Wilson, K. L. "Black-White Differences in Educational Attainment." *American Sociological Review,* 1976, *41,* 414-501.

Proshansky, H., and Newton, P. "The Nature and Meaning of Negro Self-Identity." In M. Deutsch, I. Katz, and A. R. Jenson

(Eds.), *Social Class, Race, Psychological Development.* New York: Holt, Rinehart and Winston, 1968.

Purkey, W. W. *Self-Concept and School Achievement.* Englewood Cliffs, N.J.: Prentice-Hall, 1970.

Reniston, E. G. "Levels of Aspiration of Black Students as a Function of Significant Others in Integrated and Segregated Schools." *Dissertation Abstracts,* 1973, *33,* 7020-7021.

Rist, R. *The Invisible Children: School Integration in American Society.* Cambridge, Mass.: Harvard University Press, 1978.

Rozelle, R. M., and Campbell, D. T. "More Plausible Rival Hypotheses in the Cross-Lagged Panel Correlation Technique." *Psychological Bulletin,* 1969, *71,* 74-80.

St. John, N. H. "The Effect of Segregation on the Aspirations of Negro Youth." *Harvard Educational Review,* 1966, *36,* 284-294.

St. John, N. H. *School Desegregation Outcomes for Children.* New York: Wiley, 1975.

Scheirer, N. A., and Kraut, R. E. "Increasing Educational Achievement by a Self-Concept Change." *Review of Educational Research,* 1979, *49,* 131-150.

Schmuck, R. A., and Luszki, M. B. "Black and White Students in Several Small Communities." *Journal of Applied Behavioral Science,* 1969, *5,* 203-220.

Schofield, J. W. "School Desegregation and Intergroup Relations." In D. Bar-Tal and L. Saxe (Eds.), *Social Psychology of Education: Theory and Research.* Washington, D.C.: Hemisphere, 1978.

Schofield, J. W., and Sagar, H. A. "Peer Interaction Patterns in an Integrated Middle School." *Sociometry,* 1977, *40,* 130-138.

Sewell, W. H., Haller, A. O., and Ohlendorf, G. W. "The Educational and Early Occupational Status Attainment Process: Replication and Revision." *American Sociological Review,* 1970, *35,* 1014-1027.

Sewell, W. H., Haller, A. O., and Portes, A. "The Educational and Early Occupational Status Attainment Process." *American Sociological Review,* 1969, *34,* 82-92.

Sewell, W. H., and Hauser, R. M. *Education, Occupation, and Earnings: Achievement in the Early Career.* New York: Academic Press, 1975.

Shaw, M. E. "Changes in Sociometric Choices Following Forced

Integration of an Elementary School." *Journal of Social Issues,* 1973, *29* (4), 143-159.

Sherif, M., and Sherif, C. W. *Groups in Harmony and Tension.* New York: Harper & Row, 1953.

Silverman, I., and Shaw, M. E. "Effects of the Sudden Mass School Desegregation on Interracial Interaction and Attitudes in One Southern City." *Journal of Social Issues,* 1973, *29,* 133-142.

Simmons, R. G. "Black and White Self-Esteem: A Puzzle." *Social Psychology Quarterly,* 1978, *41,* 54-57.

Slavin, R. E., and Madden, N. A. "School Practices That Improve Race Relations." *American Educational Research Journal,* 1979, *16,* 169-180.

Spenner, K. I., and Featherman, D. L. "Achievement Ambitions." *Annual Review of Sociology,* 1978, *4,* 373-420.

Stephan, W. G. "School Desegregation: An Evaluation of Predictions Made in Brown vs. Board of Education." *Psychological Bulletin,* 1978, *85,* 217-238.

Stephan, W. G., and Rosenfield, D. "Effects of Desegregation on Race Relations and Self-Esteem." *Journal of Educational Psychology,* 1978, *70,* 670-679.

Stephan, W. G., and Rosenfield, D. "Black Self-Rejection: Another Look." *Journal of Educational Psychology,* 1979, *71,* 708-716.

Taylor, N. C., and Walsh, E. J. "Explanations of Blacks' Self-Esteem: Some Empirical Tests." *Social Psychology Quarterly,* 1979, *42,* 242-252.

U.S. Commission on Civil Rights. *Racial Isolation in the Public Schools.* Washington, D.C.: U.S. Government Printing Office, 1967.

Webb, E. J., Campbell, D. T., Schwartz, R. D., and Sechrest, L. *Unobtrusive Measures.* Chicago: Rand McNally, 1966.

Weinberg, N. "The Relationship Between School Desegregation and Academic Achievement: A Review of the Research." *Law and Contemporary Problems,* 1975, *39,* 240-270.

White, K., and Knight, J. H. "School Desegregation, SES, Sex, and the Aspirations of Southern Negro Adolescents." *Journal of Negro Education,* 1973, *42,* 71-78.

Williams, T. H. "Educational Aspirations: Longitudinal Evidence on Their Development in Canadian Youth." *Sociology of Education,* 1972, *45,* 107-133.

Williams, T. H. "Educational Ambition: Teachers and Students." *Sociology of Education,* 1975, *48,* 432-456.

Williams, T. H. "Teacher Prophecies and the Inheritance of Unequality." *Sociology of Education,* 1976, *49,* 223-236.

Wilson, A. B. "Residential Segregation of Social Classes and Aspirations of High School Boys." *American Sociological Review,* 1959, *24,* 843-854.

Wilson, J. Q. "On Pettigrew and Armor: An Afterword." *Public Interest,* 1973, *24,* 132-134.

Wilson, K. L. "The Effects of Integration and Class on Black Educational Attainment." *Sociology of Education,* 1979, *52,* 84-98.

Wilson, W. "Rank Order of Discrimination and Its Relevance to Civil Right Priorities." *Journal of Personality and Social Psychology,* 1970, *15,* 188-224.

Wylie, R. C. *The Self-Concept: Theory and Research on Selected Topics.* (Rev. ed.) Vol. 2. Lincoln: University of Nebraska Press, 1979.

Yarrow, M. R., Campbell, J. D., and Yarrow, L. J. "Acquisition of New Norms: A Study of Racial Desegregation." *Journal of Social Issues,* 1958, *14,* 8-28.

Donald T. Campbell

Comment: Another Perspective on a Scholarly Career

Editors' Note: The following autobiographical commentary succeeds as no ponderous essay could in conveying the spirit of Donald Campbell's blind-variation-and-selective-retention theme in the context of his own career. In contrast to the picture of a well-ordered system of ideas portrayed in the introductory sections of the book, this chapter provides a glimpse of the fits, starts, and blind alleys that characterize the life of the working scientist and intellectual. As editors we have taken responsibility—sometimes above Campbell's objections—for shortening this chapter from the original manuscript, and any resulting omissions of persons and events are likely due to our cuts. However, the informal tone and occasional quixotic self-revelation have escaped editorial interference because they provide a vehicle for conveying the breadth and richness of experiences and influences that have shaped Campbell's epistemology and world view.

Marilynn Brewer and Barry Collins have persuaded me to add to the volume they have instigated and integrated. They have tied together the scattered threads of my career so ably, have made its diversity seem so much like the deliberate application of a single central core, that I am loath to add to their comments for fear of spoiling a beautiful but fragile portrait. Yet they have known me in frustrating blind alley explorations as well as in success, and in depressed and overwhelmed periods as well as in buoyant and overexpansive ones, and they have nonetheless persisted in their request for a biographical comment.

The honor of such a volume should lower the defensiveness of one's self-presentation, reduce the need to put up a brave front of continual competence, and enable one to talk about intelligent hypotheses that proved wrong, stupid investments of research energy that were never worth undertaking even in anticipation, and promising lines of research that were dropped before fruition due to faintheartedness or doing too many things at once. This sort of review of my research career is what I aim at here. Several uses, I hope, can be achieved by this awkward counterpoint to the book's thematic content. It will fill out a picture that is otherwise extremely selective. It will enable me to exemplify the blind-variation-and-selective-retention epistemology—now one of the core concepts in my theory of science—by illustrating with my own career the inevitable wastefulness of scientific exploration, the chancy indirectness of discovery, and the further chanciness of recognition. I hope it will also give heart to younger colleagues when they find that they are at age thirty-three by the time of their first publication, or have wasted five years on an unpublishable exploration, or a year or so due to depression, or have bitten off too much and have had to default on promises. Science is a wasteful and chancy process, but a big blind alley exploration or prolonged absence need not wipe one out.

Undergraduate Research Interests

I finished high school in 1934 knowing that I wanted to be a scientist, but I had not chosen a field. Before going on to college,

I worked for a year for $40 a month plus room and board on a turkey ranch near Victorville, California. The offer grew out of a summer job. Accepting it was supported by a family ideology favoring such broadening experiences. During that year I chose psychology as my field of science, inspired primarily by an article on the psychological novel, read in a neighbor's old issue of *Harper's* or the *Atlantic Monthly,* or perhaps the *New Yorker.* The article focused on Stendahl and Dostoevsky. (I have not succeeded in relocating it.) My notion of science was already of the experimental physics sort, whereas the article was solely about humanistic psychology and literature. How I put the two together I do not know. (I later read many of Stendahl's works but found Dostoevsky too threatening to complete, although my memory now is too vague to specify why.) In my career as a psychologist, the scientific part of the commitment turned out to be stronger than the personality-motivational-social-situational psychology of the psychological novel.

I returned from that year to live at home and go to San Bernardino Valley Junior College, where I spent my freshman and sophomore years. My teacher for psychology and philosophy was an ex-missionary, ordained but with no Ph.D. degree. My zoology, physics, and political science teachers had Ph.D.s, and the zoologist was a squirrel hunter who published articles on valley-to-valley differences in squirrel fur color relevant to evolutionary theory. From them, and from laboratory instructors at Berkeley later, I got my image of the worker scientist, and the knowledge that I could enjoy the scholarly life they were leading at the junior college level, a lower edge on a broad span of aspiration that has often comforted me and that I have often wished specific friends could share.

In the fall of 1937, I went to the University of California at Berkeley as a junior and was lucky to get a job as one of the "readers" or examination graders for the large (1,000) introductory psychology class. Later I became a reader for Robert Tryon's introductory statistics and C. W. Brown's advanced statistics. This made me socially a part of the ten or fifteen career-oriented psychology majors and the similar number of graduate teaching and research assistants. I was a full-time professional psychologist from then on.

I took almost all the courses that the department had to offer, with the result that probably 75 percent of my course load was psychology. This array included child psychology lectures and laboratory, personality, clinical and abnormal psychology, experimental psychophysics and animal learning, statistics and experimental design. There was no social psychology, although I defined myself as in that field and served as Robert Tryon's assistant or reader when he first offered such a course in a summer session (probably 1939). While I intend to report here on research activities rather than the course work, I should mention the unpolarized breadth of what was regarded as acceptable psychology in that great department. Freudian theory may well have been the most ubiquitous content, showing up sympathetically presented not only in courses taught by the psychoanalysts Erik Erikson and Nevitt Sanford and in Jean Macfarlane's clinical psychology but also in the teachings of Harold Jones (a scientifically conservative, quantitative child psychologist who, I later learned, had had six months or so of personal psychoanalysis), Edward Tolman, Egon Brunswik, and others. I added course work in anthropology and sociology—partly through the encouragement of Harold Jones. Perhaps his advice also influenced my taking genetics and human genetics in the biology department.

My undergraduate research experiences were varied. Under Harold Jones's supervision, I did Parten time-sample observations of social behavior in the Institute of Child Welfare's nursery school, plus still and motion pictures with telephoto lens. Under Robert Tryon's direction, I did a cluster analysis of the items on M. H. Harper's old liberalism-conservatism questionnaire. In the second semester of Warner Brown's experimental psychology course, we had to choose our own problem. Fascinated with how *New Yorker* cartoons could convey character in so few lines, I compromised on transient emotion, sought out in the comics a character expressing numerous emotions, and ended with a dozen drawings of Olive Oyl from the Popeye series, judged in and out of context on thirty trait terms by some fifty undergraduates. This and the cluster analysis were well enough done to be on the lower edge of publishability, though the outcomes were not memorable and no one suggested I submit them.

During the summer of 1938, after my junior year, I had a

job at the Institute of Child Welfare doing simple, multiple, and partial correlations for Else Frenkel-Brunswik's monograph on latent and manifest needs. This job enabled me to spend the summer in Berkeley and further my full professionalization. I also worked on data analysis for Robert Tryon's project on the inheritance of maze-running ability in rats. The aspect I worked most on involved dice-rolling simulations of breeding experiments (genes plus error), trying to determine how many gene loci might be involved in the maze-running ability. (Even with twenty-five genes, the dice rats separated brights from dulls in many fewer generations than had the live rats.)

Graduate Research Experience

During my junior and senior years at Berkeley, I developed the intention to continue on for the Ph.D. degree, and at Berkeley. I do not remember considering other places. The scholarly life was an exciting one. I was already self-supporting and could look forward to increased income as a graduate assistant. Since the Depression was still on, we were not being lured away by other careers. As my June 1939 graduation approached, I was offered two assistantships, one from Edward Tolman running rats and one from Harold Jones at the Institute of Child Welfare, working on a study of adolescents. I chose the latter, as being human social psychology. (In spite of this choice, Tolman, Brunswik, and Tryon have been the most influential of my teachers. At times I have suspected myself of seeking autonomy through avuncular rather than paternal identifications, perhaps also shown later in my selection among job offers.) My graduate career began that summer. Kurt Lewin was visiting; and, in addition to taking his advanced undergraduate courses on personality and child psychology, I was permitted to attend a small seminar on Lewin's theories, in which Tolman and Brunswik were active participants.

The Adolescent Study was nearing the close of its data collection activities, and numerous analysis opportunities were available. I was given a relatively free hand in designing and executing my own project. John Gardner, in his just-completed dissertation research, had given most of the adolescents a level of aspiration experiment, in the Lewin-Frank tradition, with artifically controlled performance feedback which should have made scores ideally comparable across individuals. I attempted to relate several such scores to the voluminous personality measures available, particularly the rich aspiration content in the self-description "Personal-Social Inventory," designing new total scores from it, generating profiles, and looking for types. None of the analyses turned up any relationship with Gardner's performance measures, thus constituting my first really thorough experience of completely negative results where relationships plausibly should have been found, and where both obvious and ingenious analyses had been performed. I am not sure that the results ever were written up. Certainly they were never submitted for publication.

For my second graduate year, I sought out and obtained a departmental teaching assistantship. Why I did not continue at the Institute of Child Welfare I do not now know. Probably I wanted to be more centrally located in the psychology department. My assistantship was in experimental psychology, under Warner Brown for the first semester and Egon Brunswik for the second. The course was predominantly laboratory, with sensory and perceptual processes dominating. The only original research effort I can remember doing that year involved an indirect test of liberalism and conservatism, based on selective information and guessing behavior on multiple-choice factual information items. This research was not carried beyond the pilot stage. Other research experiences included interviewing and projective test administration in the clinical psychology practicum and, for Erik Erikson and Alfred Kroeber's seminar on the Yurok, coding Yurok myths as though they were projective test protocols.

At the end of two years of graduate work, I had no research products—in part because a master's degree was not required of students aiming at and approved for the Ph.D. degree. (Subsequently I have always required a master's thesis of my students, even when the department did not. It improves the Ph.D. dissertation, making it more realistic and modest. It also furthers a tendency to complete projects, and it often produces a pre-Ph.D. publication.)

The University of California had in those years "traveling fellowships." For my third year, I applied for and received one, choosing to spend it at Harvard with H. A. Murray. This year,

1941-42, was interrupted after a very intense three months. Murray was organizing an ambitious study; and there were long, intense weekly seminars devoted to its planning. Murray, Gordon Allport, and the anthropologist Clyde Kluckhohn were running an equally intense Morale Seminar, which was developing background papers on social psychology relevant to a potential war effort. I extended my financial support by being a discussion section leader in Edwin Boring's introductory psychology course. I took Allport's course on the history of social psychology. For my own research, I attempted to devise two types of structured indirect tests of civilian morale: multiple-choice information tests and estimates of public opinion.

Graduate Studies Interrupted

Registration, draft, and military training were already under way in response to the war in Europe. Educational deferments were not expected to last long. My Berkeley professor Robert Tryon was putting together a Social Psychology section in the Research and Analysis Branch of the Coordinator of Information (COI), a new defense agency in Washington, D.C. I withdrew from graduate school and joined him on December 1, 1941—just before the December 7 bombing of Pearl Harbor. COI was a foreign intelligence agency, soon to change its name to the Office of Strategic Services. The Research and Analysis Branch was organized into numerous area specialty divisions, plus a few topical divisions—economics, cartography, and psychology. We did studies of social attitudes, competing propaganda themes, and accessibility of the population to radio broadcasts for potentially focal regions like Portugal, Spain, Morocco, and Algiers, using all available sources, classified and public.

I spent a lot of time devising and applying a propaganda-coding scheme, combining Harold Lasswell's world attention survey and ethnocentrism theory. The exposure to German, Russian, and Italian home propaganda made a lasting impression on me. The social science treatments of the Nazi ideology, and a reading of Hitler's *Mein Kampf,* had led me to expect an internal presentation of Germany's war effort in terms of the rights of the strong, *Lebensraum,* aggression as justified, pity portrayed as a weakness, and the like. Instead, for its own people, Germany's war was presented as a defensive necessity. There seemed to me to be no difference in the values invoked among the home propagandas of any of the allied or axis nations (except for the ethnocentric choice of own nation as the moral and peace-loving one).

I applied for and received a commission in the U.S. Naval Reserve and was on active training duty by February 15, 1943, and within a few months was in charge of a U.S. Navy gun crew of twenty-six enlisted men aboard a civilian merchant marine ship on the United States–England run, in which service I remained for two and a half years. During this period I allowed professional content to drop out of my life completely. (This period, like my earlier turkey ranch year, provided a thorough immersion in popular cultures which I would otherwise have missed.) After the surrender of the Japanese in the summer of 1945, I was assigned for a few months to San Francisco as Personnel Classification Interview Officer, helping to administer the demobilization point system (a great achievement of applied social science measurement under very complex circumstances), inform veterans about the Veterans Administration educational support program, and do a modicum of vocational counseling. By the end of January 1946, I had enough demobilization points myself and was released in time to return to graduate school in Berkeley by the spring semester.

Graduate School Postwar

Harold Jones had kept in correspondence with me during the war and provided me with a research assistantship to supplement my VA support. (I had by that time my wife, Lola, and son, Thomas, with Martin on the way.) Else Frenkel-Brunswik and Jones had funds for an adolescent version of the big Authoritarian Personality Study (which was just completing its data collection). Milton Rokeach, who had been a fellow graduate student before the war, Murray Jarvik, and I were among the research assistants busily engaged in designing new attitude and personality tests appropriate for the fifth through the twelfth grades. Jones also had me apply for a Social Science Research Council fellowship, especially designed for war veterans. When it came through in September, I continued to help with the project in making contacts with schools and administering group tests. Quantitative scores were collected on some 1,500 children ages eleven to sixteen, with five separate component scores on attitudes toward Jews, Negroes, Japanese, Mexicans, and outgroups in general; liberalism-conservatism items dealing with domestic and foreign policy; and F-scale opinions. To Milton Rokeach's and my continued regret, the completed statistical analyses of these data were never published.

If I had the option (and I probably did), I probably should have written my dissertation on parts of those data for which I had designed the instruments. Instead, I chose to do a separate project, also in the ethnic attitude area, but still more self-consciously orderly in the use of attitude measures. Finished in September 1947, my dissertation, "The Generality of a Social Attitude," measured attitudes toward five groups (Negro, Japanese, Jewish, Mexican, and English) on five topics (Social Distance, Blaming, Capability, Morality, and Affection), each tapped with a five-item scale. The 125 resulting items were meticulously counterbalanced over five pages that were assembled in five different orders. Data were collected on 150 college and 239 high school students, and analyzed in a dozen different ways, both obvious and subtle. Apparatus factors (response sets) were worried about, response quality deterioration from page 1 to page 5 was examined, hetero-outgroup–heterotopic correlations and scale analyses were used as a comparison base, and topical differences in the preference ordering of outgroups were examined. I was sure then as now that the study was full of publishable material. Over the next few years, I did some supplementary analyses that seemed required; but, with new data collections and coauthored research taking priority, I have not yet submitted a single article based on it. (Boyd McCandless used my scales along with those of the Authoritarian Personality Study in a heroic two full days of testing of 150 randomly selected students at San Francisco State College in 1945, and we did publish a compact article on my scales in 1951.)

Four of us completed our work in social psychology in 1947: Milton Rokeach, Daniel Levinson, Fred Glixman, and myself. I first received job offers in child psychology from Stanford University and the University of Minnesota; I finally accepted one in social psychology from Ohio State University, arriving on the job that September with no teaching experience.

Ohio State University, 1947-1950

We had a nine-hour teaching load on a quarter system. In addition to social psychology, undergraduate and graduate, I alternated with Harold Burtt in teaching his course on advertising research and Sidney Pressey in his course on individual differences. On top of this, I began one of my worst periods of overinitiation of research, of biting off more than I could chew.

Public opinion-polling techniques had not been a part of my training, but I immediately plunged into this area. Harold Burtt turned over to me the local supervision of the Psychological Corporation's Brand Barometer Survey. I introduced polling projects as an alternative to term papers in large social psychology and advertising research courses. I organized colleagues in sociology, political science, and speech into an "Ohio State Interdepartmental Public Opinion Research Project," and we conducted a 1948 election panel study from May to October, using the class project efforts of our students. Although there was much of value in the data from this heroic effort (particularly in citizens' estimates of local and national opinion), and although I kept large boxes of the data until my move from Northwestern University in 1979, all that was ever written up was a minor methodological note, which Dale Wyatt and I published in 1950, and a brief comment in a footnote (in 1951).

From the summer of 1948 on, including one summer a few years after I had left Ohio State, I worked on well-funded research with Carroll Shartle and Ralph Stogdill and the Ohio State Leadership Studies. My main activity was planning and executing a study of the officers and crews of a squadron of ten submarines stationed at New London, Connecticut. Data collection was probably complete by 1949. In addition to the Shartle-Stogdill leader behavior description instruments, I introduced enlisted men's morale ballots, the task of estimating group responses on each morale item, measures of group reputation, and administrative records of ship efficiency. By 1953 I had finally produced a 200-page report, which Ralph Stogdill edited (without taking credit for it) as my monograph *Leadership and Its Effects upon the Group* (Campbell, 1956c). The rich and theoretically provocative results from the opinion estimation task were never written up but long remained in high priority for me. (My minor 1955 paper "The Informant in

Quantitative Research" is my only other publication from this project.)

A miscellany of influences contributed to what eventually became central foci in my scholarship. The OSU department was very behavioristic, in the Hull-Spence tradition. Arguments with D. D. Wickens, Richard Littman, Julian Rotter, and David Bakan kept alive my commitment to Tolman's purposive behaviorism, by now combined with my growing interest in cybernetics. (My 1954 paper reflects this discussion, especially with Wickens.) These arguments were enriched with logical positivist philosophy of science. I was already partially aware of the difficulties of operationally defining theoretical variables. A philosophy of science table at the faculty club each Friday, hosted by Eliseo Vivas and Vergil Hinshaw, furthered this side of my education. Kurt Wolff was a regular participant and introduced me to the sociology of knowledge. John Bennett in anthropology and Melvin Seeman in sociology extended my social science commitments. My social psychology texts (such as David Krech and Richard S. Crutchfield's *Theory and Problems of Social Psychology,* 1948) were Gestaltish, and in lectures I tried to translate these principles into the locally prevailing behavioristic ones and to integrate both with a list of bias tendencies in human cognition, centered on Francis Bacon's "idols" (see Wyatt and Campbell, 1951; Campbell, 1959b).

During the same period, an ambivalent interest persisted in Else Frenkel-Brunswik's, Nevitt Sanford's, and Milton Rokeach's authoritarian personality, tolerance of ambiguity, and rigidity syndrome. I had a student write opposites for F-scale items and collect a set of data on direction of wording effects (never published, but see Chapman and Campbell, 1957b, 1959b; Small and Campbell, 1960). My first Ph.D. student, John Francis Michael, did his dissertation (completed in 1952) on "An Attempt to Check and Extend the Ethnocentrism and Rigidity Hypothesis" (negative results, never published).

Indirect Attitude Measurement

In spite of all the above, my central research focus while I was at Ohio State, shown especially in my involvement with students, was on structured, quantitative, disguised measures of social attitudes. Since I now regard this interest as fundamentally mistaken, since it persisted another dozen years or so, and since it is the topic with the largest number of my publications, it seems best to break out from the chronological format and treat it as a whole.

My first published article ("The Indirect Assessment of Social Attitudes," 1950) reviewed the literature on this topic. In it I included reports on several Ohio State master's theses; and other such theses were under way, some of which eventually got published (Parrish and Campbell, 1953; Rankin and Campbell, 1955; Campbell and Tyler, 1957; and, if we include other uses of estimation of group opinion, Hites and Campbell, 1950).

At the University of Chicago, funds from Louis Wirth's Committee on Education, Research, and Training in Race Relations supported continuation of work on bias-diagnostic information tests of attitudes toward Negroes, resulting in massive waves of testing and test revision in 1951, 1952, and 1953 with cooperating introductory psychology students, black and white, from Baton Rouge, Nashville, Boulder, and Chicago. Although Louise Kidder and I devoted several pages of our 1970 article to the results, and Warner Wilson and I have a rejected manuscript on his 1969 resurvey of some of the colleges, the results of this massive data collection and analysis have never been published. William Kruskal in 1951 developed for me a measure of nonrandomness in seating patterns, which we eventually published with some repeat observations made in 1963 by William Wallace (see Campbell, Kruskal, and Wallace, 1966), resulting in the only product of this great investment of effort. Again, greater priority to coauthorship affecting the careers of students and premature investment in new data collection are to blame. Certainly it was not from lack of pride in the research. When I was interviewed by Dean Simeon Leland in 1953 for my job at Northwestern, I told him that most of my publications for the next few years would be on these data. None ever got out.

In a later study of "superior versus subordinate orientation," more than half of the measurement efforts were in the indirect attitude area with a wide variety of measures (see Campbell and Burwen, 1956; Burwen and Campbell, 1957; Campbell and Damarin, 1961; Campbell and Mehra, 1958; Campbell and Shanan, 1958; Campbell and McCormack, 1957; Campbell and Chapman,

1957; Chapman and Campbell, 1957a). Several more papers fell in this area (Campbell, 1957b; Westfall, Boyd, and Campbell, 1957; Maher, Watt, and Campbell, 1960; Renner, Maher, and Campbell, 1962)—all in all, some nineteen or twenty publications.

When G. F. Summers asked permission to reprint my 1950 review in his collection on *Attitude Measurement,* I took the occasion to provide an updated revision, joined by Louise Kidder. Our opening two paragraphs express my present perspective:

> This is an ambivalent updating of a twenty-year-old review covering *disguised* and *projective* measures of social attitudes (Campbell, 1950). Many of the reasons for positive evaluation of indirect attitude measures still remain. At their best, they are admirably ingenious. They utilize and illustrate psychological laws to a greater degree than direct attitude tests, and are thus more characteristic of measurement in the successful sciences wherein yesterday's crucial experiments are today's routine measurement procedure. And even if not better, they are different, thus fitting in with multiple operationalism, which attempts by using multiple methods of, hopefully, independent biases to curb the inevitable biases of single methods.
>
> The sources of negative valence do today loom much larger than they did in 1950, and it seems well to review these briefly, raising issues which the reader can keep in mind as we discuss specific problems. Four points can be examined: (1) the invasion of privacy issue, (2) the deceptive-deprecatory-exploitative attitudes of psychologists toward subjects, (3) the failure to do the research implied in the introduction of indirect tests, and (4) the disappointing nature of the research results [Kidder and Campbell, 1970, p. 333].

University of Chicago: 1950-1953

I turned down an associate professorship at Ohio State University to accept an assistant professorship at the University of Chicago, beginning in September 1950. Although I have often regretted this decision, my Chicago years made an important con-

tribution to my intellectual development. My inaugural colloquium to the department was "On the Psychological Study of
Knowledge." It foreshadowed my subsequent investments in epistemological processes but differed from most of these publications
in being oriented toward the sociology of knowledge rather than
the philosophy of science. Unlike many of my colloquia presentations, it was fully written out. It is still available and seems even
now publishable. Why I never submitted it I do not know. It does
a better job than any paper since of integrating my whole research
program, interpreting the indirect information-attitude tests as
studies of the sociology of popular knowledge. (I had tooled up
for this presentation by teaching a new course on "The Psychology of Knowledge" the summer before at Ohio State.) At Chicago
I supervised two dissertations in this area: a 1954 dissertation by
Kanwal Mehra, comparing experimental and theoretical physicists,
and a 1955 dissertation by Elsa Whalley, entitled "Individual Life-
Philosophies in Relation to Personality and to Systematic Philosophy."

Immediately on arriving at Chicago, I was made a part of
the Committee on Education Research and Training in Race Relations, of whom Louis Wirth, Everett Hughes, Sol Tax, and Philip
Hauser were the leaders. Their support of my indirect attitude
assessment research has already been noted. My involvements went
far beyond this, however; and overall, until his death in 1952, I
worked more closely with Louis Wirth than with any other person
on campus, performing chores such as facilitating empirical research by the gifted and prejudiced Gustav Ichheiser and supervis-
ing a cooperative four-person sociology master's thesis on deliberate segregation in the assignment of children in Chicago's public
schools.

For the year 1951-52, Chicago had a Ford Foundation grant
for the development of interdisciplinary "behavioral science." In
the anthropologist Robert Lowie's courses at Berkeley, I had acquired a great admiration of the University of Chicago sociologists
Robert Park and W. I. Thomas, and in particular for their concept
of the marginal man. This topic was already a staple in my first
social psychology course at Ohio State, and I combined it with
Else Frenkel-Brunswik's concept of "tolerance of ambiguity," predicting that intolerant people achieve a superficial resolution of

their marginality by an exclusive overidentification with one of the cultures to which they have been exposed, whereas people who are tolerant of ambiguity and have greater ego strength creatively accept multiple cultural identifications. Louis Wirth and I were provided financial support for a year-long interdisciplinary seminar on the topic.

Out of this seminar came two sociology dissertations which I directed. Irwin Rinder's 1953 dissertation, "Jewish Identification and the Race Relations Cycle," sought out adult children of Eastern European immigrants, scaled the degree of their Jewish identification from monolithic through mixed to rejecting, and found both extremes higher on the F scale than the middle. Thurstone's perceptual closure factors produced no relationship. Eugene Uyeki's 1953 dissertation studied adults of the Nisei generation and found only the monolithic-to-mixed part of the continuum—which made sense, since "passing" was not a feasible alternative. (Both Rinder and Uyeki published articles based on their dissertations.) My only publication in this area was an essay with Rinder (see Rinder and Campbell, 1952).

In the spring of this same year began the Superior-Subordinate Orientation Project. Undoubtedly this creative and administrative burden was a mistake, leading to the neglect of activities I would now (and probably even then) regard as more important. Pressure or encouragement from the department was a factor. Although I had worked on funded projects, I had not yet raised any funds, and I was proud to be able to do so. The conceptualization of the problem seemed to me both original and appropriate, and to effectively wed psychology and sociology. Modern social life places most of us at intermediate places in a hierarchy, with competing pressures and rewards from superiors and subordinates. Early family settings and other personality or attitude determinants might well give individuals persisting tendencies to favor superiors over subordinates, or vice versa. Stipulating (probably erroneously, although our respondents were predominantly in military training) that there was such a general dimension of consistent individual differences, which would show up among occupants of parallel loci sharing the same organization structure, we set out to measure it in a dozen different ways.

The project was well enough funded to hire Leroy S. Burwen as full-time research director for the three-year period, plus five or so research assistants from clinical and social psychology. We all designed instruments like mad, sparing no costs in getting real people and personality ratings for the Photo Judgment test booklets. (Despite these efforts, the results of the project were generally frustrating, although well published, in part through using unrefereed journals; only six of the eighteen research reports that were written up are unpublished as of this date.)

During my final year at Chicago, 1952-53, I was privileged to be relieved of all teaching duties to participate with six or eight others in James G. Miller's "Committee on Behavioral Sciences," focused on cybernetics, information theory, and general systems theory. During this year, and the summer of 1954, I worked on relating W. Ross Ashby's cybernetics and other natural selection analogues to learning theory and perception (see Campbell, 1956a, 1956d). I also developed a large mimeographed opus on the integration of the many social science and psychological terms referring to the residues of past experience as they influence behavior, with special focus on translating between behavioristic and phenomenological concepts (Campbell, 1963b). Overall, this "Behavioral Science" year greatly furthered my commitment to a kind of theorizing that transcends social psychology, encompassing learning, perception, and cognition—and, eventually, biological evolution, social evolution, and theory of science. It is here for the first time that my scholarly activities overtly show one of the major thematic cores of this volume.

A second thematic core, quasi-experimental methodology, was also near the surface at this time. My graduate methods course, "Social Attitude Research," at Ohio State and at Chicago included not only concepts of reliability, validity, and pitfalls but also experimental design in social settings extended to correlational causal inference. My planning documents for the OSU Leadership Studies recognized the assertion of leadership as a causal hypothesis for which experimental design is appropriate, and explored how the transfer of officers—combined with group measures in each billet before, during, and after a given officer's tour of duty—could generate quasi-experimental evidence of transsituational individual differences in leadership effectiveness.

The third key theme, epitomized by the multitrait-multimethod matrix, grew out of the methodology concerns discussed in lectures as apparatus factors (from Robert Tryon's course on individual differences at Berkeley in 1938-39, based on his rat research), "halo effects" in ratings, and response sets in questionnaires. Large correlational matrices crossing different methods and topics were in my dissertation, the Submarine Leadership Study, and the Superior-Subordinate Orientation Study. In the latter two, the methodological triangulation theme appears. But while Donald Fiske was friend and colleague during all three of my Chicago years, we did not start our classic paper and do the centerpiece analysis of the Kelly-Fiske clinical assessment data until I had been at Northwestern for several years.

It is obvious from this recounting of my three Chicago years that I had great opportunities and took advantage of them, that I was highly approved and given favored treatment. Why, then, the net regret about the Chicago experience, which I regularly share with friends when job choices and tenure problems are being discussed? Most obvious to me is that I found the scholarly life much less enjoyable, overall, than I had found it at Ohio State. At Chicago the tenure pressures were extreme. There were twelve untenured assistant professors between thirty and thirty-five years of age (I came in at thirty-four), with promotion room for only two or three. Publications were everything. The last departmental head had been retired at the associate professor level after some two decades of chairing. Worse still, we were explicitly told that our publications had to be works of genius. This requirement reduced productivity for all; and several colleagues—for whom Chicago was their first job, who worked long hours on research, and who would have had productive scientific careers at any average teaching institution offering graduate study—published nothing at all during a six-year period. I ended my three years at Chicago with only five papers (including one in press) in career-relevant journals. My peers and administrators regarded my other seven papers as handicaps, spoiling my image as the kind of scholar that Chicago or any other proud university would want. I was hired at Chicago (and later at Northwestern) on promise, in spite of low productivity. I had definite depressions during my second and third winters at Chicago, and survived Chicago only because of the self-esteem and enjoyment of the scholarly life that Berkeley and

Ohio State had nourished. While some (both persons more self-assured and persons more modest than I) survived well in this atmosphere, I know that my reactions were not at all unique.

Tenure policies such as those prevailing at the University of Chicago are (in my judgment) very wasteful of talented new Ph.D.s. Norman Bradburn (when chairman of behavioral sciences at Chicago) and I once planned and came close to implementing a retrospective quasi-experimental study of the effect of pressure on quality, which would go back in the records to that period when the faculty of the University of Chicago College was organizationally discrete from the Graduate Divisions, and when College faculty members earned their living by teaching—in contrast to the Divisions, where publications were all that counted. In our causal consideration of examples in the social sciences, it is the College, rather than the Divisions, that has produced eminent people (such as David Riesman and Edward Shils) from among new Ph.D. appointees, and yet any selection advantage in hiring would run the opposite way. We both feel, too, that many more career tragedies afflicted new Ph.D.s who went into the Divisions. My observations, from greater distance, of other universities that deliberately hire many more assistant professors than they can promote confirms my belief that this practice both wastes scholarly talent and reduces the joy of scholarship.

During my second summer, I seriously considered an applied job in Air Force research at Montgomery Field, Alabama, and I used this offer to explore my prospects at Chicago with Dean Ralph Tyler. He responded with a raise. That same summer Carl Duncan (from Northwestern) and I were cochairmen of local arrangements for the national convention of the American Psychological Association. When Northwestern's offer of a five-year term as associate professor came through, I saw in it (in general, correctly) an opportunity to return to a mutually affirming scholarly atmosphere, where worker scientists were honored and where I would earn my living through teaching. (By the time I became aware that publications were necessary at Northwestern, I had overcome the log jam, still shamelessly using unrefereed journals.) Dean Tyler was willing to offer me another three-year term as assistant professor but agreed that, unless my publication record (both volume and quality) dramatically changed, I stood no chance of tenure. After my departure I continued to be in contact with my Chicago colleagues—Donald Fiske and William Kruskal in particular, but also a half-dozen others—and I developed new colleagueships, as with William Wimsatt in philosophy and evolutionary epistemology. Needless to say, when these friends promoted an honorary degree for me in 1978, it was a very healing and rewarding experience. Chicago behaved rationally and fairly in letting me go, both on record and promise. (It would have taken immediate tenure to keep me.) Even now, I am convinced that I would not have earned that honorary degree or this honorary volume (assuming for the sake of argument that I have earned them) had I remained. This is, of course, a statement about me fully as much as it is a statement about the University of Chicago setting.

Northwestern University, 1953-1979

When, urgently needing a new setting, I resigned from Northwestern at age sixty-three and a half, I quixotically asked for emeritus status with this statement: "I remain tremendously proud of Northwestern University and of my twenty-six-year-long association with it. While I enjoyed six academic years as a faculty member before coming to Northwestern, and thus spent only 79 percent of my scholarly career there, some 99 percent of my published pages bear the 'Northwestern University' signature, and at least as high a proportion of my citations.... I cannot imagine that my pages or citations will ever fall below 90 percent in Northwestern identification. We are thus permanently and inextricably co-identified. I am glad this is so, as well as sad that personal problem solving has led to the present official severance of our ties. Pondering on this, I have decided that if it were to fall within a permissible interpretation of the regulations, I would greatly appreciate the status of 'Professor Emeritus, Northwestern University.' " (However irregular the action, emeritus status was granted.)

Enhancement of Contrast and Adaptation Level. A central concept in my teaching of social psychology was Muzafer Sherif's "frame-of-reference" concept, translated by Harold Helson into adaptation levels producing contrast illusions and, when combined with categorizing cues, enhancing the contrast. In my first year at Northwestern, Robert Clarke did an M.A. thesis demonstrating this bias in pupils' judgments of the examination scores of black and white classmates (Clarke and Campbell, 1955). I spent the spring of 1955 visiting at Yale, handling Leonard Doob's courses and working with Carl Hovland's great Communication and Attitudes Project. There Marshall Segall and I did two studies on the effect of adaptation range and level of heard opinion on opinion expressed, asking for responses in "your own words" and "illustrating" what we were interested in by playing five tape-recorded samples. The samples were carefully designed with spliced tapes and, in pretransistor technology, carried by Marshall from door to door with a twenty-five-pound tape recorder. Although the results were utterly frustrating, they should have been published nonetheless, along with somewhat more challenging results from the door-to-door interviews of the social psychology class. (In spite of this frustrating collaboration, Segall transferred his graduate studies back to Northwestern.)

Back at Northwestern, these interests were fused with those of William Hunt (my departmental chairman) and Nan Agricola Lewis, who was working on Hunt's Office of Naval Research project on clinical judgment (Campbell, Hunt, and Lewis, 1957, 1958; Campbell, Lewis, and Hunt, 1958). This work used tasks ranging from judgments of the degree of disturbance evidenced in schizophrenics' definitions of words to judgments of the pitch of electronic notes using the piano keyboard as the response scale. The capstone of this successful program was David Krantz's M.A. thesis (Krantz and Campbell, 1961). The achievement for my social psychology was to make clear that the frame-of-reference, adaptation-level, contrast effects were genuinely "illusory," occurring without respondent awareness, rather than merely linguistic artifacts of the laboratory-specific judgment language dominating the psychological tradition. It was my opinion that these contributions should have been incorporated into the experimental psychology of judgment. While three of the crucial papers appeared in the *Journal of Experimental Psychology,* these contributions were not picked up by that tradition. I did not persist. I had satisfied my own curiosity and was on to other things. (My contributions to Hicks and Campbell, 1965, and Brickman and Campbell, 1971, also comes from this interest.)

David Krantz's Ph.D. dissertation (in experimental psychology, not social) was on esthetic judgment and was also supervised by me. This crossing of subdepartmental lines was representative of the Northwestern University psychology department of those years, as well as of my own collaborative style. As David's adviser I approved courses in art, philosophy, and religion, as well as in readings in the psychology of art, which I knew nothing about. (Later, at nearby Lake Forest College, David's research interests shifted to history and sociology of science, as exemplified in psychology, and he became one of my tutors on these matters as well as a member of Northwestern's theory of science community. My "Tribal Model" paper—Campbell, 1979d—describes our later collaboration.)

Phenomenological Behaviorism. The central core of my graduate and undergraduate teaching in social psychology at Northwestern was a long list of "principles" drawn mainly from learning theory but also presented in "view of the world" or phenomenological translation. My strong feeling that these two orientations should be interrelated was not widely felt then or since. Had it been and had the data from the projection project come out right, I might have ended up a major theorist rather than a methodologist. My projected social psychology textbook was under contract for some fifteen years, and the duplicated sixty-five pages of 124 principles and graphs is still used in my teaching. The articles that came out of this integrative effort (Campbell, 1956b, 1961a; 1963; 1967b) would nearly make a book, although marred by the ridiculous "periscope diagram," well known to all my students, and by the out-of-date, albeit cognitively interpreted, Hullian symbols.

This theory led to a well-funded research grant on "Varieties of Projection in Trait Attribution," obtained from the National Institutes of Health. Norman Miller had arrived the year previously and had done a classic M.A. thesis (Miller and Campbell, 1959). He became the backbone of the project that first year, helping to design the instruments, making the contacts, doing the five hours of testing, and serving as paymaster for the 450 respondents involved. The respondents came from nineteen different residential groups; and each person judged the other residents in his or her particular residential group, plus thirty persons known only through their photographs, on twenty-seven traits. The basic paradigm was to correlate reputational personality (ratings received) with attributions of personality (ratings given to others). Edward O'Connell joined the next year and became the project's statistical superego (though he allowed me to override his advice on the "double-standardized" scores) and computer manager. Jacob Lubetsky also joined us, taking over all the "assimilative projection" analyses (relating self-descriptions to descriptions of others) for his 1960 dissertation. We were a high-morale team, working on the best integration of theory and data collection I have ever achieved, with an extensive enough sample size to be definitive.

The results were crushingly negative and threw me, at least, into a temporary depression. My research report writeup began: "This thorough, tedious, expensive, and disappointing study . . ." Eventually Norman cleared out some of the pathological symptoms from the writeup and got it accepted as a *Psychological Monograph* (Campbell, Miller, Lubetsky, and O'Connell, 1964), and the team squeezed two more publications out of the project data (Miller, Campbell, Twedt, and O'Connell, 1966; Lubetsky and Campbell, 1963).

Cross-Cultural Research and the Ethnocentrism Project. Soon after I arrived at Northwestern, I was in regular conversation with Melville Herskovits. Perhaps the fact that I had studied with Robert Lowie and Alfred Kroeber, fellow Boasians, helped our rapport. My social psychology preoccupation with unconscious bias made me receptive to much of his cultural relativism. (For an appreciation, see Campbell, 1972b.) In any event, we were soon planning a renewal of W. H. R. Rivers's 1901 work on cultural differences in optical illusions, with new methods optimizing task communication. As a summer job on Program of African Studies funds, Marshall Segall designed and got printed the classic 1956 edition of our field manual, which Herskovits disseminated to students and friends all over the world (Herskovits, Campbell, and Segall, 1956). Interesting cultural differences in susceptibility to illusions were immediately found, and Marshall analyzed the whole while at Columbia and Iowa (Segall, Campbell, and Herskovits, 1963, 1966; Campbell, 1964).

By the time the optical illusion data were coming in, Richard Snyder and Harold Guetzkow had joined Northwestern's political science department. Snyder's programmatic grants in International Relations and Comparative Politics, which were spent in interdepartmental ventures (including philosophy), added on to the Program of African Studies base to create a pervasive interdisciplinary thrust for cross-cultural studies. Robert A. LeVine was a new assistant professor, with an appointment split between anthropology and political science. We were both admirers of the Yale Human Relations Files (which distressed Herskovits but did not remove his support). Because we believed that the field-manual approach could get similar data of more uniform quality on topics that traditional ethnographies usually neglected, we developed and distributed an ethnocentrism field manual (Campbell and LeVine, 1961), which Segall pretested on three groups in Uganda. This effort was supported by the grant money from African Studies, Comparative Politics, and International Relations. The latter two supported Paul Rosenblatt for a year's study of theories of ethnocentrism and nationalism (which he published in 1964), even though he was a psychology student.

Our field-manual schedule took a great deal of time to carry out and was rarely used fully. But it was a ready-made research proposal, and with it we applied to the Carnegie Corporation for funds to subsidize anthropologists already in the field to spend an extra few months collecting such data. In the spring of 1962, we received $250,000 (with no university overhead) for a five-year grant, which we spent over an eleven-year period. Marilynn Brewer soon became principal research assistant, did her dissertation on the data, worked as a full-time research associate during 1968-1970, and for several years continued to be LeVine's and my major collaborator on the project.

About this time LeVine moved to the University of Chicago but continued to be codirector of the project. The field manual was appropriately revised and distributed to cooperating anthropologists, along with our analysis of the relevant competing theories. LeVine was officially codirector of the project from the beginning, but he became in practice senior director at about this time, since locating fellow anthropologists was the main activity. But more than this, I became overcommitted, so that he and Marilynn ran the project most of the time.

The "Cooperative Cross-Cultural Study of Ethnocentrism" (as we called it, if not Carnegie) supported some twenty-five anthropologists in fieldwork for three months to one year. In several cases ethnographers returned to research sites studied years earlier. It is very difficult to finance anthropological research, and our paying travel expenses for about half of them, as well as living expenses and salary, thus was a valuable side product. Of the twenty-two who eventually turned over field notes for us, twenty made use of much of our field manual. Marilynn spent one full year editing and reorganizing these notes, so that the field manual could serve as an index, and we subsidized their archiving in the Human Relations Area Files (see LeVine and Campbell, 1972b). These data are still, however, underanalyzed, although Marilynn has pilot-tested several coding schemes. One of us should seek additional funding to get these done, and set aside time to do it.

Overall, however, the project has more published pages than the average research project funded at that level for that many years. There are two books (LeVine and Campbell, 1972a, of which Marilynn wrote most of Chapters Eleven and Thirteen, as well as helped edit; and Brewer and Campbell, 1976). Seven other publications of mine are footnoted to the project (Campbell, 1965a, 1967b; Campbell and LeVine, 1968, 1970; Werner and Campbell, 1970; Brewer, Campbell, and LeVine, 1971; Campbell, 1972a). Both LeVine and Brewer have several solo-authored papers so footnoted, as do many of the cooperating anthropologists.

The fact that there is a great unevenness in the ratio of invested time and money to publications is well illustrated in our ethnocentrism project. In 1965 LeVine designed a public opinion interview schedule, using the social distance and stereotype elicitation content from our field manual. With an amount of money equivalent to two anthropologists for three months, Marco Surveys collected interviews from fifty persons in each of thirty ethnic groups in East Africa. Brewer and Campbell (1976) and several journal articles report these data. (LeVine kept his name off the title page for fear of losing access for child development research due to local resentment of published tribal comparisons, as had happened in Nigeria.) In published pages per dollars, this survey project has paid off five times as well as the major study.

The ethnocentrism grant was only one involvement with cross-cultural work. Robert Winch of sociology, Francis Hsu of anthropology, and I promoted an interdisciplinary training grant, which brought Raoul Naroll to anthropology in an initially soft-money position (he already had collaborative ties to Richard Snyder of political science). My coauthorships and coanthologizing with all three are thus to be explained, although these understate our collaboration. Similar explanation holds for the otherwise anomalous series of publications by William Crano on cross-cultural method.

Cross-cultural and interdisciplinary studies at Northwestern reached their high point with the funding of the Council for Intersocietal Studies, promoted by a consortium involving almost all the faculty actors mentioned in this section, plus Lee Sechrest in psychology, and headed by Richard Schwartz. This history is also a part of my history. I had met Schwartz when I was at Yale, where he was on the faculty in sociology, in the spring of 1955. The Northwestern sociology department hired him around 1958, and our collegial and personal friendship further developed, a partnership underrepresented in publications, but evidenced most recently in his collaboration with Marshall Segall in doing the background work behind my distinguished and privileged appointment at Syracuse in 1979.

Overcommitment as an Occupational Hazard. I feel the need to break topical continuity here for a discussion of overcommitment, overwhelmedness, and students and colleagues who have bailed me out. Most of my friends, including many who know me well, will be surprised at the amount of depressive content in this chapter. Most will have known me as uniformly good natured, optimistic, expansive, with the only pathology (perhaps unnoted as such) being the delusion that I had time to engage fully each one of my shared intellectual interests. Probably that *has* been my mood during 90 percent of the scholarly career I am reporting. This chapter is indeed giving disproportionate weight to the blues. But I somehow feel it would be less honest and less useful to others if I left them out.

A few months after the ethnocentrism grant had been awarded, I was approached by the Educational Media Branch of the U.S. Office of Education and offered a three-year contract to support my work on research methods. The work on field experiments (Campbell, 1957a, plus the quasi-experimental design paper in widely lectured and mimeographed form) and the multitrait-multimethod matrix (Campbell and Fiske, 1959) had attracted them. They tempted me with an annual spendable budget equal to the Carnegie grant (plus full overhead to Northwestern). I was in an expansive mood, had lots of things of this sort I wanted to do and thought I could delegate, and accepted the offer without consulting my friends at the Carnegie Corporation.

Things always take longer than anticipated. Fluencies in writing and thinking are highly variable. By the spring of 1963, with the addition of personal problems, I was seriously depressed. Barry Collins on very short notice took over my big social psychology course, which he had taken as an undergraduate (Barry has always been a closer colleague than our coauthorship record reflects). Marilynn Brewer took over the ethnocentrism project management. The following year Robert LeVine and I went off to Kenya for three months of fieldwork to collect one set of data among the Gusii, with whom he earlier had worked for two years. There I sampled anthropological fieldwork of an old-fashioned sort ("veranda anthropology") as LeVine and I interviewed the grandfathers about precolonial intertribal relations, using their grandsons as interpreters.

Barry, Marilynn, and Bob were my rescuers during this time. But I have usually been blessed in my overwhelmed periods by having a graduate student or another associate who "carried" me: before ethnocentrism, Ralph Stogdill on Navy Leadership; Roy Burwen on the Air Force project, Marshall Segall on optical illusions cross-culturally; Norman Miller and Edward O'Connell on the projection project; in quasi-experimentation, to be discussed below, Robert Boruch, David Kenny, Thomas Cook, and Charles Reichardt. No doubt there are others whose asymmetrical contributions do not at the moment come to mind, for I would not be human if I did not share the universal bias of tending to exaggerate one's own contribution in interdependencies. Granting such a bias, I have, I believe, balanced out the larger account by generously "carrying" numerous students more overwhelmed in their time than I was. On the other hand, there have been a few students whose careers were harmed by their knowing me only in one of my depressed periods.

Quasi-Experimental Design and Social Measurement. Sometime around 1960 William Hunt had used a Carnegie grant designed to bring education and psychology closer together to fund me for a quarter off to complete my essay on quasi-experimental design. At that time, or earlier, I had sought out a statistician partner to help with the project. My friends William Kruskal and John Cotton were too busy at the moment. Julian Stanley and I were not acquainted but were in correspondence about his earlier research on the covariance analysis of raters and rating targets, in which he had used concepts like those in the multitrait-multimethod matrix. Our paper on quasi-experimental design eventually became a chapter in Nathan Gage's *Handbook of Research on Teaching* (see Campbell and Stanley, 1963) and was published separately in 1966. This must be for each of us our most cited publication, rivaled for me by the Campbell and Fiske (1959) multitrait-multimethod matrix.

The U.S. Office of Education contract of 1962-1965 provided full funding for my methodological work. Beginning in 1966, through three five-year continuation grants, the National Science Foundation has continued this level of support up to the present day. Through this and training grants, the Northwestern social psychology program was able to establish the custom of providing full-time summer research employment for each of its graduate students. These often led to master's and doctoral research efforts much more extensive than the support, increasing the apparent productivity of the projects.

A thick part of the final report of the USOE project was a duplicated manuscript entitled "Other Measures: A Survey of Unconventional and Cooperation-Free Measures for Social Science." The final report goes on to say: "This survey has been circulated widely among persons thought to be experts in this field, and has met with both specific suggestions for addition and generous praise. In considerably revised form, with more illustrations, with, sad to say, fewer efforts at humor, . . . and with a new title, possibly *Unobtrusive Measures,* it will be published by Rand McNally early in 1965 [see Webb, Campbell, Schwartz, and Sechrest, 1966]. While the only acknowledgment of financial support will be to this contract, it should be noted that no personal support was provided Webb, Schwartz, or Sechrest. . . . In the opinions of some, this book will be the most important contribution of this contract." Whoever those "some" were, their opinions were prescient. This book is still selling well and must be close to the top of our most-cited publications. (Lee Sechrest has revised it for 1981 publication.) No doubt the USOE project did facilitate the effort. Perhaps without the project money, it would never have been published. But the book itself grew out of no project initiative; instead, it emerged from our weekly "Social Psych Sack Lunch," held in the social psychology space on the top floor of Kresge Hall, in which it became a game to come up with novel odd-ball methods. Gene Webb, whose appointment then was in advertising research in the journalism school, was the major wit and author. I wrote the pessimistic parts of the methodological first chapter. (But as to humor, Donald Fiske, Nan Lewis, and I are the "Anonymoi" of the reference list, whose study of haircut length and hardheadedness in research is reported on page 116 and 117. There are real data. I think I still have them. We long intended a full writeup.)

My need for statistically gifted collaborators in methodological work must be noted. I lack such training; and, while I may have good mathematical intuition on some issues, my most stubborn insights have often proven wrong. Kruskal, Stanley, O'Connell, and Cotton have been mentioned. I have several times tempted mathematically gifted graduate students into developing new statistical procedures that went far beyond their training, not always for the good of their own careers. They have been my teachers, as well as the statistical superegos and unofficial dissertation advisers of their fellow graduate students working with me. Joyce Sween, Edward Kepka, Stanley Rickard, David Kenny, Leslie McCain, and Charles Reichardt have played such roles, supported by their teachers of advanced psychological statistics, especially Albert Erlebacher (our joint paper of 1970 is tangential to this more important, unacknowledged, indirect collaboration). Robert Boruch and I jointly prepared the National Science Foundation grant proposal of 1971, funding him as a full-time research associate, from which he transferred to the psychology department budget after a few years. He continued to provide statistical expertise for the project long after budgetary ties had ceased and the bulk of his own efforts had turned to policy issues in program evaluation.

The Civil Rights Movement, Vietnam, and the Experimenting Society. My account of research endeavors for 1965-1975 cannot be complete without some mention of the guiding impact of my students' and colleagues' commitments during those years. At Northwestern—in contrast to other contemporaneous social science communities—the leftward politicization of our ablest students was *not* accompanied by an antiscientistic, antiquantitative, humanistic turn. Instead, they continued to make contributions to better quantitative methods for real-world hypothesis testing as well as atheoretical ameliorative program evaluations. Essential in achieving this student involvement was (and is) my partnership with Andrew Gordon of urban affairs and sociology. We began our collaboration with an interdisciplinary seminar on urban experiments. From that seminar experience I designed a proposal to the Russell Sage Foundation with the flamboyant title of "Methods for the Experimenting Society," which had the more modest goal of providing summer funding for laboratory experimental psychology students to affiliate themselves as volunteer methodological assistants to action programs. With my National Science Foundation funds, similar activities were sometimes already being funded, but the opportunities had to promise more methodological novelty and rigor. For the Russell Sage grant, the social action component was an advantage or carried greater weight. After the four years of that program (approximately 1970-1974), Andrew Gordon and the Center for Urban Affairs have continued to provide this real-world outlet for the commitments and energies of social psychology students with laboratory experimental backgrounds. With Russell Sage's explicit permission, we were able to provide four months' transition support for Ricardo Zuniga of the Catholic University of Santiago to write up his experiences as a Catholic Marxist applied social scientist supporting the experimenting society of the Unidad Popular in Chile prior to the coup of the rightist junta.

My best-known unpublished papers, widely distributed and occasionally cited, are footnoted to this Russell Sage grant, as well as to the National Science Foundation. They are "Methods for the Experimenting Society" (Campbell, 1971a) and "Qualitative Knowing in Action Research" (Campbell, 1974c). (As soon as certain incoherences are removed and the scotch-tape marks smoothed over, I will get them published. Every year or so, I have spent a month trying to revise and update them, though no new revisions have been duplicated, nor are the papers really improved. I will get them out soon, even if not updated.)

Respect for Tradition and Evolutionary Theory

This brief chapter is intended only to be a research autobiography, and to avoid personal history except as it is relevant to the content and course of my scholarly career. But for many social scientists, their personal relationship to religious tradition does have such relevance; and, since I have conspicuously gone into print on such issues, providing some personal context seems in order. My father had grown up in a rural fundamentalist protestant church in southwestern Pennsylvania, but he did not participate in church activities as an adult. My mother had been reared in a liberal protestant church and taught Sunday school during my childhood, taking us along. My parents treated children's opinions with respect, responding to immature arguments with no pressure but with reasoned arguments at most. By high school or early college, I had drifted away from whatever belief in God I had had as a child, with no personal or familial crisis, no resentment of the ethical beliefs with which I had been reared or the means by which they had been taught, and considerable respect for the quality of lives led by those devout believers who combined subtle understanding of themselves with high ethical standards.

On the other hand, my early period as an observer in the Berkeley Institute of Child Welfare nursery school had convinced me that the recipes by which psychiatrists and psychologists were rearing their children were less adequate than traditional ones, rather than better. My increasing understanding of what it would take to be scientific about the important choices in life furthered a double standard, in which I was committed to helping create a thoroughly scientific social psychology but unwilling to jettison traditional wisdom about how to live life and rear children.

My own commitment to materialist theories of biological and social evolution was augmented by the mechanistic analysis of individual and organizational purpose provided by cybernetics. This analysis supported a commitment to atheism, sometimes openly and dogmatically expressed when such issues came up in my courses on social psychology and naturalistic epistemology. On the other hand, strong belief in a blind-variation-and-selective-retention version of social evolution gave me scientific grounds for considerable respect for religious tradition, at least on issues where the world adapted to had not changed. From such a background came my grumpy elder statesman's presidential address to the American Psychological Association in 1975 (see Campbell, 1975b). This address and an earlier paper (Campbell, 1972a) also made me an accepted participant in the sociobiology movement, even though my emphasis on the power of culture in conflict with behavioral dispositions produced by biological evolution is not widely accepted.

Theory of Knowledge

During all my years at Northwestern, I offered a seminar or course called "Knowledge Processes" at least every second year. Logical positivism, ably represented by Herbert Hochberg of philosophy, dominated graduate education from political science to psychology. Accepting the logical positivists' delineation of the proper spheres of philosophy and science, if not the philosophy itself, I was careful to distinguish between what I was doing and philosophy of science, as exemplified in my "Methodological Suggestions from a Comparative Psychology of Knowledge Processes" (Campbell, 1959a). Today we would call it "naturalistic epistemology," and, although still a minority activity, it is not mainly being pursued in philosophy departments.

Usually I had a philosophy coteacher, who doubled as my tutor: William Todd, Arnold Levison, Hugh Petrie over the longest period of years, Kenneth Seeskin, Carl Kordig, Edward Sankowski, and John Heffner, plus cybernetician William Powers and historian of science Donald Moyer. William Wimsatt frequently came up from Chicago as a guest lecturer or for private discussions, joined in the late 1970s by Robert Richards. The course tended to recruit some of the most able intellectuals among the undergraduates, who also became my tutors and induced me to study existentialism, phenomenology, and hermeneutics. Some years an undergraduate, such as W. J. Dowling, was my closest collaborator in this area. Before the expansion of the social psychology program, most of my social psychology students took the course, including those who have edited and participated in this volume, although, I am proud to say, it was never a required or even a recommended part of any curriculum and included evolutionary biology, cybernetics, and visual perception, but at that time no social psychology.

While in later years the products of this program have often acknowledged research grants, and legitimately so insofar as typing, duplicating, and payment of translators was involved, yet it would not be out of place to characterize the effort as using time stolen from data collection projects, or as an avocational indulgence. Reading, writing, essay writing, and visiting in this area brought me personal acquaintance with Karl Popper, Michael Polanyi, W. V. Quine, and Konrad Lorenz. My published essays in this area are spread out over a span of twenty-three years, from 1956 to 1979, culminating in the William James Lectures cited extensively in this volume.

In the spring of 1968-69, while I was Fulbright Professor at Oxford University, Stephen Toulmin and I jointly taught such a course, listed as "Evolutionary Epistemology." In the spring term of 1977, visiting at Harvard, I offered a version of this course under the title of "Descriptive Epistemology" and was lucky enough to have in addition to the twelve enrolled graduate and undergraduate students an equal number of volunteer participants and coteachers from the Cambridge community. Peter Skagestad was one of those, and our debates as to whether Charles Sanders Peirce or I had the truer evolutionary epistemology, shared with the seminar in both oral and written form, exemplified the vigor of seminar exchanges. It was during this term, too, that Abner Shimony and I, fellow evolutionary epistemologists, renewed our acquaintance and clarified our agreements and differences.

While I must plead innocent to selecting or recruiting these philosophers and historians of science (or, for that matter, any of the other participants), I am of course very flattered by their generous contributions to this volume. That five of the fifteen chapters come from this area is very disproportionate—as to my research effort, my sources of citations, and my national reputation. However, it is not disproportionate in terms of my current and planned future scholarly investment.

As a sample of that future, I conclude with plans for the June 1981 Syracuse Conference on Epistemologically Relevant

Internalist Sociology of Science, codirected by Syracuse philosopher of science Alex Rosenberg and funded through the State of New York's support of the Schweitzer Chair. The invitation to this conference concludes with this statement of my own position and predicament: "Campbell is a social psychologist turned research methodologist and lately theorist of science. He accepts epistemological relativism but stops short of ontological nihilism. His agenda for this conference will be to extend his evolutionary epistemology via social evolution into the social processes of science. His program falters here because it attributes all 'fit' (adaptation, validity) to selection, and because we know enough about a wide variety of selective processes impinging on scientific belief assertions to see that many selectors are tangential or inimical to improved accuracy in the description of nature." Let's hope my program can get beyond that faltering.

List of Honors

1965   Fellow, Center for Advanced Study in the Behavioral Sciences, Stanford, California

1968   Fulbright Lecturer and Visiting Professor in Social Psychology, University of Oxford

1970   Distinguished Scientific Contribution Award, American Psychological Association

1973   Member, National Academy of Sciences
Fellow, American Academy of Arts and Sciences

1974   Kurt Lewin Memorial Award, Society for the Psychological Study of Social Issues

1977   William James Lecturer, Harvard University
Hovland Memorial Lecturer, Yale University
Myrdal Prize in Science, the Evaluation Research Society

1980   Award for distinguished Contribution to Research in Education, American Educational Research Association

Honorary Degrees

1974   Doctor of Laws, University of Michigan
1975   Doctor of Science, University of Florida
1978   Doctor of Social Sciences, Claremont Graduate School
Doctor of Humane Letters, University of Chicago

1979   Doctor of Science, University of Southern California

Book Dedications

Kiesler, C. A., Collins, B. E., and Miller, N. *Attitude Change: A Critical Analysis of Theoretical Approaches.* New York: Wiley, 1969.

Crano, W. D., and Brewer, M. B. *Principles of Research in Social Psychology.* New York: McGraw-Hill, 1973.

Powers, W. T. *Behavior: The Control of Perception.* Chicago: Aldine, 1973.

Glass, G. V., Willson, V. L., and Gottman, J. M. *Design and Analysis of Time-Series Experiments.* Boulder: Colorado Associated University Press, 1975.

Kidder, L. H., and Stewart, V. M. *The Psychology of Intergroup Relations: Conflict and Consciousness.* New York: McGraw-Hill, 1975.

Cook, T. D., Del Rosario, M., Hennigan, K. M., Mark, M. M., Trochim, W. M. K. (Eds.), *Evaluation Studies Review Annual.* Vol. 3. Beverly Hills, Calif.: Sage, 1978.

McCleary, R., and Hay, R. A., Jr. *Applied Time Series for the Social Sciences.* Beverly Hills, Calif.: Sage, 1980.

Chronological Bibliography of Donald T. Campbell

1947

Campbell, D. T. "The Generality of a Social Attitude." Unpublished doctoral dissertation, Department of Psychology, University of California, Berkeley, 1947.

1950

Campbell, D. T. "The Indirect Assessment of Social Attitudes." *Psychological Bulletin,* 1950, *47,* 15-38.

Campbell, D. T., and Mohr, P. J. "The Effect of Ordinal Position upon Responses to Items in a Checklist." *Journal of Applied Psychology,* 1950, *34,* 62-67.

Hites, R. W., and Campbell, D. T. "A Test of the Ability of Fraternity Leaders to Estimate Group Opinion." *Journal of Social Psychology,* 1950, *32,* 95-100.

Wyatt, D. F., and Campbell, D. T. "A Study of Interviewer Bias as

Related to Interviewers' Expectations and Own Opinions." *International Journal of Opinion and Attitude Research,* 1950, *4,* 77-83.

1951

Campbell, D. T. "On the Possibility of Experimenting with the 'Bandwagon' Effect." *International Journal of Opinion and Attitude Research,* 1951, *5,* 251-260.

Campbell, D. T., and McCandless, B. R. "Ethnocentrism, Xenophobia, and Personality." *Human Relations,* 1951, *4,* 185-192.

Wyatt, D. F., and Campbell, D. T. "On the Liability of Stereotype or Hypothesis." *Journal of Abnormal and Social Psychology,* 1951, *46,* 496-500.

1952

Campbell, D. T. "The Bogardus Social Distance Scale." *Sociology and Social Research,* 1952, *36,* 322-326.

Rinder, I. D., and Campbell, D. T. "Varieties of Inauthenticity." *Phylon,* 1952, *13,* 270-275.

1953

Campbell, D. T. "Operational Delineation of 'What Is Learned' via the Transportation Experiment." *Psychological Review,* 1953, *61,* 167-174.

Parrish, J. A., and Campbell, D. T. "Measuring Propaganda Effects with Direct and Indirect Attitude Tests." *Journal of Abnormal and Social Psychology,* 1953, *48,* 3-9.

1954

Campbell, D. T. "A Rationale for Weighting First, Second, and Third Sociometric Choices." *Sociometry,* 1954, *17,* 242-243.

1955

Campbell, D. T. "An Error in Some Demonstrations of the Superior Social Perceptiveness of Leaders." *Journal of Abnormal and Social Psychology,* 1955a, *51,* 694-696.

Campbell, D. T. "The Informant in Quantitative Research." *American Journal of Sociology,* 1955b, *60,* 339-342.

Clarke, R. B., and Campbell, D. T. "A Demonstration of Bias in Estimates of Negro Ability." *Journal of Abnormal and Social Psychology,* 1955, *51,* 585-588.

Kidd, J. S., and Campbell, D. T. "Conformity to Groups as a Function of Group Success." *Journal of Abnormal and Social Psychology,* 1955, *51,* 390-393.

Rankin, R. E., and Campbell, D. T. "Galvanic Skin Response to Negro and White Experiments." *Journal of Abnormal and Social Psychology,* 1955, *51,* 30-33.

1956

Campbell, D. T. "Adaptive Behavior from Random Response." *Behavioral Science,* 1956a, *1,* 105-110.

Campbell, D. T. "Enhancement of Contrast as Composite Habit." *Journal of Abnormal and Social Psychology,* 1956b, *53,* 350-355.

Campbell, D. T. *Leadership and Its Effects upon the Group.* Ohio Studies in Personnel, Bureau of Business Research Monograph No. 83. Columbus: Ohio State University, 1956c.

Campbell, D. T. "Perception as Substitute Trial and Error." *Psychological Review,* 1956d, *63,* 330-342.

Campbell, D. T., and Burwen, L. S. "Trait Judgments from Photographs as a Projective Device." *Journal of Clinical Psychology,* 1956, *12,* 215-221.

Burwen, L. S., Campbell, D. T., and Kidd, J. "The Use of a Sentence Completion Test in Measuring Attitudes Toward Superiors and Subordinates." *Journal of Applied Psychology,* 1956, *40,* 248-250.

Herskovits, M. J., Campbell, D. T., and Segall, M. "Materials for a Cross-Cultural Study of Perception." Program of African Studies, Northwestern University, 1956.

1957

Campbell, D. T. "Factors Relevant to the Validity of Experiments in Social Settings." *Psychological Bulletin,* 1957a, *54,* 297-312.

Campbell, D. T. "A Typology of Tests, Projective and Otherwise." *Journal of Consulting Psychology,* 1957b, *21,* 207-210.

Campbell, D. T., and Chapman, J. P. "Testing for Stimulus Equivalence Among Authority Figures by Similarity in Trait Description." *Journal of Consulting Psychology,* 1957, *21,* 253-256.

Campbell, D. T., Hunt, W. A., and Lewis, N. A. "The Effects of Assimilation and Contrast in Judgments of Clinical Materials." *American Journal of Psychology,* 1957, *70,* 347-360.

Campbell, D. T., and McCormack, T. H. "Military Experience and Attitudes Toward Authority." *American Journal of Sociology,* 1957, *62,* 482-490.

Campbell, D. T., and Tyler, B. B. "The Construct Validity of Work-Group Morale Measures." *Journal of Applied Psychology,* 1957, *41,* 91-92.

Burwen, L. S., and Campbell, D. T. "A Comparison of Test Scores and Role-Playing Behavior in Assessing Superior Versus Subordinate Orientation." *Journal of Social Psychology,* 1957a, *46,* 49-56.

Burwen, L. S., and Campbell, D. T. "The Generality of Attitudes Toward Authority and Nonauthority Figures." *Journal of Abnormal and Social Psychology,* 1957b, *54,* 24-31.

Chapman, L. J., and Campbell, D. T. "An Attempt to Predict the Performance of Three-Man Teams from Attitude Measures." *Journal of Social Psychology,* 1957a, *46,* 277-286.

Chapman, L. J., and Campbell, D. T. "Response Set in the F Scale." *Journal of Abnormal and Social Psychology,* 1957b, *54,* 129-132.

Cotton, J. W., Campbell, D. T., and Malone, R. D. "The Relationship Between Factorial Composition of Test Items and Measures of Test Reliability." *Psychometrika,* 1957, *22,* 347-357.

Westfall, R. L., Boyd, H. W., and Campbell, D. T. "The Use of Structured Techniques in Motivation Research." *Journal of Marketing,* 1957, *22,* 134-139.

1958

Campbell, D. T. "Common Fate, Similarity, and Other Indices of the Status of Aggregates of Persons as Social Entities." *Behavioral Science,* 1958, *3,* 14-25.

Campbell, D. T., and Gruen, W. "Progression from Simple to Complex as a Molar Law of Learning." *Journal of General Psychology*, 1958, *59*, 237-244.

Campbell, D. T., Hunt, W. A., and Lewis, N. A. "The Relative Susceptability of Two Rating Scales to Disturbances Resulting from Shifts in Stimulus Context." *Journal of Applied Psychology*, 1958, *42*, 213-217.

Campbell, D. T., and Kral, T. P. "Transposition away from a Rewarded Stimulus Card to a Nonrewarded One as a Function of a Shift in Background." *Journal of Comparative and Physiological Psychology*, 1958, *51*, 592-595.

Campbell, D. T., Lewis, N. A., and Hunt, W. A. "Context Effects with Judgmental Language That Is Absolute, Extensive, and Extra-Experimentally Anchored." *Journal of Experimental Psychology*, 1958, *55*, 220-228.

Campbell, D. T., and Mehra, K. "Individual Differences in Evaluations of Group Discussions as a Projective Measure of Attitudes Toward Leadership." *Journal of Social Psychology*, 1958, *47*, 101-106.

Campbell, D. T., and Shanan, J. "Semantic Idiosyncracy as a Method in the Study of Attitudes." *Journal of Social Psychology*, 1958, *47*, 107-110.

1959

Campbell, D. T. "Methodological Suggestions from a Comparative Psychology of Knowledge Processes." *Inquiry*, 1959a, *2*, 152-182.

Campbell, D. T. "Systematic Error on the Part of Human Links in Communication Systems." *Information and Control*, 1959b, *1*, 334-369.

Campbell, D. T., and Fiske, D. W. "Convergent and Discriminant Validation by the Multitrait-Multimethod Matrix." *Psychological Bulletin*, 1959, *56*, 81-105.

Chapman, L. J., and Campbell, D. T. "Absence of Acquiescence Response Set in the Taylor Manifest Anxiety Scale." *Journal of Consulting Psychology*, 1959a, *23*, 465-466.

Chapman, L. J., and Campbell, D. T. "The Effect of Acquiescence Response-Set upon Relationships Among the F Scale, Ethnocentrism, and Intelligence." *Sociometry*, 1959b, *22*, 153-161.

Miller, N., and Campbell, D. T. "Recency and Primacy in Persuasion as a Function of the Timing of Speeches and Measurements." *Journal of Abnormal and Social Psychology*, 1959, *59*, 1-9.

1960

Campbell, D. T. "Blind Variation and Selective Retention in Creative Thought as in Other Knowledge Processes." *Psychological Review*, 1960a, *67*, 380-400.

Campbell, D. T. "Blind Variation and Selective Survival as a General Strategy in Knowledge Processes." In M. C. Yovits and S. H. Cameron (Eds.), *Self-Organizing Systems*. Oxford, England: Pergamon Press, 1960b.

Campbell, D. T. "Recommendations for APA Test Standards Regarding Construct, Trait, or Discriminant Validity." *American Psychologist*, 1960c, *15*, 546-553.

Campbell, D. T., Miller, N., and Diamond, A. L. "Predisposition to Identify Instigating and Guiding Stimulus as Revealed in Transfer." *Journal of General Psychology*, 1960, *63*, 69-74.

Maher, B. A., Watt, N., and Campbell, D. T. "Comparative Validity of Two Projective and Two Structured Attitude Tests in a Prison Population." *Journal of Applied Psychology*, 1960, *44*, 284-288.

Small, D. O., and Campbell, D. T. "The Effect of Acquiescence Response-Set upon the Relationship of the F Scale and Conformity." *Sociometry*, 1960, *23*, 69-71.

Thistlethwaite, D. L., and Campbell, D. T. "Regression-Discontinuity Analysis: An Alternative to the Ex Post Facto Experiment." *Journal of Educational Psychology*, 1960, *51*, 309-317.

1961

Campbell, D. T. "Conformity in Psychology's Theories of Acquired Behavioral Dispositions." In I. A. Berg and B. M. Bass (Eds.), *Conformity and Deviation*. New York: Harper & Row, 1961a.

Campbell, D. T. "The Mutual Methodological Relevance of Anthropology and Psychology." In F. L. K. Hsu (Ed.), *Psychological Anthropology: Approaches to Culture and Personality.* Homewood, Ill.: Dorsey Press, 1961b.

Campbell, D. T., and Clayton, K. N. "Avoiding Regression Effects in Panel Studies of Communication Impact." *Studies in Public Communication,* 1961, *3,* 99-118.

Campbell, D. T., and Damarin, F. L. "Measuring Leadership Attitudes Through an Information Test." *Journal of Social Psychology,* 1961, *55,* 159-175.

Campbell, D. T., and LeVine, R. A. "A Proposal for Cooperative Cross-Cultural Research on Ethnocentrism." *Journal of Conflict Resolution,* 1961, *5,* 82-108.

Jacobs, R. C., and Campbell, D. T. "The Perpetuation of an Arbitrary Tradition Through Several Generations of a Laboratory Microculture." *Journal of Abnormal and Social Psychology,* 1961, *62,* 649-658.

Krantz, D. L., and Campbell, D. T. "Separating Perceptual and Linguistic Effects of Context Shifts upon Absolute Judgments." *Journal of Experimental Psychology,* 1961, *62,* 35-42.

1962

Renner, K. W., Maher, B. A., and Campbell, D. T. "The Validity of a Method for Scoring Sentence-Completion Responses for Anxiety, Dependency, and Hostility." *Journal of Applied Psychology,* 1962, *46,* 285-290.

1963

Campbell, D. T. "From Description to Experimentation: Interpreting Trends as Quasi-Experiments." In C. W. Harris (Ed.), *Problems in Measuring Change.* Madison: University of Wisconsin Press, 1963a.

Campbell, D. T. "Social Attitudes and Other Acquired Behavioral Dispositions." In S. Koch (Ed.), *Psychology: A Study of a Science.* Vol. 6: *Investigations of Man as Socius.* New York: McGraw-Hill, 1963b.

Campbell, D. T., and Stanley, J. C. "Experimental and Quasi-Experimental Designs for Research on Teaching." In N. L. Gage (Ed.), *Handbook of Research on Teaching.* Chicago: Rand

McNally, 1963. (Reprinted as *Experimental and Quasi-Experimental Designs for Research,* Rand McNally, 1966.)

Harvey, O. J., and Campbell, D. T. "Judgments of Weight as Affected by Adaptation Range, Adaptation Duration, Magnitude of Unlabeled Anchor, and Judgmental Language." *Journal of Experimental Psychology,* 1963, *65,* 12-21.

Lubetsky, J., and Campbell, D. T. (Eds.). "Age and Sex as Sources of Stimulus Equivalence in Judgments of Photos and Peers." *Journal of Clinical Psychology,* 1963, *19,* 502-505.

Segall, M. H., Campbell, D. T., and Herskovits, M. J. "Cultural Differences in the Perception of Geometric Illusions." *Science,* 1963, *139,* 769-771.

Watson, R. I., and Campbell, D. T. (Eds.). *E. G. Boring, History, Psychology, and Science: Selected Papers.* New York: Wiley, 1963.

1964

Campbell, D. T. "Distinguishing Differences of Perception from Failures of Communication in Cross-Cultural Studies." In F. S. C. Northrop and H. H. Livingston (Eds.), *Cross-Cultural Understanding: Epistemology in Anthropology.* New York: Harper & Row, 1964.

Campbell, D. T., Miller, N., and Lubetsky, J., and O'Connell, E. J. "Varieties of Projection in Trait Attribution." *Psychological Monographs,* 1964, *78* (entire issue 592).

Spear, N. E., Hill, W. F., and Campbell, D. T. "Effect of Unconsumed Reward on Subsequent Alternation of Choice." *Psychological Reports,* 1964, *15,* 407-411.

1965

Campbell, D. T. "Ethnocentric and Other Altruistic Motives." In D. Levine (Ed.), *The Nebraska Symposium on Motivation 1965.* Lincoln: University of Nebraska Press, 1965a.

Campbell, D. T. "Variation and Selective Retention in Socio-Cultural Evolution." In H. R. Barringer, G. I. Blanksten, and R. W. Mack (Eds.), *Social Change in Developing Areas: A Reinterpretation of Evolutionary Theory.* Cambridge, Mass.: Schenkman, 1965b.

Hicks, J. M., and Campbell, D. T. "Zero-Point Scaling as Affected by Social Object, Scaling Method, and Context." *Journal of Personality and Social Psychology,* 1965, *2,* 793-808.

1966

Campbell, D. T. "Pattern Matching as an Essential in Distal Knowing." In K. R. Hammond (Ed.), *The Psychology of Egon Brunswik.* New York: Holt, Rinehart and Winston, 1966.

Campbell, D. T., Kruskal, W. H., and Wallace, W. P. "Seating Aggregation as an Index of Attitude." *Sociometry,* 1966, *29,* 1-15.

Campbell, D. T., and Tausher, H. "Schopenhauer (?), Séquin, Lubinoff, and Zehender as Anticipators of Emmert's Law: With Comments on the Uses of Eponymy." *Journal of the History of the Behavioral Sciences,* 1966, *2,* 58-63.

Miller, N., Campbell, D. T., Twedt, H., and O'Connell, E. J. "Similarity, Contrast, and Complementarity in Friendship Choice." *Journal of Personality and Social Psychology,* 1966, *3,* 3-12.

Segall, M. H., Campbell, D. T., and Herskovits, M. J. *The Influence of Culture on Visual Perception.* Indianapolis: Bobbs-Merrill, 1966.

Webb, E. J., Campbell, D. T., Schwartz, R. D., and Sechrest, L. B. *Unobtrusive Measures: Nonreactive Research in the Social Sciences.* Chicago: Rand McNally, 1966.

1967

Campbell, D. T. "Administrative Experiments, Institutional Records, and Nonreactive Measures." In J. C. Stanley and S. M. Elam (Eds.), *Improving Experimental Design and Statistical Analysis.* Chicago: Rand McNally, 1967a.

Campbell, D. T. "Stereotypes and the Perception of Group Differences." *American Psychologist,* 1967b, *22,* 817-829.

Campbell, D. T., and O'Connell, E. J. "Methods Factors in Multitrait-Multimethod Matrices: Multiplicative Rather than Additive?" *Multivariate Behavioral Research,* 1967, *2,* 409-426.

Campbell, D. T., Siegman, C. R., and Rees, M. B. "Direction of Wording Effects in the Relationships Between Scales." *Psychological Bulletin,* 1967, *68,* 293-303.

1968

Campbell, D. T. "A Cooperative Multinational Opinion Sample Exchange." *Journal of Social Issues,* 1968a, *24,* 245-258.

Campbell, D. T. "Experimental Design: Quasi-Experimental Design." In D. L. Sills (Ed.), *International Encyclopedia of the Social Sciences.* Vol. 5. New York: Macmillan and Free Press, 1968b.

Campbell, D. T., and LeVine, R. A. "Ethnocentrism and Intergroup Relations." In R. P. Abelson, E. Aronson, W. J. McGuire, T. M. Newcomb, M. J. Rosenberg, and P. H. Tannenbaum (Eds.), *Theories of Cognitive Consistency: A Sourcebook.* Chicago: Rand McNally, 1968.

Campbell, D. T., and Ross, H. L. "The Connecticut Crackdown on Speeding: Time-Series Data in Quasi-Experimental Analysis." *Law and Society Review,* 1968, *3,* 33-53.

1969

Campbell, D. T. "Definitional Versus Multiple Operationalism." *et al.,* 1969a, *2,* 14-17.

Campbell, D. T. "Ethnocentrism of Disciplines and the Fish-Scale Model of Omniscience." In M. Sherif and C. W. Sherif (Eds.), *Interdisciplinary Relationships in the Social Sciences.* Hawthorne, N.Y.: Aldine, 1969b.

Campbell, D. T. "A Phenomenology of the Other One: Corrigible, Hypothetical and Critical." In T. Mischel (Ed.), *Human Action: Conceptual and Empirical Issues.* New York: Academic Press, 1969c.

Campbell, D. T. "Prospective: Artifact and Control." In R. Rosenthal and R. Rosnow (Eds.), *Artifact in Behavior Research.* New York: Academic Press, 1969d.

Campbell, D. T. "Reforms as Experiments." *American Psychologist,* 1969e, *24,* 409-429.

Rozelle, R. M., and Campbell, D. T. "More Plausible Rival Hypotheses in the Cross-Lagged Panel Correlation Technique." *Psychological Bulletin,* 1969, *71,* 74-80.

Winch, R. F., and Campbell, D. T. "Proof? No. Evidence? Yes. The Significance of Tests of Significance." *American Psychologist,* 1969, *4,* 140-143.

1970

Campbell, D. T. "Considering the Case Against Experimental Evaluations of Social Innovations." *Administrative Science Quarterly,* 1970a, *15,* 110-113.

Campbell, D. T. "Natural Selection as an Epistemological Model." In R. Narroll and R. Cohen (Eds.), *A Handbook of Method in Cultural Anthropology.* New York: Natural History Press/Doubleday, 1970b.

Campbell, D. T., and Erlebacher, A. "How Regression Artifacts in Quasi-Experimental Evaluations Can Mistakenly Make Compensatory Education Look Harmful." In J. Hellmuth (Ed.), *Compensatory Education: A National Debate.* Vol. 3: *Disadvantaged Child.* New York: Brunner/Mazel, 1970.

Campbell, D. T., and Frey, P. W. "The Implications of Learning Theory for the Fade-Out of Gains from Compensatory Education." In J. Hellmuth (Ed.), *Compensatory Education: A National Debate.* Vol. 3: *Disadvantaged Child.* New York: Brunner/Mazel, 1970.

Campbell, D. T., and LeVine, R. A. "Field-Manual Anthropology." In R. Naroll and R. Cohen (Eds.), *A Handbook of Method in Cultural Anthropology.* New York: Natural History Press/Doubleday, 1970.

Brewer, M. B., Campbell, D. T., and Crano, W. D. "Testing a Single-Factor Model as an Alternative to the Misuse of Partial Correlations in Hypothesis-Testing Research." *Sociometry,* 1970, *33,* 1-11.

Kidder, L., and Campbell, D. T. "The Indirect Testing of Social Attitudes." In G. F. Summers (Ed.), *Attitude Measurement.* Chicago: Rand McNally, 1970.

Raser, J. R., Campbell, D. T., and Chadwick, R. W. "Gaming and Simulation for Developing Theory Relevant to International Relations." In A. Rapoport (Ed.), *General Systems: Yearbook of the Society for General Systems Research.* Vol. 15. Ann Arbor, Mich.: Society for General Systems Research, 1970.

Ross, H. L., Campbell, D. T., and Glass, G. V. "Determining the Social Effects of a Legal Reform: The British 'Breathalyser' Crackdown of 1967." *American Behavioral Scientist,* 1970, *13,* 493-509.

Werner, O., and Campbell, D. T. "Translating, Working Through Interpreters, and the Problem of Decentering." In R. Naroll and R. Cohen (Eds.), *A Handbook of Method in Cultural Anthropology*. New York: Natural History Press/Doubleday, 1970.

1971

Campbell, D. T. "Methods for the Experimenting Society." Paper presented to the Eastern Psychological Association, New York City, April 17, 1971, and to the American Psychological Association, Washington, D.C., Sept. 5, 1971a.

Campbell, D. T. "Temporal Changes in Treatment-Effect Correlations: A Quasi-Experimental Model for Institutional Records and Longitudinal Studies." In G. V. Glass, (Ed.), *Proceedings of the 1970 Invitational Conference on Testing Problems*. Princeton, N.J.: Educational Testing Service, 1971b.

Brewer, M. B., Campbell, D. T., and LeVine, R. A. "Cross-Cultural Test of the Relationship Between Affect and Evaluation." In *Proceedings, 79th Annual Convention*. Washington, D.C.: American Psychological Association, 1971.

Brickman, P., and Campbell, D. T. "Hedonic Relativism and Planning the Good Society." In M. H. Appley (Ed.), *Adaptation-Level Theory: A Symposium*. New York: Academic Press, 1971.

1972

Campbell, D. T. "On the Genetics of Altruism and the Counterhedonic Components in Human Culture." *Journal of Social Issues*, 1972a, *28*, 21-37.

Campbell, D. T. "Herskovits, Cultural Relativism, and Metascience." In M. J. Herskovits, *Cultural Relativism*. New York: Random House, 1972b.

Campbell, D. T. "Measuring the Effects of Social Innovations by Means of the Time-Series." In J. M. Tanur, F. Mosteller, W. H. Kruskal, R. F. Link, R. S. Pieters, and G. R. Rising (Eds.), *Statistics: A Guide to the Unknown*. San Francisco: Holden-Day, 1972c.

Crano, W. D., Kenny, D. A., and Campbell, D. T. "Does Intelligence Cause Achievement?: A Cross-Lagged Panel Analysis." *Journal of Educational Psychology*, 1972, *63*, 258-275.

LeVine, R. A., and Campbell, D. T. *Ethnocentrism: Theories of Conflict, Ethnic Attitudes, and Group Behavior.* New York: Wiley, 1972a.

LeVine, R. A., and Campbell, D. T. *The Gusii of Kenya.* Vols. 1 and 2. New Haven, Conn.: Human Relations Area Files, 1972b.

1973

Campbell, D. T. "Ostensive Instances and Entitativity in Language Learning." In W. Gray and N. D. Rizzo (Eds.), *Unity Through Diversity.* Pt. 2. New York: Gordon & Breach, 1973a.

Campbell, D. T. "The Social Scientist as Methodological Servant of the Experimenting Society." *Policy Studies Journal,* 1973b, *2,* 72-75.

Webb, E. J., and Campbell, D. T. "Experiments on Communication Effects." In I. de S. Pook, W. Schramm, F. W. Frey, N. Maccoby, and E. B. Larker (Eds.), *Handbook of Communications.* Chicago: Rand McNally, 1973.

1974

Campbell, D. T. " 'Downward Causation' in Hierarchically Organized Biological Systems." In F. Ayala and T. Dobzhansky (Eds.), *Studies in the Philosophy of Biology.* London: Macmillan, 1974a.

Campbell, D. T. "Evolutionary Epistemology." In P. A. Schilpp (Ed.), *The Philosophy of Karl Popper.* La Salle, Ill.: Open Court, 1974b.

Campbell, D. T. "Qualitative Knowing in Action Research." Kurt Lewin Award Address, Society for the Psychological Study of Social Issues, presented at 82nd annual meeting of the American Psychological Association, New Orleans, Sept. 1, 1974c.

Campbell, D. T. "Unjustified Variation and Selective Retention in Scientific Discovery." In F. Ayala and T. Dobzhansky (Eds.), *Studies in the Philosophy of Biology.* London: Macmillan, 1974d.

Riecken, H. W., Boruch, R. F., Campbell, D. T., Caplan, N., Glennan, T. K., Pratt, J., Rees, A., and Williams, W. *Social Experimentation: A Method for Planning and Evaluating Social Intervention.* New York: Academic Press, 1974.

1975

Campbell, D. T. "The Conflict Between Social and Biological Evolution and the Concept of Original Sin." *Zygon,* 1975a, *10,* 234-249.

Campbell, D. T. "On the Conflicts Between Biological and Social Evolution and Between Psychology and Moral Tradition." *American Psychologist,* 1975b, *30,* 1103-1126.

Campbell, D. T. " 'Degrees of Freedom' and the Case Study." *Comparative Political Studies,* 1975c, *8,* 178-193.

Campbell, D. T. "Reintroducing Konrad Lorenz to Psychology." In R. I. Evans (Ed.), *Konrad Lorenz: The Man and His Ideas.* New York: Harcourt Brace Jovanovich, 1975d.

Campbell, D. T., and Boruch, R. F. "Making the Case for Randomized Assignment to Treatments by Considering the Alternatives: Six Ways in Which Quasi-Experimental Evaluations in Compensatory Education Tend to Underestimate Effects." In C. A. Bennett and A. Lumsdaine (Eds.), *Evaluation and Experiments: Some Critical Issues in Assessing Social Programs.* New York: Academic Press, 1975.

1976

Campbell, D. T. "Focal Local Indicators for Social Program Evaluation." *Social Indicators Research,* 1976a, *3,* 237-256.

Campbell, D. T. "Reprise." *American Psychologist,* 1976b, *31,* 381-384.

Brewer, M. B., and Campbell, D. T. *Ethnocentrism and Intergroup Attitudes: East African Evidence.* New York: Halsted Press, 1976.

Cook, T. D., and Campbell, D. T. "The Design and Conduct of Quasi-Experiments and True Experiments in Field Settings." In M. D. Dunnette (Ed.), *Handbook of Industrial and Organizational Psychology.* Chicago: Rand McNally, 1976.

1977

Campbell, D. T. "Descriptive Epistemology: Psychological, Sociological, and Evolutionary." William James Lectures, Harvard University, 1977a.

Campbell, D. T. "Discussion Comment on 'The Natural Selection Model of Conceptual Evolution.' " *Philosophy of Science*, 1977b, *44*, 502-507.

Campbell, D. T. "Keeping the Data Honest in the Experimenting Society." In H. W. Melton and D. J. H. Watson (Eds.), *Interdisciplinary Dimensions of Accounting for Social Goals and Social Organizations*. Columbus, Ohio: Grid, 1977c.

Campbell, D. T., Boruch, R. F., Schwartz, R. D., and Steinberg, J. "Confidentiality-Preserving Modes of Access to Files and to Interfile Exchange for Useful Statistical Analysis." *Evaluation Quarterly*, 1977, *1*, 269-299.

1978

Campbell, D. T. "Response to Dyer." *American Psychologist*, 1978, *33*, 770-772.

Campbell, D. T., and Cecil, J. S. "Protection of the Rights and Interests of Human Subjects in the Areas of Program Evaluation, Social Experimentation, Social Indicators, Survey Research, Secondary Analysis of Research Data, and Statistical Analysis of Data from Administrative Records." In National Commission for the Protection of Human Subjects of Biomedical and Behavioral Research, *The Belmont Report: Ethical Principles and Guidelines for the Protection of Human Subjects of Research*. Washington, D.C.: U.S. Government Printing Office, 1978.

1979

Campbell, D. T. "Assessing the Impact of Planned Social Change." *Evaluation and Program Planning*, 1979a, *2*, 67-90.

Campbell, D. T. "Comnents on the Sociobiology of Ethics and Moralizing." *Behavioral Science*, 1979b, *24*, 37-45.

Campbell, D. T. "Darning up the Gaps in the Seamless Web of Scholarship." *Syracuse Scholar*, Winter 1979-1980c, *1*, 73-77.

Campbell, D. T. "A Tribal Model of the Social System Vehicle Carrying Scientific Knowledge." *Knowledge*, 1979d, *2*, 181-201.

Cook, T. D., and Campbell, D. T. *Quasi-Experimentation: Design and Analysis for Field Settings*. Chicago: Rand McNally, 1979.

Name Index

Subject Index

515